NITRATION

Methods and Mechanisms

ORGANIC NITRO CHEMISTRY SERIES

Managing Editor
Dr. Henry Feuer
Purdue University
West Lafayette, Indiana 47907
(USA)

EDITORIAL BOARD

Hans H. Baer
Ottawa, Canada

Robert G. Coombes
London, England

Leonid T. Eremenko
Chernogolovka, USSR

Milton B. Frankel
Canoga Park, CA, USA

Nathan Kornblum
West Lafayette, IN, USA

Philip C. Myhre
Claremont, CA, USA

Arnold T. Nielsen
China Lake, CA, USA

Wayland E. Noland
Minneapolis, MN, USA

George A. Olah
Los Angeles, CA, USA

Noboru Ono
Kyoto, Japan

C.N.R. Rao
Bangalore, India

John H. Ridd
London, England

Glen A. Russell
Ames, IA, USA

Dieter Seebach
Zurich, Switzerland

François Terrier
Rouen, France

Heinz G. Viehe
Louvain-la-Neuve, Belgium

Also in the Series:

NITROAZOLES
The C-Nitro Derivatives of Five-Membered
N- and N,O-Heterocycles
by Joseph H. Boyer

**NITRILE OXIDES, NITRONES, AND NITRONATES
IN ORGANIC SYNTHESIS**
Novel Strategies in Synthesis
by Kurt B. G. Torssell

NITRATION

Methods and Mechanisms

George A. Olah
Ripudaman Malhotra
Subhash C. Narang

George A. Olah
Loker Hydrocarbon Research Institute
University of Southern California
Los Angeles, CA 09989

Ripudaman Malhotra and
Subhash C. Narang
S.I.R. International
Menlo Park, CA 94025

Library of Congress Cataloging-in-Publication Data

Olah, George A. (George Andrew), 1927-
   Nitration : methods and mechanisms / George A. Olah, Ripudaman Malhotra, Subhash C. Narang.
      p.   cm. — (Organic nitro chemistry series) 3
   Includes bibliographical references.
   ISBN 0-89573-144-4
   1. Nitration. I. Malhotra, Ripudaman. II. Narang, Subhash C.
   III. Title. IV. Series: Organic nitro chemistry.
QD281.N5033 1989   89-16582
547'.041 — dc20   CIP

British Library Cataloguing in Publication Data

Olah, George A. (George Andrew *1927-*)
   Nitration : methods and mechanisms.
   1. Aromatic compounds. Nitration
   I. Title II. Malhotra, Ripudaman III. Narang, Subhash C.
   IV. Series
   547'.604593

ISBN 0-89573-144-4

© 1989 VCH Publishers, Inc.
This work is subject to copyright.
All rights are reserved, whether the whole or part of the material is concerned, specifically those of translation, reprinting, re-use of illustrations, broadcasting, reproduction by photocopying machine or similar means, and storage in data banks.
Registered names, trademarks, etc., used in this book, even when not specifically marked as such, are not to be considered unprotected by law.

Printed in the United States of America
ISBN 0-89573-144-4 VCH Publishers
ISBN 3-527-26698-4 VCH Verlagsgesellschaft

Printing History:
10 9 8 7 6 5 4 3 2 1

Published jointly by:

VCH Publishers, Inc.
220 East 23rd Street
Suite 909
New York, New York 10010

VCH Verlagsgesellschaft mbH
P.O. Box 10 11 61
D-6940 Weinheim
Federal Republic of Germany

VCH Publishers (UK) Ltd.
8 Wellington Court
Cambridge CB1 1HW
United Kingdom

Series Foreword

In the organic nitro chemistry era of the fifties and the early sixties, a great emphasis of the research was directed towards the synthesis of new compounds which would be useful as potential ingredients in explosives and propellants.

In recent years, the emphasis of research has been directed more and more toward utilizing nitro compounds as reactive intermediates in organic synthesis. The activating effect of the nitro group is exploited in carrying out many organic reactions, and its facile transformation into various functional groups has broadened the importance of nitro compounds in the synthesis of complex molecules.

It is the purpose of the Series to review the field of organic nitro chemistry in its broadest sense by including structurally related classes of compounds such as nitroamines, nitrates, nitrones and nitrile oxides. It is intended that the contributors who are active investigators in the various facets of the field will provide a concise presentation of recent advances which have generated a renaissance in nitro chemistry research.

Henry Feuer
Purdue University

Mookie, the Olahs' cocker spaniel, enlightens an otherwise blank page.

(*photo by Mark Sassaman*)

Preface

It is not our purpose to try to give an all encompassing review of nitration which, anyhow, would be impossible in a small volume. Instead, we emphasize synthetic methods and the present understanding of the mechanisms. The book inevitably reflects some of our own preferences in selecting the topics and their presentation. We have tried, however, to give a balanced presentation of sometimes still differing views and their significance. Recent years have seen significant progress not only in the development of new and improved methods of nitration, but also in our understanding of their mechanisms. Whereas no clear consensus of the latter has yet emerged, differences have narrowed to the point where it seemed timely to review the field.

Henry Feuer is thanked for not only inviting us to write this book, but also for his patience and understanding during inevitable delays in completing the manuscript. In the process of writing the book, we benefited by numerous discussions and exchanges with colleagues and friends which helped to bring into focus many aspects. We wish to particularly acknowledge fruitful discussion and comments by Edward M. Arnett, Gernot Frenking, David M. Golden, Rudy Marcus, Donald F. McMillen, Peter Politzer, John Ridd, David S. Ross, Paul Schleyer, Michael Zerner and Lennart Eberson. Philip Myhre and G. K. Surya Prakash are thanked for reading the manuscript and for their suggestions and criticisms. Dr. Chandra Rao helped with the task of references and Joseph Bausch with the figures and indexing.

Ms. Michele Dea and Mrs. Reiko Choy expertly typed the manuscript.

We thank the following copyright owners for permission to reproduce material from their publications in Tables and Figures, where the individual references are given: Acta Chemica Scandinavica, American Chemical Society (Accounts of Chemical Research, Journal of the American Chemical Society, Journal of Organic Chemistry); National Research Council of Canada (Canadian Journal of Chemistry); Royal Society of Chemistry (Journal of the Chemical Society, Perkin II) and Thieme Verlag (Synthesis).

The Olah's group research on nitration chemistry over the years was helped by support by the U. S. Army Office of Research and more recently by the Office of Naval Research. G.A.O. also thanks the Guggenheim Foundation for a Fellowship which helped to complete the writing of the book.

December, 1988

George A. Olah
Ripudaman Malhotra
Subhash C. Narang

*In memory of Sir Christopher Ingold,
who transformed nitration chemistry into an exact science.*

CONTENTS

SERIES FOREWORD

PREFACE

CHAPTER 1. *Introduction and General Aspects*

Definition	1
Historical	1
Scope	2
Mechanistic Classification	4
Industrial Use and Research Significance	5
References	7

CHAPTER 2. *Reagents and Methods of Aromatic Nitration*

I. ACID-CATALYZED ELECTROPHILIC NITRATION 9
- 2.1 Nitric Acid ... 11
- 2.2 Nitric Acid-Graphite (Graphite Nitrate) ... 13
- 2.3 Nitric-Sulfuric Acid (Mixed Acid) ... 13
- 2.4 Nitric Acid-Oleum ... 15
- 2.5 Nitric-Phosphoric (Polyphosphoric) Acid ... 15
- 2.6 Nitric-Perchloric Acid ... 18
- 2.7 Nitric Acid-Hydrogen Fluoride ... 19
- 2.8 Nitric Acid-Hydrogen Fluoride-Boron Trifluoride ... 20
- 2.9 Nitric Acid-Boron Trifluoride ... 20
- 2.10 Nitric-Trifluoroacetic Acid ... 21
- 2.11 Nitric-Methanesulfonic Acid ... 22
- 2.12 Nitric-Trifluoromethanesulfonic (Triflic) Acid ... 22
- 2.13 Nitric-Fluorosulfuric Acid ... 25
- 2.14 Nitric-Magic Acid (FSO_3H-SbF_5) ... 26
- 2.15 Nitric Acid-Solid Acid Catalysts ... 26
- 2.16 Nitric Acid-Supported Acid Catalysts ... 28
- 2.17 Monodentate Metal Nitrates ... 29

	2.18	Alkyl Nitrates	36
	2.19	Acetone Cyanohydrin Nitrate	40
	2.20	Trimethylsilyl Nitrate	42
	2.21	Acyl Nitrates	43
	2.21.1	Acetyl Nitrate	43
	2.21.2	Benzoyl Nitrate	44
	2.21.3	Trifluoroacetyl Nitrate	44
	2.22	Nitryl Halides	45
	2.22.1	Nitryl Chloride	45
	2.22.2	Nitryl Bromide	47
	2.22.3	Nitryl Fluoride	48
	2.23	Nitrogen Oxides	48
	2.23.1	Dinitrogen Trioxide	49
	2.23.2	Dinitrogen Tetroxide (Nitrogen Dioxide)	50
	2.23.3	Dinitrogen Pentoxide	55
	2.24	Nitronium Salts	57
	2.24.1	Nitronium Tetrafluoroborate	61
	2.24.2	Nitronium Hexafluorophosphate	67
	2.24.3	Nitronium Trifluoromethanesulfonate (Triflate)	67
	2.25	Transfer Nitrating Agents	68
	2.25.1	Nitronium Tetrafluoroborate-Crown Ether Complexes	68
	2.25.2	N-Nitropyridinium and N-Nitroquinolinium Salts	69
	2.25.3	Pyridine-N-Oxide- and Dimethylsulfoxide-Nitrosonium Tetrafluoroborate	70
	2.25.4	N-Nitropyrazole	72
	2.25.5	9-Nitroanthracenium Ion	73
	2.25.6	Nitrohexamethylbenzenium Ion	74
	2.26	Oxidative Nitration with Nitrosonium Salts	75
	2.27	Nitration via Metallation	75
	2.27.1	Nitration via Mercuration	76
	2.27.2	Nitration via Palladation	78
	2.27.3	Nitration via Thallation	79
	2.27.4	Nitro-Demetallation with Other Metal Compounds	80
	2.27.5	Desilylative Nitration	80
II.	HOMOLYTIC (RADICAL) NITRATION		83
	2.28	Nitrogen Dioxide (Dinitrogen Tetroxide)	83
	2.28.1	Solution Nitration	83
	2.28.2	Gas-Phase Nitration	88
	2.29	Photochemical Nitration with Nitrogen Oxides	89
	2.30	Nitration with $NO/O_2/N_2O_4$	92
	2.31	Metal Complex Promoted Nitration with N_2O_4	93
	2.32	Bidentate Metal Nitrations	94
III.	NUCLEOPHILIC NITRATION		96
	2.33	Nitro-Dehalogenation	96
	2.34	Nitro-Dediazoniation	97
	2.35	Nitrolysis of Diarylhalonium Ions	98
IV.	CONCLUSIONS		102
	References		103

CHAPTER 3. *Mechanisms of Aromatic Nitration*

I. ELECTROPHILIC NITRATION 117
   3.1   The Ingold-Hughes Mechanism 117
      3.1.1 Nitration by the Nitronium Ion 117
      3.1.2 Acidity Dependence of the Nitric Acid/Nitronium Ion
            Equilibrium 124
      3.1.3 Effect of Nitrous Acid 129
   3.2   Olah's Modified Mechanism of Two Separate Intermediates ... 134
      3.2.1 Studies with Preformed Nitronium Salts 135
      3.2.2 The Necessity for Two Intermediates and Their Nature 136
      3.2.3 The Nature of the Nitronium Ion in Nitrations 144
   3.3   Criticism of the Two Intermediate Mechanism 152
   3.4   Schofield's Encounter Pair as the First Intermediate 154
   3.5   *Ipso* Attack and Nitration 158
   3.6   Electron Transfer in Nitration 164
      3.6.1 Early Suggestions 164
      3.6.2 Perrin's Renewal of the ET Concept 165
      3.6.3 Probing the Role of Electron Transfer
            in Aromatic Nitrations 166
          3.6.3.1 Eberson's Electrochemical Studies 167
          3.6.3.2 Use of Chemical Oxidants 169
          3.6.3.3 Gas-Phase Ion-Molecule Nitration 173
          3.6.3.4 Radiolytic Gas-Phase Nitration 177
          3.6.3.5 Theoretical Calculations 180
          3.6.3.6 Kochi's Studies of Charge Transfer 184
          3.6.3.7 Application of Marcus Theory 187
          3.6.3.8 Relationship between Ionic and ET Mechanisms 189
   3.7   Our Present Understanding of the Mechanism of Electrophilic
         Aromatic Nitration 197
      3.7.1 The Nitronium Ion Mechanism and Its Modification for
            Reactive Aromatics 197
      3.7.2 Is the Nitronium Ion Capable of Oxidizing Aromatic
            Compounds? 199
      3.7.3 Can the Aromatic Radical Cation Couple with NO_2 to
            Give the Nitroaromatic Product? 199
      3.7.4 Are the Rates of the Individual Steps Consistent with
            Observed Rates of Nitration? 200
      3.7.5 Is the Product Distribution of the Reaction of $ArH^{\ddagger}$
            with NO_2 the Same as that of Mixed Acid
            or Nitronium Salt Nitration? 200
      3.7.6 How Do the Polar and ET Mechanisms Relate? 200
II. FREE-RADICAL NITRATION 201
III. NUCLEOPHILIC NITRATION 205
IV. CONCLUSIONS 207
    References 207

CHAPTER 4. *Aliphatic Nitration*

- 4.1 General Aspects 219
- 4.2 Nitration of Alkanes, Cycloalkanes and Arylalkanes 220
  - 4.2.1 Free-Radical Nitration 220
    - 4.2.1.1 Liquid-Phase Nitration with Dilute Nitric Acid and Nitrogen Oxides (Konovalov Reaction) 221
    - 4.2.1.2 Gas-Phase Nitration with Nitrogen Dioxide 225
  - 4.2.2 Electrophilic Nitration 229
    - 4.2.2.1 Active Methylene Compounds 229
    - 4.2.2.2 Nitration with Nitronium Salts 233
    - 4.2.2.3 Nitroalkanes via Nitro-Desilylation and Nitro-Destannylation 241
- 4.3 Nitration of Alkenes 243
  - 4.3.1 Nitric Acid 244
  - 4.3.2 Nitric-Sulfuric Acid and Nitric Acid-Anhydrous Hydrogen Fluoride 247
  - 4.3.3 Acetyl Nitrate 247
  - 4.3.4 Dinitrogen Tetroxide/Nitrogen Dioxide 250
  - 4.3.5 Dinitrogen Trioxide 255
  - 4.3.6 Dinitrogen Pentoxide 257
  - 4.3.7 Nitryl Halides 257
  - 4.3.8 Nitroalkenes via Nitro-Mercuration and Nitro-Selenation .. 260
  - 4.3.9 Tetranitromethane and Hexanitroethane 260
  - 4.3.10 Nitronium Salts 261
  - 4.3.11 Nitroalkenes via Nitro-Destannylation and Nitro-Desilylation 265
- 4.4 Nitration of Alkynes 266
- 4.5 Nitration at Heteroatoms 269
  - 4.5.1 Nitration at Oxygen 269
  - 4.5.2 Nitration at Nitrogen 275
    - 4.5.2.1 Nitric Acid and Derivatives 275
    - 4.5.2.2 Nitronium Salts 278
  - 4.5.3 Nitration at Sulfur 286
  - 4.5.4 Nitration at Phosphorus 286
- 4.6 Nucleophilic Aliphatic Nitration 287
  - 4.6.1 Alkyl Halides with Silver Nitrite (Victor Meyer Reaction) 287
  - 4.6.2 Alkyl Halides with Alkali Metal Nitrites 290
  - 4.6.3 Kornblum Modification 290
  - 4.6.4 ter Meer Reaction 292
  - 4.6.5 Kaplan-Shechter Reaction 293
  - 4.6.6 Nitrolysis of Dialkylhalonium and Trialkyoxonium Ions 295

Outlook 296
References 297
Author Index 313
Subject Index 325

CHAPTER 1.
Introduction and General Aspects

Definition

Nitration is defined[1] as the reaction between an organic compound and a nitrating agent (generally nitric acid or its derivatives) to introduce a nitro group on to a carbon atom (C-nitration) or to produce nitrates (O-nitration) or nitramines (N-nitration).[1]

$$-\overset{|}{\underset{|}{C}}-H + HNO_3 \longrightarrow -\overset{|}{\underset{|}{C}}-NO_2 + H_2O$$

$$-\overset{|}{\underset{|}{C}}-OH + HNO_3 \longrightarrow -\overset{|}{\underset{|}{C}}-ONO_2 + H_2O$$

$$\underset{/}{\overset{\backslash}{N}}-H + HNO_3 \longrightarrow \underset{/}{\overset{\backslash}{N}}-NO_2 + H_2O$$

The nitro group most frequently substitutes a hydrogen atom, however, other atoms or groups can also be substituted (e.g., halogen atoms). Nitro compounds can also be formed by addition of nitric acid or nitrogen oxides to unsaturated compounds (olefins, acetylenes).

Historical

Faraday was probably the first to perform nitration of benzene with nitric acid. Although he did not recognize it, his preserved experimental notes show that when he added nitric acid to benzene an almond-scented substance was formed which was indeed nitrobenzene.[2] Mitscherlich in 1834 first reported[3] nitrobenzene (mirbane oil) by the reaction of benzene with fuming nitric acid. From early on nitration with nitric acid was found to be applicable to a variety of aromatic compounds, and such nitrations were facilitated by the addition of sulfuric acid.

$$\text{ArH} + \text{HNO}_3 \longrightarrow \text{ArNO}_2 + \text{H}_2\text{O}$$

Mansfield suggested that the use of sulfuric acid was in binding to the water that was liberated during nitration, which otherwise would dilute the nitric acid. Mixed acid nitration became the most common method for nitration of aromatics and was first employed on an industrial scale in 1847 by Mansfield.[4] Nitration of aliphatic compounds was first recorded by Beilstein and Kurbatov in 1880.[5] However, it was only after the advent of the petrochemical industry that aliphatic nitrations also gained importance. Hass developed the gas phase nitration of hydrocarbons for the manufacture of nitroalkanes that were to be used primarily as solvents.[6]

Scope

The purpose of this book is to give an overview of nitration chemistry with regard to both synthetic methods and mechanisms. This small volume cannot give a comprehensive survey of the literature. More than 6000 studies on nitrations were reported in Chemical Abstracts in the period between 1970 and 1985 alone indicating a great upsurge of interest in nitration. We have included only relatively brief accounts of the background and well-reviewed chemistry and have focused on the more recent developments of preparative methods and mechanisms of nitration. Our review is intended to present the reader with a picture of the present status of the field.

Many books and reviews have been written on the subject of nitration. Elucidation of the mechanism of electrophilic aromatic nitration has been described by Ingold in his classic book, *Structure and*

Mechanism in Organic Chemistry.[7] Olah has given reviews of acid-catalyzed nitration methods.[8,9] Books by Schofield[10] give a detailed discussion of the mechanistic aspects of aromatic nitrations. Two volumes (and supplements) in Patai's series *Chemistry of the Functional Groups*[11] are devoted to nitro and nitroso groups and contain excellent reviews on aliphatic and aromatic nitrations. Earlier aspects of aliphatic nitration are further discussed in the works by Asinger[12] and Urbański.[13]

The readers are advised to consult these and other reviews for more details and topics not included in our discussion. We have attempted throughout the book to give the reader an overview of nitration with some emphasis on our own experience in the field, but at the same time hoping to provide a balanced perspective.

Aromatic and aliphatic compounds alike undergo nitration, although frequently the former are considered predominant. It is our hope that by discussing both areas a proper perspective will emerge emphasizing the scope and diversity of aromatic and aliphatic nitrations. Aliphatic and aromatic nitrations are of equal importance, although because of their key role in establishing mechanistic fundamentals the latter generally are more extensively discussed.

We have not limited our discussion to C-nitration. Nitrations on the heteroatoms N, O, P, and S are also included. O-Nitration is generally considered esterification, but mechanistically it is not significantly different from other nitrations since most nitrations involve the substitution of hydrogen for a nitro group. However, substitution of various other atoms and groups such as metals, organometallic groups, and halogens by nitro groups are also known. These are frequently of substantial use for achieving unusual positional selectivities.

We limited the scope of our discussion to only those reactions in which nitro groups are directly introduced into a given molecule (i.e., direct nitrations). Reactions in which nitrated synthons are attached to molecules, such as the nucleophilic displacement of tertiary halogens by alkylnitronate anions or oxidation of amino compounds, constitute important ways of preparing nitro compounds but these reactions were considered outside the scope of our review. An ever-increasing number of synthetic methods allow the chemist to find the most suitable way to introduce the nitro group into molecules with increasing emphasis on selectivity.

Mechanistic Classification

Nitration reactions (and we emphasize the plural on purpose to point out the variety of differing reactions which can achieve the introduction of nitro groups into organic compounds) in a broad sense can be divided into *ionic, radical ion* and *free radical* reactions. Within ionic nitrations we can differentiate the more predominant *electrophilic* nitrations (proceeding through the nitronium ion, NO_2^+, or some of its polarized $^{\delta+}NO_2^{\delta-}$-X carriers) and *nucleophilic* nitrations (displacement reactions of suitable leaving groups by the nitrite ion, NO_2^-). Consequently, in terms of discussing mechanisms (and we again emphasize the plural) we will deal with aspects of both ionic and free-radical reactions. Questions that will be examined in some depth include the nature of the interactions of the electrophilic nitrating agent (i.e., nitronium ion, NO_2^+) with aromatics. The possible role of *single electron transfer* in nitration is also probed. It is becoming evident that in addition to conventional two-electron-transfer nitration single-electron-transfer can also play a significant role. At the same time there are clear limitations to the systems where electron transfer may be operative. Electrophilic nitrations retain their significance in the plurality of aromatic nitrations. The relationship of ionic and electron transfer nitrations is one of the more fascinating of recently emerging aspects in the study of the mechanism of nitration.

From the perspective of historical development it is interesting to note that whereas inorganic chemistry is replete with electron-transfer reactions, indeed its progress was often guided by the concept that chemical reactions are a result of a sequence of single-electron transfers, organic chemistry until recently was mostly devoid of consideration of electron-transfer reactions. Only a few reactions such as electrochemical redox and dissolving metal reductions were seriously considered to involve electron transfer. The concept dominating the field of organic chemistry has been that of "a faithful couple: the electron pair."[14] The situation is, however, changing. The significance of single electron transfer in organic reactions is increasingly recognized. The theoretical underpinnings of the Marcus theory and Taube's concepts of electron transfer in inorganic systems are slowly reaching organic chemistry. A reevaluation is clearly indicated for many organic reactions including nitration. Eberson reviewed the role of electron transfer reactions in organic chemistry[15] tracing how the concept of electron transfer became "normal science". Many more reactions are now being proposed to proceed via electron transfer. Examples include hydride reductions, Grignard reactions, some Wittig reactions, and many nucleophilic substitutions to name a few. Pross[16] and Shaik[17] have advocated a formal basis for considering whether reactions involve single-electron shifts and have elaborated the relationship between polar and electron transfer pathways. According to Pross, electron transfer reactions and polar reactions represent just the two poles of the mechanistic spectrum. In the polar pathway the bond-formation process occurs simultaneously with the electron shifts. In Taube's terminology the polar pathway constitutes inner-sphere electron transfer. At the other extreme, the outer-sphere electron transfers, the electron shift is complete before there is any bond formation. While the inorganic chemists have focused on the electronic changes, the organic chemists have emphasized the atomic changes that have taken place. However, the analogy between the processes is clear.

While discussing the significance of a possible electron transfer mechanism we must emphasize that acid-catalyzed nitration of most benzenoid aromatics, such as that of benzene and toluene, proceed via an electrophilic, formally two-electron-transfer mechanism. There is no single mechanism of nitration, but a continuum of reactions depending on specific systems and conditions. Pross[16] and Shaik[17] emphasized the general role of single electron shifts in electrophilic reactions. Indeed, in a strict sense electrons are indistinguishable and always move individually; not in pairs, as used so extensively by organic chemists in their formulation of curved-arrow electron-pushing mechanisms. However, single electron shifts do not necessarily lead to identifiable radical ions, as fast molecular reorganization following electron shifts can merge the electron transfer and polar pathways. We discuss these questions in some detail in Chapter 3.

Industrial Use and Research Significance

Nitration has been an active area of industrial chemistry for over a century.[18] The continued interest is a testimony to the importance of nitration as a process. By far the most common industrial nitration is the sulfuric acid catalyzed reaction with nitric acid. This process is used for the production of many large-volume chemicals such as (1) nitrobenzene, which is used as a solvent and in the manufacture of aniline; (2) dinitrobenzene, nitrotoluenes, and nitrochlorobenzenes, which are used as intermediates for dyes, pharmaceuticals, and perfumes; and (3) dinitrotoluenes, which are converted into toluene diisocyanates (TDI) for the manufacture of polyurethanes. Acid catalyzed nitrations are also used in the manufacture of high explosives such as trinitrotoluene (TNT), glyceryl trinitrate (nitroglycerin), cellulose nitrate,cyclo-1,3,5-trimethylenetrinitramine (RDX), cyclo-1,3,5,7-tetramethylenetetranitramine (HMX), and related nitramines. Ever improving reactor designs have brought the technology from batchwise processes to continuous flow processes using advanced tubular mixing reactors. Gas phase nitrations, such as the production of nitroalkanes via nitration of alkanes with nitrogen dioxide, are also increasingly used.

Academic and industrial research emphasizes both methods and mechanisms. Aromatic nitrations have also played an important pedagogic role. Some of our understanding of the principles of physical organic chemistry stems from the study of aromatic nitration. Just to mention a few examples:

1. Recognition of activating and deactivating substituents based on the effect they have on the rates of reactions.
2. Directing effect of substituents.
3. Correlation between the activating/deactivating effect of substituents and their directing ability and subsequent refinement of the concept in terms of quantitative structure/activity relationships.
4. The theory of "molecular electronics," (i.e., the electronic theory of organic chemistry) as developed by Robinson[20] and Ingold[21] which describes the course of organic reactions in terms of movement of electrons has as its keystone in the study of aromatic nitration and related substituent effects.

Ingold[7] has pointed out some of the main reasons why aromatic nitration has been particularly useful in studying reaction mechanisms: the wide applicability of the reactions, the fact that introduction of a nitro group into the molecule deactivates the system sufficiently so that further nitration does not take place readily, the generally irreversible nature of the reaction, and that, under most reaction conditions, nitrations are affected by the same reactive species: the nitronium ion. To this list we should add the fact that the products are generally easy to separate and analyze. Furthermore, unlike other electrophilic aromatic substitutions, nitrations generally take place without concurrent or subsequent isomerizations. This feature allows one to relate directly isomeric product yields to the reactivity of the particular position. Complications, however, do arise with highly reactive and/or hindered substrates. Further, initial attack of the nitronium ion at the *ipso*-position often results in the intramolecular migration of the nitro group to the adjacent positions.

Continued interest in nitration is thus well justified and is a testimony to its importance in both preparative as well as mechanistic chemistry.

References

1. Albright, L. F. in *"Encyclopedia of Chemical Technology"*, 3rd ed.; Kirk, Orthmer, eds., Wiley: New York, 1981, Vol. **15**, p. 841.

2. *Faraday's Diary*, T. Martin, ed. London, 1932, Vol. **1**, p. 221, quoted by Schofield, K. in ref. 10b, p. 1.

3. Mitscherlich, E. *Ann. Phys. Chem.* **1934**, *31*, 625; *Ann. Pharm.* **1934**, *12*, 305.

4. For an account see Ward, E. R. *Chem. In Britain* **1979**, *15*, 297.

5. Beilstein, F., Kurbatov, A., *Ber.* **1880**, *13*, 1818, 2029.

6. Haas, H. B. *et al. Ind. Eng. Chem.* **1936**, *28*, 339.

7. Ingold, C. K. *"Structure and Mechanism in Organic Chemistry,"* 2nd ed., Cornell University Press:Ithaca, New York, 1969.

8. Olah, G. A., Kuhn, S. J. in *"Friedel-Crafts and Related Reactions"*, Olah, G. A., ed., Wiley-Interscience:New York, Vol. **2**, 1964, p. 1393.

9. Olah, G. A., *American Chemical Society Symposium Series,* Vol. **22** Albright, F. A., ed., Washington, D.C. *1976*, Chapter 1, pp. 1-47.

10. a) Hoggett, J. G., Moodie, R. B., Penton, J. R., Schofield, K. *"Nitration and Aromatic Reactivity"*, Cambridge University Press:London, 1971; b) Schofield, K. *"Aromatic Nitration"*, Cambridge University Press:London, 1980.

11. a) Feuer, H., ed. *"The Chemistry of the Nitro and Nitroso Groups"*, Vols. **I-II**, in the series of *"The Chemistry of Functional Groups"*, Patai, S., ed., Wiley-Interscience: New York, 1969; b) Patai, S. ed., "Supplement F: The Chemistry of Amino, Nitroso and Nitro Compounds and Their Derivatives," Wiley:New York, 1982.

12. Asinger, F. *"Paraffins"*, Pergamon Press:London, New York, 1968, p. 364.

13. Urbański, T. *"Chemistry and Technology of Explosives"*, Vols. **1-2**, MacMillan Co.:New York, 1964-1965.
14. Salem, L. *J. Chem. Ed.*, **1978**, *55*, 344.
15. Eberson, L. *"Electron-transfer Reactions in Organic Chemistry"*, Springer: Berlin-Heidelberg, 1987.
16. Pross, A. *Acc. Chem. Res.*, **1985**, *18*, 212.
17. a) Shaik, S. S. *J. Am. Chem. Soc.* **1981**, *103*, 3692; *Prof. Phys. Org. Chem.* **1985**, *15*, 197; b) Pross, A. *Adv. Phys. Org. Chem.* **1985**, *21*, 99.
18. Kuhn, L. P., Taylor, N. H., Groggin, P. H. In: *"Unit Processes in Organic Chemistry"* 5th ed. Groggins, P. H. ed., McGraw Hill Book Co., Inc.: New York, 1958, Chapter 5.
19. Albright, L. F. *"Nitration"* In: "Encyclopedia of Chemical Technology", 3rd ed., Kirk-Othmer, eds. Wiley:New York, 1981, Vol. **15**, p. 841.
20. Robinson, R. *"Two Lectures on An Outline of An Electrochemical (Electronic) Theory of the Course of Organic Reactions"*, Institute of Chemistry: London, 1932.
21. Ingold, C. K. *Chem. Rev.* **1934**, *15*, 225.

CHAPTER 2.
Reagents and Methods of Aromatic Nitration*

I. Acid-Catalyzed Electrophilic Nitration

Electrophilic nitrations are carried out by acid-catalyzed reactions of HNO_3 and its derivatives. Nitrating agents are of the general formula NO_2-X, which serve as sources of the nitronium ion, NO_2^+, the effective nitrating agent. Ingold termed the NO_2-X compounds "carriers of the nitronium ion."[1] From the ease of X-elimination he derived a relative sequence of nitrating activity for different nitrating agents as follows: nitronium ion, (NO_2^+) > nitracidium ion $(NO_2^+-OH_2)$ > nitryl chloride, (NO_2-Cl) > dinitrogen pentoxide (NO_2-NO_3) > acetyl nitrate, $(NO_2-O(CO)CH_3)$ > nitric acid, (NO_2-OH) > methyl nitrate (NO_2-OCH_3).

However, the range of nitrating agents, by now is much wider[2] (Table 1). Our discussion in this chapter will review reagents and methods available for aromatic nitration.

Warning: Preparation and manipulation of nitro and particularly of polynitro compounds can potentially lead to explosions. Proper safety precautions are therefore essential particularly when working with acyl nitrates (or acid anhydrides plus nitric acid), alkyl nitrates, nitramines, nitroalkanes and their polynitro compounds. Many nitro compounds are toxic or carcinogenic and are readily absorbed through the skin.

Table 1. Electrophilic Nitrating Agents

NO_2^+ Carrier	Acid Catalyst	NO_2^+ Carrier	Acid Catalyst
HNO_3	H_2SO_4 (mixed acid)	$(CH_3)_2C(CN)ONO_2$	
	$H_2SO_4SO_3$	$(CH_3)_3SiONO_2$	
	H_3PO_4	$RC(O)ONO_2$	
	PPA(polyphosphoric acid)	NO_2F	BF_3
	$HClO_4$	NO_2F	BF_3, PF_5, AsF_5
	HF	NO_2Cl	HF, $AlCl_3$
	$HF-BF_3$		$TiCl_4$
	BF_3	N_2O_3	BF_3
	CH_3SO_3H	N_2O_4	H_2SO_4
	CF_3SO_3H		$AlCl_3, FeCl_3$
	R_FSO_3H		BF_3
	FSO_3H		SbF_5, AsF_5, IF_5
	solid acids (Nafion-H, polystyrene-sulfonic acid)		
		N_2O_5	BF_3
$AgNO_3, NaNO_3$	$FeCl_3, BF_3, AlCl_3, CF_3CO_2H$	$NO_2^+BF_4^-, NO_2^+PF_6^-$ and other salts	
KNO_3, NH_4NO_3,			
$Ti(NO_3)_4$		N-Nitropyridinium salts	
$RONO_2$		N-Nitropyrazole	
$C_2H_5ONO_2$	H_2SO_4, BF_3	9-Nitroanthracene	
	$AlCl_3, SnCl_4, SbCl_5, FeCl_3$		
CH_3ONO_2	BF_3		
		$C_6(CH_3)_6NO_2^+$	HF-TaF_5, Nafion-H

Acid-Catalyzed Electrophilic Nitration

The reactivity of the nitrating agent is also affected by the solvent used. A decreasing activity is observed in the series of solvents of increasing nucleophilicity: sulfuric acid > nitric acid > nitromethane > acetic acid > 1,4-dioxane > water.[3]

The nitronium ion, NO_2^+, as established by Ingold's studies in the 1940s[1] following an early suggestion by Euler,[4] is the reactive nitrating agent in electrophilic nitrations. Forty years of subsequent studies have not changed this picture. (Chapter 3 gives a more detailed discussion of these and other aspects of the mechanism of nitration).

Table 1 summarizes the most frequently used electrophilic nitrating agents.

2.1 Nitric Acid

By far the most common method for nitrating aromatic compounds is with HNO_3 in the presence of acid catalysts.[5] Nitrations can be conducted with neat HNO_3 (*vide infra*), but the presence of a strong acid accelerates the reaction.

Anhydrous HNO_3, best prepared in pure form from N_2O_5 with equimolar water, was shown to attain the equilibrium[5]

$$2HNO_3 \rightleftharpoons NO_2^+ + NO_3^- + H_2O$$

Anhydrous HNO_3 is an effective nitrating agent. Aromatic hydrocarbons as well as their nitrated derivatives have appreciable solubility in 100% HNO_3 and nitration proceeds homogeneously. Benzene and toluene, when rapidly nitrated at 0°C also give the dinitro products with no detectable trinitro products. Ciaccio and Marcus studied the kinetics of nitration of nitrobenzene with 100% HNO_3 at −13°C.[6] The second-order rate constant for nitration was roughly ten times larger than that extrapolated for nitration in 90% H_2SO_4 from data at higher temperatures, probably as a result of medium effects that are not fully understood. The amount of phenolic byproducts are also considerably less than those formed under the typical mixed acid nitrations. These features of HNO_3 nitration may make it attractive for production of nitroaromatics. The chief drawback of using 100% HNO_3 is that its reactivity drops substantially as water is produced during the nitration reaction. Ridd has studied the nitration of a number of aromatic substances in aqueous HNO_3 systems of strengths ranging from 64% to 100%.[7] At lower acidities, reactive substrates, such as phenol, anisole, and mesitylene, and at higher acidities quaternary anilinium salts were nitrated efficiently. The desirability for a nitration process that is not dependent on H_2SO_4 regeneration/disposal prompted several reports for nitrations with neat HNO_3. Central to the application of these nitrations is the question of regeneration of the spent HNO_3. The use of N_2O_4 followed by air–oxidation has been suggested.[8] Harrar and Pearson have described a system for anodic oxidation of N_2O_4 in anhydrous HNO_3 to generate N_2O_5.[9] When coupled with the nitration reaction, this system offers a means for nitration in anhydrous HNO_3 using N_2O_4 as the nitrogen source to reconvert the water that is formed into HNO_3. Azeotropic removal of water from nitric acid nitration over solid superacid catalyst is discussed in Section 2.15.

Nitroaromatics are soluble in concentrated HNO_3 making them difficult to recover on a practical scale. Distillation is one possible approach. Carr *et al.* described a method[10] for recovering nitroaromatics from the spent acid by decomposing it with NO.

$$2HNO_3 + NO \longrightarrow 3NO_2 + H_2O$$

HNO_3 is decomposed only to the extent necessary to cause the nitroaromatics to separate and the off–gases are reacted with oxygen to regenerate anhydrous HNO_3.

When HNO_3 is used in acetic anhydride solution, intermediate formation of acetyl nitrate ($CH_3CO(O)NO_2$) becomes a contributor to the nitration reactions.

$$(CH_3CO)_2O + HNO_3 \rightleftharpoons CH_3CO(O)NO_2 + CH_3CO_2H$$

Subsequent ionization gives the nitronium ion

$$CH_3CO(O)NO_2 \rightleftharpoons NO_2^+ + CH_3CO_2^-$$

Acetyl nitrate, like benzoyl nitrate and other acyl nitrates, was also prepared in pure form and used in nitration of aromatics. Due to their instability great care is needed when using acyl nitrates in isolated form; therefore, *in situ* preparation, such as from acyl chlorides and silver nitrate, is preferred. Nitrations with acyl nitrates are discussed subsequently (see Section 2.21).

Trifluoroacetic acid has a much higher acidity than acetic acid and is comparable to that of HNO_3. It thus can protolytically ionize HNO_3, in a manner similar to other strong acids. HNO_3 in trifluoroacetic anhydride was used by Tedder in nitrating aromatics.[11] The reaction presumably proceeds through formation of trifluoroacetyl nitrate, but has not gained wide use.

$$(CF_3CO_2)_2O + HNO_3 \rightleftharpoons CF_3CO(O)NO_2 + CF_3CO_2H$$

HNO_3 in trifluoromethanesulfonic (triflic) anhydride was found to be an exceedingly reactive nitrating system.[12] In all probability it forms the nitronium ion according to

$$(CF_3SO_2)_2O + HNO_3 \longrightarrow CF_3SO_2(O)NO_2 + CF_3SO_3H$$
$$\updownarrow$$
$$NO_2^+ CF_3SO_3^-$$

2.2 Nitric Acid-Graphite (Graphite Nitrate)

The intercalation of HNO_3 into graphite gives "graphite nitrate" $(C_{12}HNO_3)_n$ which was reported to be a nitrating and oxidizing agent under heterogeneous conditions.[13] Alkylbenzenes, anisole, and phenol are nitrated in moderate yield giving products with usual isomer distribution. This would indicate that nitration is not taking place in the graphite layers but at the surface.

2.3 Nitric-Sulfuric Acid (Mixed Acid)

Nitration of aromatic compounds with nitric-sulfuric acid, the so-called "mixed acid" is the most frequently used method of nitration of aromatic compounds. H_2SO_4 aids the ionization of HNO_3 to the nitronium ion, the *de facto* nitrating agent, and also binds the water formed in the reaction. A convenient way of altering the nitrating activity is by changing the concentration of the H_2SO_4. A number of excellent reviews are available.[14] Therefore, we will not review in detail this extensive field except to emphasize some aspects. The reader is referred to the reviews for further information and details.

It has been shown that the regioselectivity of nitration of aromatics by "mixed acid" can be changed by selecting a suitable cosolvent.[15] Studies have also shown that the regioselectivity of nitration can be altered by altering the concentration of H_2SO_4.

Table 2 shows the effect of solvent and acid concentration on the ortho/para isomer ratio of nitro products in the nitration of toluene, ethylbenzene, isopropylbenzene, and *tert*-butylbenzene.[16] For comparison data for nitration with neat HNO_3 are also included.

Table 2. Isomer Distribution in Acid Catalyzed Nitration of Alkylbenzenes at 25°C

Compound	Reagent	Percent Isomer Distribution			½ o/p ratio
		ortho	meta	para	
Toluene	100% HNO_3	57.5	4.6	37.9	0.76
	HNO_3-CH_3NO_2	61.5	3.1	35.4	0.87
	HNO_3-AcOH	56.9	2.8	40.3	0.71
	HNO_3-sulfolane	61.9	3.5	34.7	0.89
	HNO_3-CF_3CO_2H	61.6	2.6	35.8	0.86
	30% of 1:1-mixed acid-sulfolane	62.0	3.4	34.6	0.89
	30% of 1:1-mixed acid-AcOH	56.5	3.1	40.4	0.70
	75% of 1:1-mixed acid-sulfolane	56.3	2.6	41.0	0.69
	75% of 1:1-mixed acid-AcOH	58.1	1.9	40.0	0.73
	Heterogeneous nitration with 1:1-mixed acid	56.4	4.8	38.8	0.73
Ethylbenzene	30% of 1:1-mixed acid in sulfolane	50.3	3.6	46.1	0.55
	75% of 1:1-mixed acid in sulfolane	44.7	2.0	53.3	0.42
	Mixed acid-CH_3NO_2	47.6	4.0	48.4	0.49
Isopropyl-benzene	HNO_3-H_2SO_4	30	7.7	62.3	0.24
	30% of 1:1-mixed acid in sulfolane	43.2	4.5	52.3	0.41
	Mixed acid-CH_3NO_2	22.9	6.4	70.7	0.16
	HNO_3-60.3% H_2SO_4	26.0	7.9	66.1	0.19
	HNO_3-66.4% H_2SO_4	25.5	7.5	66.9	0.19
	HNO_3-74.5% H_2SO_4	25.9	6.6	67.5	0.19
tert-Butyl-benzene	Mixed acid-CH_3NO_2	11.1	10.4	78.5	0.07
	HNO_3-CH_3NO_2	12.2	8.2	79.6	0.07

Slightly differing ortho/para ratios most probably reflect the solvation of the reactive NO_2^+ and thus somewhat varying steric requirements.

Manufacture of mononitrotoluenes on an industrial scale is conducted in 58% H_2SO_4. At this acidity no appreciable amounts of the dinitro products are formed. The production of dinitrotoluenes is carried out in 65% H_2SO_4. These reactions are heterogeneous in nature as the aromatic has only a limited solubility in the mineral acid mixture.[17] The rate of nitration in commercial reactors is therefore often limited by diffusion of the aromatic into the acid phase, or at least to the interface. Rys has studied the regiochemistry of mixed-acid nitrations and found it to be strongly influenced by mixing effects.[18]

One of the largest industrial applications of aromatic nitration is in the production of 2,4- and 2,6-dinitrotoluenes which are then converted to toluene diisocyanates for the manufacture of polyurethane foam.[19] Any meta-nitrotoluene formed during the first stage of nitration leads to 3,5-dinitrotoluene, which is an undesired isomer. Hence, a considerable effort has been extended to study means of suppressing meta-nitration of toluene.

2.4 Nitric Acid-Oleum

Nitric acid-oleum is an extremely active nitrating agent. Its use, however, is limited to the nitration of highly deactivated aromatics since oleum otherwise can also cause sulfonation and oxidation of more reactive aromatics. 1,3-Dinitrobenzene is nitrated to 1,3,5-trinitrobenzene at 110°C during prolonged heating with anhydrous HNO_3 and 60% oleum in 71% yield.[20]

2.5 Nitric-Phosphoric (Polyphosphoric) Acid

Phosphoric acid is a weaker acid than H_2SO_4 and thus higher weight percentages of H_3PO_4 compositions are needed to bring about comparable nitration in aqueous solutions. The acidity functions of >80% H_3PO_4 are not known and can only be extrapolated.[21] Higher concentration of H_3PO_4 solutions, generally prepared by adding P_2O_5 to H_3PO_4, are referred to as polyphosphoric acid (PPA). PPA is considered to be a mixture of polymers of the composition

$$\text{HO} - \overset{\overset{\displaystyle O}{\|}}{\underset{\underset{\displaystyle OH}{|}}{P}} - O {\left[\overset{\overset{\displaystyle O}{\|}}{\underset{\underset{\displaystyle OH}{|}}{P}} - O \right]}_n \overset{\overset{\displaystyle O}{\|}}{\underset{\underset{\displaystyle OH}{|}}{P}} - \text{OH}$$

where n is usually 1–7 or higher. Commercial PPA has a H_3PO_4 equivalent of 82–84%. PPA is a syrupy, viscous liquid which does not form homogeneous systems with HNO_3. It is also frequently used as a solid supported acid.[22]

The regioselectivity of the nitration of toluene can be altered by using increasingly stronger H_3PO_4 resulting in decreasing ortho over para nitration as compared with H_2SO_4[23] (Table 3).

The reason in all probability is the polymeric nature of H_3PO_4 at higher concentrations causing steric hindrance to ortho nitration.

Tsang *et al.* carried out nitrations by continuously removing the byproduct water with P_2O_5.[24] Nitration of toluene with HNO_3 and P_2O_5

Table 3. Regioselectivity of the Nitration of Toluene with HNO_3–H_3PO_4[23]

Reagent	Percent isomer distribution			½ o:p
	ortho	meta	para	
HNO_3-89.9% H_3PO_4	46.0	4.6	39.3	0.58
HNO_3-91.8% H_3PO_4	43.6	4.8	41.4	0.53
HNO_3-96.7% H_3PO_4	42.8	4.7	43.2	0.49
HNO_3-55% H_2SO_4	61.0	3.7	30.9	0.98

Table 4. Nitration of Toluene with Nitric Acid and Phosphorous Pentoxide at Room Temperature[24]

Solvent	C_7H_8-HNO_3 (moles)	P_2O_5-HNO_3 (moles)	Time (days)	o-p ratio	m-$C_7H_7NO_2$ (percent)	Yield (percent)
$CHCl_3$	3	1.0	2	0.84	3.7	85
$CHCl_3$	3	0.5	1	0.99	2.6	49
$CHCl_3$	12	7.9	3	0.82	2.4	78
$CHCl_3$	4	3.2	3	1.07	–	84
CH_3NO_2	3	1.6	1	1.30	–	73

led to moderate to high yields of nitrotoluenes with low ortho–para ratio. The role of solvent, nature of the phosphorous species, surface area of the dehydrating agent, and stoichiometry was examined. It was found that the ortho–para ratio of nitrotoluenes is dependent upon the polarity of the solvent, being lower in chloroform than in the more polar nitromethane (Table 4).

PPA is also responsible for the change in regioselectivity (Table 5). The ratio of toluene and nitric acid has no effect on the regioselectivity.

At least an equimolar amount of P_2O_5 is necessary to obtain low ortho–para ratios. Increasing the surface area of P_2O_5 by ball milling increases the ortho–para ratio from 0.84 to 1.07. It is interesting to note that nitration with N_2O_5 provides an ortho–para ratio of 1.04, as the reaction of HNO_3 with P_2O_5 is known to yield N_2O_5. A P_2O_5 content of 86.5% in PPA was found to be optimum for obtaining the lowest ortho–para ratio and highest yield of nitrotoluenes (Table 6).

No clear mechanistic conclusions can be drawn from these nitrations in the highly viscous, heterogeneous reaction media where NO_2^+ is strongly solvated (probably as a tight ion pair) by an increasingly bulky polyphosphate.

It is also not clear what possible role the intercalation of toluene

Table 5. Nitration with Nitric Acid and Polyphosphoric Acid[24]

C_7H_8-HNO_3 (moles)	Solvent	Time (days)	o-p ratio	Yield (percent)
3	$CHCl_3$	1	1.16	42
3	None	2	0.86	45
2	None	1	0.87	44

Table 6. Effect of Phosphoric Acid Polymer Species on o-p Ratio[24]

Weight perecent of P_2O5 in H_3PO_4	Polymer species (percent)	o-p Ratio	Yield (percent)
None	None	1.8	0.7
76.2	Monomer, 49; dimer, 42	1.10	31.7
80.0	Monomer to tetramer, 92	1.04	56.6
83.0	Dimer to heptamer, 79	1.01	86.5
86.5	Decamer and Higher, 68	1.03	89.5

or interaction with the heterogeneous surface of PPA has in controlling the regiochemistry of the reaction. However, irrespective of the significance of various factors, the regioselectivity of nitration can be altered by changing the nature (physical and/or chemical) of the nitrating agent. In spite of the poorly understood nature of the nitrating system, Tsang's work[24] further shows that a remarkable change in regioselectivity can be obtained by changing the nature of the nitrating agent. In fact, by changing from HNO_3 to bulkier alkyl nitrates (such as neopentyl nitrate), the ortho-para ratio can be further changed from 1.30 to 0.49. Alkyl nitrate nitrations are discussed in more detail in Section 2.18.

Pearson *et al.* have reported[26] the nitrating ability of a mixture of sodium nitrate, trimethyl phosphate, and P_2O_5 for the nitration of anthracene and 9-methylphenanthrene. Unfortunately, they did not carry out further studies to demonstrate the utility of this system and its effect on regioselectivity.

2.6 Nitric-Perchloric Acid

Aqueous $HClO_4$ has limited use in catalyzing HNO_3 nitration. It was used in nitrating reactive aromatics such as 1,3,5-trimethoxybenzene and 3,5-dimethoxytoluene.[27] A disadvantage is the increasing danger of explosion acompanying use of $HClO_4$ in concentrations greater than 72%.

HNO_3-concentrated $HClO_4$ played a significant role in the development of our knowledge of acid catalyzed nitrations.[28] Hantzsch in his pioneering studies used this system in an attempt to identify the reactive nitrating agent.[29] From mixtures of appropriate proportions of anhydrous HNO_3 and $HClO_4$ he isolated two salts whose structure he proposed as

$$HNO_3 + HClO_4 \longrightarrow H_2NO_3^+ ClO_4^-$$

$$HNO_3 + 2HClO_4 \longrightarrow H_3NO_3^{2+} 2ClO_4^-$$

The salts had high electrical conductivity and it was claimed that the values of the molar conductances at infinite dilution indicated formation of binary and ternary electrolytes, respectively.

Ingold and his associates reinvestigated Hantzch's work and found that the isolable perchlorate salt of the composition $H_3NO_3^{2+}2ClO_4^-$ could be separated into nitronium perchlorate ($NO_2^+ClO_4^-$) and hydronium perchlorate ($H_3O^+ClO_4^-$) by fractional recrystallization from nitromethane solution.[30] The structure of the pure nitronium salt was established by X-ray crystallography.[31] Raman spectroscopic studies established unequivocally the formation of the NO_2^+ in strongly acidic solutions of HNO_3.[32]

Ingold and co-workers at the same time were unable to obtain the compound of the composition $H_2NO_3^+ClO_4^-$, i.e., nitracidium perchlorate. To this date, protonated HNO_3 (nitracidium ion) has not been observed spectroscopically and no salt has been isolated. A weakly bound hydrate of the nitronium ion, i.e., $O_2^+N \cdot \cdot H_2O$, can, however, form. It was suggested that because HNO_3 is difficult to remove from nitronium perchlorate, Hantzsch in fact may have obtained a mixture of the observed composition by chance. Since these earlier observations numerous nitronium salts have been obtained and used as nitrating agents (see Section 2.24).

As anhydrous HNO_3-$HClO_4$ mixtures are also strong oxidizing agents for organic compounds, they never gained wide use and extreme caution is needed. Aqueous $HClO_4$ is also little used and still dangerous. Nitronium perchlorate itself is a potentially explosive compound (as it forms an equilibrium with the covalent perchlorate, see Section 2.24). One of us remembers Ingold's recollection that a sample of the isolated salt exploded overnight causing substantial damage and ending his studies with nitronium perchlorate.

2.7 Nitric Acid-Hydrogen Fluoride

Gillespie and Millen showed that ionization of HNO_3 to NO_2^+ takes place in anhydrous HF.[33]

Simons, Passino and Archer studied the nitration of benzene by HNO_3 in anhydrous HF.[34] At 0°C benzene gave 85% nitrobenzene, but nitrobenzene was not nitrated further at this temperature. Sodium nitrate in anhydrous HF is a convenient way to carry out HNO_3-HF nitrations. Sodium nitrate in HF nitrates benzene at 0°C.[35] Mesitylene is efficiently nitrated in liquid anhydrous HF with sodium nitrate at 5°C to 2,4-dinitro-1,3,5-trimethylbenzene.[36] 2,4-Difluoro-1,3,5-trimethylbenzene is also nitrated very smoothly under similar conditions.[37]

2.8 Nitric Acid-Hydrogen Fluoride-Boron Trifluoride

Realizing the difficulties involved with HNO_3-$HClO_4$ and the limited nitrating ability of the HNO_3-anhydrous HF system, Olah and Kuhn in 1956 introduced HNO_3-anhydrous HF-BF_3 as an extremely effective and safe nitrating agent.[38] HNO_3 ionizes with HF-BF_3 according to

$$HNO_3 + HF + 2BF_3 \longrightarrow NO_2^+BF_4^- + BF_3 \cdot H_2O$$

Nitronium tetrafluoroborate can be isolated as stable salt (see Section 2.24) and used as such as nitrating agent, or the system can be used for *in situ* nitration of aromatics.

$$ArH + HNO_3 + HF + BF_3 \longrightarrow ArNO_2 + H_3O^+BF_4^-$$

BF_3 can be readily regenerated from its hydrate by distilling it from H_2SO_4 and thus the reaction can be made catalytic with recycling of the acid.

Instead of BF_3 other Lewis acid fluorides such as PF_5, TaF_5, NbF_5, and SbF_5 can be also used effectively in related nitrations (see Section 2.22).

2.9 Nitric Acid-Boron Trifluoride

Hennion *et al.* first nitrated aromatics, including nitrobenzene to dinitrobenzene, with HNO_3-BF_3.[39]

$$ArH + HNO_3 + BF_3 \longrightarrow ArNO_2 + BF_3 \cdot H_2O$$

Csürös *et al.* nitrated *o*-nitrotoluene to dinitrotoluenes with the system.[40] It was established that $BF_3 \cdot H_2O$ formed in the reaction (and also present by complexing any water initially present in HNO_3) promotes the reaction. Thus BF_3 by itself does not directly ionize HNO_3

$$HNO_3 + BF_3 \rightleftharpoons NO_2^+BF_3OH^-$$

but it is $BF_3 \cdot H_2O$ (or $BF_3 \cdot 2H_2O$) which protonates HNO_3 leading to formation of NO_2^+ ion, similar to other strong protic acids.

$$HNO_3 + BF_3 \cdot H_2O \rightleftharpoons H_2NO_3^+BF_3OH^- \xrightarrow{BF_3} NO_2^+BF_3OH^- \text{ (or } BF_4^-\text{)} + BF_3 \cdot H_2O$$

$$ArH + NO_2^+BF_3OH^- \longrightarrow ArNO_2 + BF_3 \cdot H_2O$$

Further, it is also very probable that small amounts of HF, which is always present as impurity in the system, can then form the ternary $HNO_3/HF/BF_3$ system.

2.10 Nitric-Trifluoroacetic Acid

Trifluoroacetic acid is a much stronger acid than acetic acid and gives reactive nitrating mixtures with HNO_3 for many aromatics. The acid strengths of HNO_3 and CF_3CO_2H are comparable.[41] The indicated ionization equilibria are

$$CF_3CO_2H + HNO_3 \xrightleftharpoons{-H_2O} NO_2^+CF_3CO_2^- \rightleftharpoons CF_3CO(O)NO_2$$

Despite the advantageous properties of the system it is only occasionally used for aromatic nitration.[42-45] Trifluoroacetyl nitrate, however, formed also in metal nitrate–trifluoroacetic anhydride nitrating mixtures is a frequently used nitrating agent (see Section 2.21.3).

2.11 Nitric-Methanesulfonic Acid

HNO_3 in methanesulfonic acid was reported as a nitrating agent of limited use.[6,46] Methanesulfonic acid, although weaker than H_2SO_4, has the advantage that aromatics are generally more soluble in it and no sulfonation or oxidative side reactions occur. Trifluoromethanesulfonic acid (CF_3SO_3H), however is a much stronger and therefore more suitable acid for nitration.

2.12 Nitric-Trifluoromethanesulfonic (Triflic) Acid

Trifluoromethanesulfonic acid (triflic acid) is one of the strongest known Bronsted acids. H_o for CF_3SO_3H is −14.5, thus comparable to that of fluorosulfuric acid. At the same time while fluorosulfuric acid, like H_2SO_4, is also a powerful sulfonating and oxidizing agent, CF_3SO_3H does not react with aromatics.[47] It is therefore a most convenient strong acid for nitration with HNO_3, which is completely ionized by it.

During an investigation on the effect of acids on the regioselectivity of nitration of toluene, Coon, Blucher, and Hill found that two equivalents of CF_3SO_3H react with 100% HNO_3 to yield a white crystalline solid, which is a mixture of nitronium trifluoromethanesulfonate and hydronium trifluoromethanesulfonate.[48]

$$2CF_3SO_3H + HNO_3 \longrightarrow NO_2^+CF_3SO_3^- + H_3O^+CF_3SO_3^-$$

The mixture of HNO_3 and CF_3SO_3H in CH_2Cl_2, CCl_4, CF_2Cl_2, $CFCl_3$, and pentane solution is an excellent nitrating agent for benzene, toluene, *m*-xylene, chlorobenzene, nitrobenzene, and benzotrifluoride (Table 7). The reactions were carried out from −110 to 30°C. Mono or dinitration of toluene can be controlled by choosing the reaction temperature. Mononitration of toluene is extremely rapid, the reaction being complete in one minute at −110°C. The dinitration is complete in 30 minutes at 0°C.

Table 7. Nitration of Aromatics with Nitric Acid–Trifluoromethanesulfonic Acid[48]

Substrate	Product	Distribution
Benzene	Nitrobenzene	98
	Dinitrobenzene	2
Toluene	2-nitrotoluene	50-62
	3-nitrotoluene	0.2-0.5
	4-nitrotoluene	37-44
Nitrobenzene	1,2-dinitrobenzene	10
	1,3-dinitrobenzene	87
	1,4-dinitrobenzene	2
Chlorobenzene	2-nitrochlorobenzene	30
	3-nitrochlorobenzene	0.1
	4-nitrochlorobenzene	70
Benzotrifluoride	2-nitrobenzotrifluoride	14
	3-nitrobenzotrifluoride	85
	4-nitrobenzotrifluoride	0.1

The regioselectivity of mononitration of toluene was studied by varying the reaction temperature. At −60°C, −90° and −110°C in halomethane solvents, mononitrotoluenes were obtained containing only 0.53, 0.36 and 0.23% meta–nitrotoluene, respectively. The ortho–para isomer ratio is also low at −110°C (Table 8).

Table 8. Mononitration of Toluene with HNO_3/CF_3SO_3H in Halomethane Solvents[48]

Solvent	Time (min)	Temperature (°C)	Yield (percent)	Isomer ratios (percent)		
				o-	m-	p-
$CFCl_3$	180	−110	>99	50.5	0.2	49.3
CH_2Cl_2	180	−90	>99	61.3	0.4	38.3
CH_2Cl_2	60	−60	>99	62.1	0.5	37.4
$CFCl_3$	1	−60	>99	61.9	0.5	37.6
$CFCl_3$	1	−110	>99	50.8	0.2	49.0

Table 9. Preparation of Dinitrotoluenes in Nitrating Mixtures Containing CF_3SO_3H[48]

Composition of nitrating mixture (weight percent)			Temperature (°C)	Yield (percent)	Isomer distribution				Total meta (percent)
CF_3SO_3H	H_2SO_4	HNO_3	H_2O			2,6	2,3-2,5	2,4	3,4	
89.0	0	11.0	0	−5	>98	15.7	0.5	82.8	1.0	1.5
45.5	45.5	6.0	3.0	−20	>98	10.2	0.3	88.7	0.9	1.0
45.5	45.5	6.0	3.0	−20	>98	14.9	0.6	83.4	1.1	1.7
22.7	68.3	6.0	3.0	−20	99.5	12.1	0.4	86.4	1.6	1.6
0	90.6	6.3	3.1	−25	99	11.8	0.5	86.5	1.3	1.8
80	0	10	10	20	99.2	16.8	1.0	1.3	2.2	—
65	0	5	30	0	100	o-/m-/p-MNT[a] = 58.86/1.96/39.18 o/p = 1.5				

[a]NMT = Mononitrotoluene

Low meta substitution allows favorable regiocontrol in the subsequent preparation of dinitrotoluenes. In general, the HNO_3–CF_3SO_3H system shows less meta substitution than other nitrating systems at comparable temperatures (Table 9).

The major factor, however, affecting low meta nitration is the use of extremely low temperatures. Solubility of the formed nitronium salt at low temperature in halomethane solutions is limited and unusual ortho/para ratios may be also a consequence of the heterogeneous nature of the reaction mixtures.

Nitronium triflate ($NO_2^+CF_3SO_3^-$) is more soluble in methylene chloride than other studied nitronium salts, and this is reflected in the isomer distribution of the nitration of toluene (Table 10).

Table 10. Nitration of Toluene with Nitronium Salts in Methylene Chloride[48]

Nitronium salt	Temperature (°C)	Time (min)	Yield (percent)	Isomer ratio (percent)		
				o-	m-	p-
$NO_2^+BF_4^-$	-65	150	70.2	56.5	0.7	42.8
$NO_2^+PF_6^-$	-65	150	88.5	46.4	0.8	52.8
$NO_2^+CF_3SO_3^-$	-60	1	>99	62.2	0.5	37.3

Olah recently found CF_3SO_3H–$B(O_3SCF_3)_3$ (triflatoboric acid) as a highly efficient new superacid for nitrating aromatics with HNO_3. Nitronium tetratriflatoborate $NO_2^+B(CF_3SO_3)_4^-$ is the reactive nitrating agent.[12]

2.13 Nitric–Fluorosulfuric Acid

Usual mixed acid nitrations give water as the byproduct in forming the NO_2^+ ion. The effectiveness of mixed acid is continuously decreased during the progress of the nitration reaction due to dilution of the acid by the water formed.

Coon et al. used fluorosulfuric acid (FSO_3H) in the nitration of toluene with HNO_3 and compared its effectiveness with that of CF_3SO_3H[48] (Table 11). They found that both acids were effective, but trifluoromethanesulfonic acid is more suitable since it does not cause oxidation or sulfonation.

Table 11. Comparison of the Nitration of Toluene
with HNO_3/CF_3SO_3H and HNO_3/FSO_3H in
CH_2Cl_2 Solution at $-60°C$[48]

Acid	Time (min)	Yield mononitrotoluene percent	Percent isomer distribution
CF_3SO_3H	60	100	o/m/p = 62/0.5/37
FSO_3H	120	89	o/m/p = 63/0.7/36

Olah *et al.* found that a mixture of HNO_3 and FSO_3H (or ternary HNO_3 + HF + FSO_3H) allows the trinitration of benzene to 1,3,5-trinitrobenzene at higher temperatures.[49] Water formed in the ionization of HNO_3 to the NO_2^+ ion reacts with FSO_3H (to H_2SO_4 and HF), and thus the nitrating system maintains high acidity. HNO_3-FSO_3H is also a very suitable strong nitrating system for other deactivated aromatics.

2.14 Nitric–Magic Acid (FSO_3H-SbF_5)

Adding Lewis acid fluorides, such as antimony, tantalum, or niobium pentafluoride to FSO_3H greatly enhances its acidity. FSO_3H-SbF_5 (magic acid) is one of the strongest known superacids.[47] Nitric–magic acid ($HNO_3-FSO_3H-SbF_5$) is an extremely effective nitrating agent for polynitration of aromatics.[12]

2.15 Nitric Acid–Solid Acid Catalysts

Nitration with HNO_3 in the presence of strong protic acids such as H_2SO_4, FSO_3H, and CF_3SO_3H or Lewis acids such as BF_3 requires subsequent separation of spent acid (due to water formed in the reaction) and neutralization of acid left in the product. One is generally left with a large amount of dilute acid for disposal, which is neutralized in the case of H_2SO_4 catalyzed nitrations to a mixture of ammonium nitrate and ammonium sulfate. By using a solid acid catalyst most of these problems can be eliminated. The solid acid catalyst is simply separated and recycled for subsequent use.

Kameo et al. reported[50a] the use of polystyrenesulfonic acid as a catalyst in the nitration of aromatics with HNO_3. Nitration of toluene with 90% HNO_3 over dried sulfonated polystyrene resin (Rohm and Haas amberlite IR-120) was also reported by Wright et al. at 65-70°C to give an ortho-para isomer ratio of only 0.68, much lower than usual ortho-para ratios in acid catalyzed nitrations.[50b] It is thought that the NO_2^+ ion is strongly ion-paired to the resinsulfonic acid. The ion-pair salt thus formed is much bulkier than the "free" NO_2^+ ion or such nitronium salts as $NO_2^+BF_4^-$.

$$HNO_3 + 2P\text{-}SO_3H \longrightarrow NO_2^+P\text{-}SO_3^- + H_3O^+P\text{-}SO_3^-$$

This method, however, is of limited use because the catalyst undergoes degradation during the reaction. Polystyrene has benzylic hydrogens that can be abstracted easily by the highly reactive species (nitrogen oxides, NO_2^+, NO^+) present in the reaction medium. The polymer therefore readily undergoes oxidative degradation. In addition, it is likely that the polymer also can undergo nitration, sulfonation, and subsequent degradation under the reaction conditions.

Nitric acid adsorbed in silica gel was found to be an effective nitrating agent for activated aromatics, such as phenols and arylmethyl ethers.[51]

When aromatics are nitrated with mixed acid, the reaction rate slows down with time, because the byproduct water dilutes the acid, thus reducing its reactivity. In preparative nitrations therefore, a large excess of acid is required, with the excess being wasted because of dilution. The disposal of spent acid also represents a significant environmental problem.

In view of these considerations and the limitations of polystyrenesulfonic acids, Olah et al. studied superacidic Nafion-H perfluorosulfonic acid resin catalyzed nitration of aromatics with HNO_3 under conditions of azeotropic removal of water (azeotropic nitration).[52] The azeotropic removal of water in Nafion-H catalyzed nitration allows utilization of HNO_3 to a significantly greater extent than under conventional conditions of nitration. Both fuming and concentrated HNO_3 are effective (Table 12).

The nitrations are carried out by heating the reaction mixture to reflux and by azeotropically distilling off the water-aromatic mixture until no HNO_3 was left in the reaction mixture. Part of the HNO_3, however, also distills over in the form of a binary or ternary azeotrope, as do some nitrogen oxides formed under the reaction conditions.

Table 12. Azeotropic Nitration of Aromatics with Nitric Acid over Nafion–H Catalyst[52]

Substrate	Yield (percent)	Isomer Distribution (percent)
Benzene	77	
Toluene	80	2-nitro (56)
		3-nitro (4)
		4-nitro (40)
o-Xylene	47	3-nitro (45)
		4-nitro (55)
m-Xylene	68	2-nitro (15)
		4-nitro (85)
p-Xylene	60	
Mesitylene	79	
Chlorobenzene	87	2-nitro (38)
		3-nitro (1)
		4-nitro (61)

2.16 Nitric Acid–Supported Acid Catalysts

Kemeo et al. have studied[50a] the nitration of toluene and other substituted benzenes with nitric acid–toluene-2,4-disulfonic acid supported on various supports. Nitrations using Celite-545 as a support provided lower ortho/para ratios than the homogeneous mixed acid nitrations under similar conditions. Nitration of o-xylene also gave increased amount of the less hindered 4-nitro–o-xylene. It was shown that other supports can also be used in the nitration yielding the desired mononitrotoluenes in moderate to high yields. Other aromatic sulfonic acids including polystyrenesulfonic acid supported on Celite-545 were also used.

Nitration of toluene with HNO_3 in the presence of calcium sulfate (Drierite) resulted in the formation of nitrotoluenes with an ortho–para ratio of 0.78, about half of that obtained under conventional mixed acid nitrations.[53] Nitration of other aromatics also showed that the nitric acid–calcium sulfate system, in general, provides increased amounts of the less hindered nitro compounds.

2.17 Monodentate Metal Nitrates

Nitrations with metal nitrates is dependent on the bonding nature of the nitrate ligand. Metal nitrates with monodentate nitrate ligands are inactive while those with bidentate nitrate ligands are extremely reactive and react with aromatics without catalysts (see Section 2.31).

Nitration with monodentate metal nitrates requires Lewis acid or Bronsted acid catalysis, because the nitrate group is present as NO_3^- and requires activation to convert it into an electrophile.

Topchiev et al. first explored the nitration of aromatic hydrocarbons with metal nitrates in the presence of Lewis acid catalysts.[54]

$$ArH \xrightarrow{MNO_3,\ AlCl_3} ArNO_2$$

They studied a number of nitrates and found the order of reactivity to be

$$AgNO_3 > KNO_3 > NaNO_3 > NH_4NO_3 > Pb(NO_3)_2 > Ba(NO_3)_2$$

The reaction is slightly exothermic, the temperature normally rising to 30–40°C. Topchiev et al. also studied the effect of different Lewis acids on product yields, and found $AlCl_3$ and BF_3 to be the best catalysts for nitration.

The Lewis acid complexes with the metal nitrate, thereby generating a nitronium salt or a polarized complex capable of nitrating aromatics

$$KNO_3 + AlCl_3 \longrightarrow NO_2^+ AlCl_3OK$$
$$\text{or}$$
$$3\ KNO_3 + 8\ AlCl_3 \longrightarrow 3\ NO_2^+ AlCl_4^- + 3\ K^+ AlCl_4^-$$

$$M\text{—}O\text{—}NO_2 + AlCl_3 \longrightarrow \underset{\delta^+}{M\text{—}O\text{—}NO_2} \overset{\overset{\delta^-}{AlCl_3}}{\uparrow}$$

Another possibility is Lewis acid catalyzed nitration by nitrogen oxides formed in the system. Indeed, Sprague *et al.* have shown the formation of a $N_2O_5 \cdot 2BF_3$ complex on reaction of KNO_3 with BF_3.[55] Any of these reactive intermediates could be responsible for the nitration of aromatics.

One of the main disadvantages of nitrating aromatics with metal nitrates is that the reaction is generally heterogeneous. It seems that the yields are dependent upon the nature of the Lewis acid and, to a significant extent, on the solubility of the nitrate in the reaction medium. An extension of Topchiev's work has shown that $TlNO_3/BF_3$ is the most efficient nitrating agent among a wide variety of nitrates examined.[56]

The solubility of silver nitrate in acetonitrile allows one to readily carry out nitrations with $AgNO_3/BF_3$.[57] Nitrations with this reagent system are homogeneous, and the silver salt can be recovered as $AgBF_4$.

$$ArH \xrightarrow{AgNO_3/BF_3} ArNO_2$$

$AgNO_3/BF_3$ has been used to achieve selective mononitration of polymethylbenzenes in high yield (Table 13) similarly to methyl nitrate–BF_3 nitration (*vide infra*).

It is very possible that $AgNO_3/BF_3$ nitrations do not proceed via the free NO_2^+ ion, but rather through a polarized complex (also indicated by ^{11}B NMR) which is more selective than $NO_2^+BF_4^-$.

Higher than usual ortho–para ratios in the nitration of anisole, fluorobenzene and chlorobenzene may also suggest "coordinative ortho nitration" with the silver ion first coordinating with oxygen or halogens.

Table 13. BF$_3$-Catalyzed Mononitration of Polymethylbenzenes[57]

Polymethylbenzene	Yield of mononitro product (percent)	
	CH$_3$NO$_3$ in CH$_3$NO$_2$	AgNO$_3$ in CH$_3$CN
pentamethylbenzene	99	92
tetramethylbenzene	95.4	86
mesitylene (1,3,5- with extra CH$_3$)	97.5	83
1,2,4-trimethylbenzene	97.0	87

The AgNO$_3$/BF$_3$ system seems to represent the only successful attempt to examine the metal nitrate nitration of aromatics under homogeneous conditions.

Nitrations with metal nitrates can also be carried out under Bronsted acid catalysis. Such nitrations have been carried out using trichloroacetic,[58] trifluoroacetic,[59] polyphosphoric,[60] and sulfuric acid.[61] Among the nitrates used were alkali metal nitrates, ammonium nitrate, and urea nitrate.

A high-temperature reaction studied is the nitration of benzene with a eutectic melt of NaNO$_3$ and KNO$_3$ containing sodium metaphosphate catalyst at 470°C.[62] Under the reaction conditions, some biphenyl is also obtained in addition to nitrobenzene, presumably because of strong oxidizing properties of the reaction mixture. Hydrated aluminum nitrate reacts with anthracene in an uncatalyzed nitration under rather harsh conditions to yield anthraquinone and 9,10-dinitroanthracene.[63]

Another method for activation of metal nitrates involves nitration in the presence of carboxylic acid anhydrides. Menke demonstrated the nitration of activated aromatics with copper(II)nitrate, ferric nitrate, and other transition metal nitrates in the presence of Ac$_2$O.[64,65] The reagents nitrated both benzenoid and nonbenzenoid aromatic hydrocarbons under mild conditions. Sondheimer et al. demonstrated the success of this nitration for annulenes, where common nitrating agents failed to give the desired products.[66] Horita et al. carried out nitration of methyl substituted [2.2] paracyclophanes.[67]

An examination of the reaction products tends to indicate that the nitration proceeds via the intermediacy of acetyl nitrate. The metal nitrate itself cannot be involved.

An unusual reaction of Cu(NO$_3$)/Ac$_2$O with 1,2-diphenylcyclopropane results in the formation of 3,5-diphenylisoxazole and the 4-nitro derivative,[68] probably through the following pathway.

Laszlo *et al.* studied aromatic nitration using metal nitrates, particularly cupric nitrate impregnated into acidic clays, as the nitrating agent.[69-71] The reaction of halobenzenes was carried out in 50–75% yield in the presence of Ac$_2$O, hexane, or methylene chloride at room temperature. The reactions are characterized by high para–preference for the incoming nitro group. Besides the strong +K>–I electronic effects of the halogens generally responsible for increased para substitution (the conjugative effect [K] remains unchanged, whereas the inductive effect [I] diminishes with distance), the dimensionality of the reaction space in the clay-supported reaction tends to further enhance the more accessible para substitution.

Nitrating toluene with cupric nitrate impregnated in K10 montmorillonite ("claycop") in the presence of Ac$_2$O and using carbon tetrachloride as the solvent for the aromatic, highly dilute solutions with long reaction times (120 hr) gave 79% para, 10% meta, and 20% ortho isomer in nearly 100% yield.[72] These interesting nitrations giving high para–regioselectivity use equimolar metal nitrate reagent. It was indicated, however that the reaction can be made catalytic.[72]

In contrast to cupric nitrate, alkali metal nitrates are not activated by Ac_2O. However, Crivello has shown[59] that alkali metal nitrates can be used with trifluoroacetic anhydride [$(CF_3CO)_2O$] and to a lesser extent with trichloro– and dichloracetic anhydride. The reaction rate was shown to be significantly affected by the extent of solubility of the inorganic nitrate in the reaction medium. Since ammonium nitrate is reasonably soluble in organic solvents it is particularly successful (*vide infra*). Many metal nitrates have been also shown to be effective nitrating agents in the presence of $(CF_3CO)_2O$. Benzene gives nitrobenzene in 90% yield in most cases. The nitration rate is highly solvent dependent, being higher in polar solvents. The reaction is quite general and successful for the nitration of aromatics that are more reactive than nitrobenzene.

Crivello, as mentioned, found that ammonium nitrate is a particularly good nitrating agent of aromatics in $(CF_3CO)_2O$.[59]

$$ArH + NH_4NO_3 + (CF_3CO)_2O \longrightarrow ArNO_2 + CF_3CO_2NH_4 + CF_3CO_2H$$

The reaction bears a close resemblance to acetyl nitrate nitration and most probably involves formation of trifluoroacetyl nitrate, which subsequently can form NO_2^+ ion.

$$(CF_3CO)_2O + NH_4NO_3 \longrightarrow CF_3C(O)ONO_2 + CF_3C(O)ONH_4$$

$$CF_3C(O)ONO_2 \longrightarrow NO_2^+ + CF_3CO_2^-$$

Attempts to isolate trifluoroacetyl nitrate by distillation led to decomposition. Acetyl nitrate is known to give high ratios of ortho–nitrated products.[73] Similar trends in orientation are seen in the product distributions obtained in $(CF_3CO)_2O-NH_4NO_3$ nitrations. However, in contrast to the rather limited range of substrates which can be nitrated by using acetyl nitrate, trifluoroacetyl nitrate is a much more

Table 14. Nitration of Aromatic Compounds at 25°C wth NH_4NO_3 $(CF_3CO)_2O^{59}$

Substrate	Reaction time (hr)	Products (percent yield)
C_6H_6	2	$C_6H_5NO_2$ (95)
$C_6H_5CH_3$	2	o-$NO_2C_6H_4CH_3$ (53), m-$NO_2C_6H_4CH_3$ (>1), p-$NO_2C_6H_4CH_3$ (35)
C_6H_5Cl	2	o-$NO_2C_6H_4Cl$ (20), p-$NO_2C_6H_4Cl$ (80)
$C_6H_5CO_2H$	2	m-$NO_2C_6H_4COOH$ (98)
$C_6H_5NH_2$	0.5	o-$NO_2C_6H_4NHC(O)CF_3$ (87)
$C_6H_5NO_2$	2	No reaction
benzothiazole	2	6-nitrobenzothiazole (63)
N-phenylmaleimide	2	ortho-nitro N-phenylmaleimide (62), para-nitro N-phenylmaleimide (37)
$C_6H_5OCH_3$	2	o-$NO_2C_6H_4OCH_3$ (55), p-$NO_2C_6H_4OCH_3$ (22)
$C_6H_5OCH_2C{\equiv}CH$	2	o-$NO_2C_6H_4OCH_2C{\equiv}CH$ (48), p-$NO_2C_6H_4OCH_2C{\equiv}CH$ (16)
$C_{10}H_8$	4	1-$NO_2C_{10}H_7$ (88), 2-$NO_2C_{10}H_7$ (trace)
$C_6H_5C_6H_5$	2	o-$NO_2C_6H_4C_6H_5$ (63), p-$NO_2C_6H_4C_6H_5$ (26), $NO_2C_6H_4C_6H_4NO_2$ (~3)
$C_6H_5OC_6H_5$	3	o-$NO_2C_6H_4OC_6H_4NO_2$-o (52), o-$NO_2C_6H_4OC_6H_4NO_2$-p (13), p-$NO_2C_6H_4OC_6H_4NO_2$-p (18)
$(C_6H_5)_3P$	2	$(C_6H_5)_3PO$

efficient and versatile nitrating agent. Results of nitrations are summarized in Table 14.

Limiting reactivity in deactivated substrates lies between nitrobenzene, which is not nitrated, and benzoic acid, which nitrates nearly quantitatively in the meta position. Compounds such as aniline, which contain a reactive amino group, undergo trifluoroacetylation prior to nitration. Substrates containing acid-sensitive acetylenic groups such as phenyl propargyl ether are smoothly converted to their nitro compounds under the mild conditions of the reaction. The method is selective, introducing only one nitro group per aromatic ring. With certain substrates, oxidation rather than nitration is observed as the major reaction. For example, $NH_4NO_3/(CF_3CO)_2O$ oxidizes cycloheptatriene to the tropylium ion and triphenylphosphine to triphenylphosphine oxide. Similarly, when phenols are treated with inorganic salts in $(CF_3CO)_2O$, nitrated compounds are not obtained as the major products; instead, quinonoid compounds resulting from carbon-oxygen and carbon-carbon couplings are produced.

The nitration of polymeric aromatics by conventional nitration methods presents two major problems: inhomogeneous nitration due to limited solubility and degradation of the polymer due to cleavage of sensitive bonds. Polymers containing aromatic groups in their backbones undergo smooth homogeneous nitration at room temperature by $NH_4NO_3-(CF_3CO)_2O$ mixtures in a non-reactive solvent for the polymer. Results are summarized in Table 15. Only in the case of hydrolytically very sensitive polycarbonate was there evidence of slight molecular weight reduction in the polymer as indicated by intrinsic viscosity measurements. In each case, only one nitro group per repeating unit was introduced as determined by infrared and NMR spectroscopy.

2.18 Alkyl Nitrates

Nitration of aromatics can be carried out with alkyl nitrates (i.e., alkyl esters of nitric acid) catalyzed by both protic or Lewis acids. Acid catalysts are assumed to form NO_2^+ ion from alkyl nitrates or strongly polarized complexes.

$$RONO_2 + H_2SO_4 \longrightarrow NO_2^+HSO_4^- + ROH$$

Table 15. Nitration of Polymers with $NH_4NO_3/(CF_3CO)_2O$ at 25°C[59]

Polymer	$[\eta]_0^a$	Reaction Time (hr)	Product	$[\eta]_0^b$
(bisphenol A carbonate polymer)	0.56	15	(nitrated product)	0.29
(polystyrene)	0.19	15	(nitropolystyrene)	0.18
(bisphenol A dipropargyl ether polymer)	0.92	4	(nitrated product)	0.86
(2,6-diphenylphenylene oxide polymer)	1.1	15	(nitrated product)	0.95

$^a[\eta]_0$, intrinsic viscosity (g/dL) in DMF before nitration
$^b[\eta]_0$, intrinsic viscosity (g/dL) in DMF after nitration

$$\text{RONO}_2 + \text{AlCl}_3 \longrightarrow \text{NO}_2{}^+\text{AlCl}_3\text{OR}^-$$

Masci[74] studied the effect of various organic solvents (CH_3NO_2, CH_3CN, sulfolane, CH_2Cl_2, $ClCH_2CH_2Cl$, $CHCl_3$, $CH_3COOC_2H_5$) on the selectivity of aromatic nitration with $Bu_4NNO_3/(CF_3CO)_2O$, a system with better solubility.

Alkyl nitrates must be prepared and stored with care, since, particularly in the presence of even traces of acids, they can become explosive.[75] They should not be stored over prolonged periods of times. This is particularly the case for nitrates of polyols. The simplest and most effective method for their preparation involves reaction of alcohols with nitronium salts in the presence of an acid–binding agent (see Section 4.5.1). Transfer–nitration with N–nitropyridium salts are particularly suited for their acid free preparation.[76]

$$\text{ROH} + \text{PyNO}_2{}^+\text{BF}_4{}^- \longrightarrow \text{RONO}_2 + \text{PyH}^+\text{BF}_4{}^-$$

Alkyl nitrates do not nitrate aromatic compounds in the absence of catalysts.[77] However, good yields of nitrated products can be obtained if aromatic compounds and alkyl nitrates react in the presence of H_2SO_4,[78-82] PPA,[83] or Lewis acid halides.[84-86]

Alkyl nitrates most frequently used are methyl nitrate and ethyl nitrate (bp. 65°C and 86–87°C, respectively).

The H_2SO_4 catalyzed nitration of benzene with ethyl nitrate at 78–80°C gave nitrobenzene in only 12% yield.[87] *p*-Fluoroacetanilide was, however, nitrated under similar conditions at 0°C with 85% yield.[88]

Recently developed strong solid acid catalysts such as Nafion–H were found active in catalyzing nitrations with alkyl nitrates.[89] Benzene and alkylbenzenes are nitrated in excellent yield with *n*-butyl nitrate at around 80°C in the presence of solid Nafion–H acid catalyst. The reaction is generally slow for preparative purposes at lower temperatures and does not proceed at room temperature. The reaction is very selective. Dinitro compounds are not formed in any significant amounts. The steric bulk of the solid acid–complexed nitrating agent seems to play a significant role in determining the isomer distribution of products. Generally, a decreased amount of the more hindered isomer (ortho) is formed compared with conventional electrophilic solution nitrations, thus increasing the selectivity of nitration at the less hindered para position (Table 16).

Table 16. Nitration of Aromatics with *n*-Butyl Nitrate over Nafion-H Catalyst[89]

Aromatic	Yield (percent)	Isomer distribution (percent)
Benzene	77	
Toluene	96	2-nitro (50)
		3-nitro (3)
		4-nitro (47)
o-Xylene	98	3-nitro (47)
		4-nitro (53)
m-Xylene	98	2-nitro (12)
		4-nitro (88)
p-Xylene	95	2-nitro (100)
Mesitylene	90	2-nitro (100)
1,2,4-Trimethylbenzene	94	3-nitro (8)
		5-nitro (92)
1,2,3,4-Tetramethylbenzene	93	
Anisole	86	2-nitro (32)
		4-nitro (68)
Chlorobenzene	15	

When secondary or tertiary alkyl nitrates are used in the nitrationreaction, competing alkylations complicate the system. Consequently methyl nitrate or ethyl nitrate are the preferred alkyl nitrates for nitration.

$AlCl_3$, $SnCl_4$, $SbCl_5$, and $FeCl_3$ catalyze the nitration of benzene with ethyl nitrate.[90]

Alkyl nitrates, particularly methyl nitrate, are very effective nitrating agents in the presence of BF_3 as catalyst.[91]

$$ArH + NO_2OCH_3 \xrightarrow{BF_3} ArNO_2 + CH_3OH \cdot BF_3$$

The reaction was found useful as a selective and mild nitration method, for example, allowing mononitration of durene and other highly alkylated benzenes, which with mixed acid usually undergo dinitration (Table 13). Methyl nitrate–boron trifluoride can also be used to achieve dinitration of tetramethylbenzenes by using two and three molar excess of methyl nitrate, respectively. Relative yields of mono– and dinitro product compositions are shown in Table 17.

Other Friedel–Crafts type catalysts can also be used, but BF_3 was found to be the most suitable. In the nitration of pentamethylbenzene,

Table 17. Boron Trifluoride Catalyzed Nitration of Tetramethylbenzenes with Excess Methyl Nitrate in Nitromethane Solution[91]

Compound	H_3C-ONO_2:Arene	Total yield (%) of nitro products	Composition	
			Dinitro product (%)	Mononitro product (%)
1,2,3,4-tetramethylbenzene	2 : 1	93	64.4	35.6
	3 : 1	90	75.0	25.0
1,2,3,5-tetramethylbenzene	2 : 1	94	89.9	10.1
	3 : 1	94	99.9	0.1
1,2,4,5-tetramethylbenzene	2 : 1	95	90.1	9.9
	3 : 1	92	100	0

aluminum trichloride and titanium (IV) chloride caused formation of significant amounts of chlorinated derivatives, (H_2SO_4 led to nitro–demethylation products).

2.19 Acetone Cyanohydrin Nitrate

Acetone cyanohydrin nitrate (ACN) was found to have enhanced reactivity compared to methyl nitrate in the preparation of various ring substituted phenylnitromethanes.[92] It was first used by Thompson and Narang[93,94a] in nitrating aromatics. The BF_3–etherate catalyzed nitration of alkylbenzenes and anisole[95] gave good yields. ACN is more reactive than ordinary alkyl nitrates because of the greater ease of O–N bond cleavage in the intermediate O– or N–coordinated (protonated) ACN.

Table 18. Nitration of Aromatics by Acetone Cyanohydrin Nitrate/BF$_3$ Etherate[95]

Compound	Yield (%)	Percent isomer distribution		
		2-Nitro	3-Nitro	4-Nitro
Toluene	77.6	59.8	4.5	35.7
o-Xylene	75.2	0	60.3	39.7
m-Xylene	78.0	15.3	0	84.7
p-Xylene	90.0	–	–	–
Mesitylene	74.1	–	–	–
Anisole	73.1	72.4	0	27.6

$$ArH + H_3C-\underset{CH_3}{\overset{CN:BF_3^{\delta-}}{C}}-O-NO_2^{\delta+} \longrightarrow ArNO_2 + (CH_3)_2\underset{CN}{C}OH \cdot BF_3$$

Use of acetone cyanohydrin nitrate has certain practical advantages over other procedures that use alkyl nitrates. It is more stable than CH_3ONO_2 and is stored easily for longer periods of time. BF_3-etherate is easier to handle than BF_3 gas used in other methods. Under similar conditions, this method provides cleaner products and higher yields than does a mixture of CH_3ONO_2 and BF_3-etherate; only small amounts of BF_3-etherate are required.

Solid superacidic catalysts can also be advantageously applied in nitration with ACN.[89]

The Nafion–H catalyzed nitration of deactivated aromatics with alkyl nitrates, such as butyl nitrate gives only very low yields.[89] Even nitration of chlorobenzene, for example, gave only 15% of chloronitrobenzenes (Table 16). Due to its greater reactivity, nitration of alkylbenzenes with ACN gives the corresponding nitro compounds in good to moderate yield (Table 19). The nitration of chlorobenzene gave 49% yield (as contrasted with 15% with butyl nitrate). The yields increase only modestly with time (the isomer ratios remain constant). This may be due to the thermal decomposition of ACN at the reaction temperatures necessary for effect catalysis by Nafion–H.

Table 19. Nitration with ACN over Nafion-H Catalyst[89]

Aromatic	Yield (percent)	Isomer distribution (percent)
Benzene	85	
Toluene	79	2-nitro (47)
		3-nitro (3)
		4-nitro (50)
o-Xylene	60	3-nitro (44)
		4-nitro (56)
m-Xylene	61	2-nitro (11)
		4-nitro (89)
p-Xylene	61	
Mesitylene	36	
1,2,3,4-Tetramethylbenzene	69	
Chlorobenzene	49	2-nitro (28)
		3-nitro (2)
		4-nitro (70)

The ortho–para ratio of nitrotoluenes obtained is lower (0.94) with ACN than with butyl nitrate (1.06), reflecting the somewhat larger bulk of the former reagent.

2.20 Trimethylsilyl Nitrate

Trimethylsilyl nitrate, $(CH_3)_3SiONO_2$, is another interesting but little studied nitrating agent.[12,93] It is prepared from chlorotrimethylsilane and silver nitrate and nitrates aromatics effectively with BF_3 as catalyst

$$2ArH + 2(CH_3)_3SiONO_2 \xrightarrow{BF_3} 2\ ArNO_2 + (CH_3)_3SiOSi(CH_3)_2 + BF_3 \cdot H_2O$$

Trimethylsilyl nitrate, however, even on standing readily decomposes according to

$$2(CH_3)_3SiONO_2 \longrightarrow 2NO_2 + \tfrac{1}{2}O_2 + [(CH_3)_3Si]_2O$$

Consequently nitrogen dioxide formed can also affect nitration in the system.

2.21 Acyl Nitrates

Acyl nitrates, the mixed anhydrides of nitric and carboxylic acids, i.e. $RC(O)ONO_2$, are reactive nitrating agents.[5]

$$ArH + RC(O)ONO_2 \longrightarrow ArNO_2 + RCOOH$$

In the isolated state, but even in solution at temperatures above 60°C, they can be extremely explosive and must be handled with great care.[96] They are more safely generated *in situ*. Their preparation can involve

$$(RCO)_2O + N_2O_5 \longrightarrow 2\ RC(O)ONO_2$$

$$RCOCl + AgNO_3 \longrightarrow RC(O)ONO_2 + AgCl$$

2.21.1 Acetyl Nitrate

Acetyl nitrate is the most widely used acyl nitrate.[97] It is readily formed in pure form from N_2O_5 (prepared by distilling HNO_3 from P_2O_5) in Ac_2O. It can be distilled under reduced pressure (bp. 22°C/70 mm) but it explosively decomposes at atmospheric pressure at 60°C. Nitration of aromatics can be carried out in carbon tetrachloride solution. It is, however, more convenient to prepare it *in situ* by addition to a solution of the aromatic and acetyl chloride to finely pulverized silver nitrate.[98]

It should be pointed out that in the extensively studied nitration with HNO_3 in Ac_2O (acetic acid) the active nitrating agent indeed is acetyl nitrate acting as the carrier for the NO_2^+ ion.

Some specific examples of nitration with acetyl nitrate are that of biphenyl to nitrobiphenyls in 87% yield, giving 58% of 2- and 42% of 4-nitro isomer[99] *p*-Quatraphenyl gave only 4-nitro and 4,4''''-dinitro-*p*-quatraphenyl[100]

$$\text{Ph-Ph-Ph-Ph} \xrightarrow{CH_3COONO_2} O_2N\text{-Ph-Ph-Ph-Ph-}(NO_2)$$

Nitration of phenol with acetyl nitrate gives 52% ortho and 48% para nitrophenol.[101]

2.21.2 Benzoyl Nitrate

Benzoyl nitrate is prepared in a manner similar to acetyl nitrate preferentially *in situ* from benzoyl chloride and silver nitrate. It has also received extensive use in nitration,[5] together with other aroyl nitrates.[129]

An interesting application of benzoyl nitrate is in the nitration of diazocyclopentandiene giving a 2:1 mixture of 2- and 3-nitro-diazocyclopentadiene.[102]

$$\text{Cp-}N_2^+ \xrightarrow{C_6H_5COCl/AgNO_3} \text{(2-NO}_2\text{-Cp-}N_2^+\text{)} + \text{(3-NO}_2\text{-Cp-}N_2^+\text{)}$$

2.21.3 Trifluoroacetyl Nitrate

Trifluoroacetyl nitrate CF_3COONO_2,[62] is also increasingly used in nitrations with acyl nitrates.

There is still no agreement about the nature of the nitrating agent in acyl nitrate nitrations.[103-110] The question is whether acyl nitrates serve simply as precursors for the NO_2^+ ion or whether they can act as specific nitrating agents in their own right.

Trifluoroacetyl nitrate, the mixed anhydride of HNO_3 and CF_3CO_2H can readily ionize in polar solvents giving nitronium ion

$$CF_3COONO_2 \rightleftharpoons NO_2^+ CF_3CO_2^-$$

At the same time unionized but polar trifluoroacetyl nitrate, when protic acid is present in the nitration system, can also directly react with aromatics through its protonated form

$$F_3C-\underset{\underset{HO^+}{\|}}{C}-O-N\underset{O}{\overset{O}{\diagup}} + ArH \longrightarrow [ArHNO_2]^+ \longrightarrow ArNO_2 + H^+$$

Whereas the linear NO_2^+ ion must bend to effect nitration, in protonated acyl nitrates a bent precursor of the incipient nitronium ion is available.[111]

2.22 Nitryl Halides

The general Friedel–Crafts acylation principle can also be applied to inorganic acid halides and anhydrides.[77] Consequently aromatic nitrations involving nitryl halides and oxides (the halides and anhydrides of HNO_3) can be considered as Friedel–Crafts type reactions.

2.22.1 Nitryl Chloride

In polar solvents at low temperatures, NO_2Cl acts as a chlorinating agent for aromatics giving only minor amounts of nitro compounds.[112] Price and Sears first carried out Friedel–Crafts nitration with NO_2Cl and HF or $AlCl_3$.[113] They found $AlCl_3$ to be the most suitable catalyst. Deactivated aromatics, however, were nitrated only with difficulty, and the method was therefore considered to be of limited value.

Isolation of various complexes formed by NO_2Cl and Lewis acid halides has been subsequently reported.[114] Olah and Kuhn have shown that aromatic compounds, including deactivated ones such as halobenzenes and benzotrifluoride, can be nitrated with ease using nitryl halides and a suitable Friedel–Crafts catalyst.[115]

$$\text{ArH} + \text{NO}_2\text{X} \xrightarrow[\text{catalyst}]{\text{Friedel-Crafts}} \text{ArNO}_2 + \text{HX}$$

(X = F, Cl, Br)

Using NO_2Cl as the nitrating agent, which in the laboratory is conveniently prepared by the reaction of HNO_3 with chlorosulfuric acid,[116] $TiCl_4$ was found to be the most suitable catalyst. $FeCl_3$, $ZrCl_4$, $AlCl_3$, and $AlBr_3$ were also effective but the reactions are more difficult. With BCl_3 only a small amount of nitrated product was obtained and considerable ring chlorination took place. SbF_5 is also an active catalyst for the NO_2Cl nitration of aromatics. BF_3 was found to be inactive as a catalyst. Typical yields obtained in the nitration of aromatics using $TiCl_4$ as catalyst are: benzene, 88%; toluene, 81.5%; ethylbenzene, 79%; fluorobenzene, 91%; chlorobenzene, 41.5%; and benzotrifluoride, 32%. There is always a certain amount of ring–chlorinated byproducts formed. Reactions are carried out either by using an excess of the aromatic as solvent ($TiCl_4$ is miscible with many aromatics) or in carbon tetrachloride solution. The amount of chlorinated byproducts can be decreased by using solvents of higher dielectric constants. Tetramethylene sulfone (sulfone) was found to be a suitable solvent for the $TiCl_4$ and also for most of the other Lewis acid–catalyzed nitrations. It has excellent solvent properties for aromatics, many catalysts, and NO_2Cl. It is superior to other solvents that can be used, such as nitromethane. Because sulfolane is completely miscible with water, the work–up of the reaction mixture after the reaction is comparably easy.

In Lewis acid–halide catalyzed nitrations with NO_2Cl the question arises whether these reactions are NO_2^+ ion nitrations according to the ionization

$$NO_2Cl + AlCl_3 \longrightarrow NO_2^+AlCl_4^-$$

or whether they are affected by a donor : acceptor complex and not by the free NO_2^+ ion.

$$\text{Cl}-\overset{\delta+}{\underset{\underset{O}{\|}}{N}}\overset{\delta-}{\underset{}{,O}}\rightarrow \text{AlCl}_3$$

Olah *et al.* carried out studies of nitration of toluene with NO_2Cl, catalyzed by Lewis acid halides.[117] Data are shown in Table 20. In excess of the aromatic as solvent, the ortho–para ratio was lower than in nitrations with mixed acid and nitronium salts, indicative of a nitryl halide–Lewis acid complex acting as the nitrating agent; a free NO_2^+ ion is not necessarily involved. However, this difference diminishes by carrying out the reactions in ionizing polar solvents, such as nitromethane (Table 20) where typical nitronium ion nitration seems to take place.

Table 20. Lewis Acid Catalyzed Friedel–Crafts Nitration of Toluene with Nitryl Chloride at 25°C[117]

Lewis acid halide	Solvent	Percent isomer distribution			o/p
		ortho	meta	para	
$AlCl_3$	Excess toluene	53.3	1.2	45.5	1.17
$TiCl_4$		53.1	1.6	45.4	1.17
BF_3		57.1	1.4	41.1	1.40
$SbCl_5$		56.4	1.4	42.4	1.34
PF_5		57.6	1.6	40.8	1.41
$AlCl_3$	Nitromethane	61.3	3.7	35.0	1.75
$TiCl_4$		61.1	3.7	35.2	1.74
PF_5		61.6	3.5	34.9	1.76

2.22.2 Nitryl Bromide

Nitryl bromide, compared with the chloride and fluoride, is quite unstable. Nitration experiments were carried out with solutions obtained by the halogen exchange of NO_2Cl with KBr (not separated from unreacted NO_2Cl and decomposition products) in sulfur dioxide at −20°C, and using $TiBr_4$ as catalyst. Yields of nitration are lower than those obtained with NO_2Cl, due to the formation of more ring-brominated products.[115] This can be attributed partly to the presence of free bromine from the decomposition of NO_2Br and to the easier homolysis of NO_2Br itself.

2.22.3 Nitryl Fluoride

Nitryl fluoride is a more powerful nitrating agent than NO_2Cl, but is more difficult to handle. Hetherington and Robinson[118] reported nitration of aromatics with NO_2F in the absence of catalysts. They suggested that, in solution, NO_2F dissociates into NO_2^+ and F^- and that the NO_2^+ ion thus formed is the active reagent in the nitrations. Less reactive aromatics such as nitrobenzene were not nitrated and considerable tar formation occurred during the reactions. Olah and Kuhn found that by using a Lewis acid fluoride catalyst such as BF_3, PF_5, AsF_5, and SbF_5, Friedel-Crafts type nitrations can be carried out with NO_2F.[115] Homolytic cleavage of NO_2F, which causes most of the side reactions, is considerably suppressed under these conditions in favor of heterolysis, yielding the NO_2^+ ion. The reactions were carried out preferably at low temperatures. Benzotrifluoride was nitrated to m-nitrobenzotrifluoride at −50°C in 90% yield using BF_3 as catalyst. Halobenzenes, including di- and polyhalobenzenes, were nitrated with ease and with yields of over 80%.

The nitrations were carried out either with an excess of the aromatic as diluent and introducing NO_2F and the Lewis acid fluoride catalyst simultaneously at low temperature into the well-stirred reaction mixture, or in a suitable solvent such as sulfolane, which can be used advantageously if the catalyst fluoride does not interact with it (SbF_5, being a strong fluorinating agent, attacks the solvent and cannot be used).

2.23 Nitrogen Oxides

Neither nitrous oxide (N_2O) nor nitric oxide (NO) alone react with aromatic hydrocarbons to give nitroaromatic compounds.

2.23.1 Dinitrogen Trioxide

Dinitrogen trioxide, the anhydride of HNO_2, shows no nitrating ability towards aromatics in H_2SO_4 solution.

Millen, while investigating the Raman spectrum of N_2O_3 in H_2SO_4 solution, found evidence for the following ionization,[119]

$$N_2O_3 + 3 H_2SO_4 \rightleftharpoons 2 NO^+ + 3 HSO_4^- + H_3O^+$$

but not for ionization giving NO_2^+ ion.

NO^+ has a much lower electrophilic reactivity, including that towards hydrolysis, than the NO_2^+ ion. This should be one of the reasons why no direct electrophilic nitrosation of benzene and other moderately reactive aromatic hydrocarbons is known (another reason is the already formed nitrosoaromatic compounds react further with excess NO^+). The absence of NO_2^+ formation explains the lack of nitration reactions with N_2O_3 in H_2SO_4.

However, Bachman and co-workers observed nitrating ability of N_2O_3 when used as its BF_3 complex, $N_2O_3 \cdot BF_3$. N_2O_3 gives a white, stable solid 1 : 1 complex with BF_3.[120] The complex is insoluble in nonpolar solvents and in most of the polar solvents with which it does not react. It sublimes at room temperature and does not melt in a

Table 21. Nitration of Aromatics with $N_2O_3 \cdot BF_3$[120]

Aromatic	Time (hr)	Temperature (°C)	Solvent	Product	Yield (%)
Benzene	42	45	–	Nitrobenzene	5
Toluene	72	60-65	Nitroethane	o-Nitrotoluene	56
				p-Nitrotoluene	1
				2,4-Dinitrotoluene	3
o-Nitrotoluene	12	85-90	Nitroethane	2,4-Dinitrotoluene	40
Naphthalene (air)	24	25	2-Nitropropane	1-Nitronaphthalene	67
Naphthalene (nitrogen)	24	25	2-Nitropropane		14
Chlorobenzene	24	85-95	Nitroethane	2-Nitrochlorobenzene	12
				4-Nitrochlorobenzene	50

sealed tube below 300°C; above this temperature it dissociates. These properties suggest that the complex is ionic in character.

The $N_2O_3 \cdot BF_3$ complex is a good nitrosating agent and only a very weak nitrating agent. However, under more forcing conditions it will nitrate aromatic nuclei.

Table 21 shows Bachman's results of nitration of aromatics with the $N_2O_3 \cdot BF_3$ complex. It will be noted that the complex was claimed to produce a different isomer ratio with toluene than is generally reported for HNO_3 nitrations. No explanation was given nor was the result subsequently reaffirmed.

However, if the data are correct they may suggest that nitration via nitrosation occurs and that the para (and meta) nitroso-isomer is selectively decomposed before its conversion to the nitro-compounds. It has been shown by Ingold that initial attack by NO^+ ion is involved in the nitration of p-nitrophenol and p-chloroanisole.

2.23.2 Dinitrogen Tetroxide (Nitrogen Dioxide)

Nitration of aromatics with N_2O_4 in the presence of strong acids (either Lewis or Bronsted) is well recognized.[77]

Pinck[121] used H_2SO_4 to catalyze the nitration of aromatics with N_2O_4. He observed that only half of the N_2O_4 was used up in the nitration, the remainder being present as nitrosylsulfuric acid. Titov[122] dissolved N_2O_4 in H_2SO_4 and used this solution as a nitrating agent. It was shown that the reaction proceeded by the ionization of N_2O_4 into nitrosonium and nitronium ion, with only the latter being the effective nitrating agent. In H_2SO_4, the initial ionization of N_2O_4 was given as:

$$N_2O_4 + H^+ \longrightarrow NO^+ + HNO_3$$

Subsequent ionization of HNO_3 proceeds only at higher acidities and, indeed, acid catalyzed nitrations with N_2O_4 generally require stronger acids in larger amounts.[123,124]

Raman spectroscopic and cryoscopic investigations of solutions of N_2O_4 in H_2SO_4 gave proof for the formation of the nitronium ion (NO_2^+)

$$N_2O_4 + 3 H_2SO_4 \longrightarrow NO_2^+ + NO^+ + H_3O^+ + 3HSO_4^-$$

together with an equimolar amount of nitrosylsulfuric acid ($NO^+HSO_4^-$). Typical yields and the required acid strengths in the nitration of aromatics with N_2O_4 and H_2SO_4 are shown in Table 22. A nearly quantitative yield in the nitration of benzene (99.8%) with N_2O_4 in 85.8% H_2SO_4 was also reported.[125]

It was Schaarschmidt[126] who first investigated the catalytic effect of $AlCl_3$ and $FeCl_3$ on the nitration of aromatics with N_2O_4, the mixed anhydride of HNO_3 and HNO_2.

Bachman used the stable $N_2O_4 \cdot BF_3$ complex, prepared as a crystalline salt from the components, in aromatic nitrations.[127] He suggested the nitronium salt structure for the complex

$$(NO_2)^+(F_3B\text{-}NO_2)^-$$

The crystalline complex, however, is only a moderate nitrating agent giving satisfactory yields of such aromatics as benzene and even naphthalene only after long reaction times (1–7 days). This result obviously is not in agreement with the suggested NO_2^+ ion structure. The $N_2O_4 \cdot BF_3$ complex also shows an ability to nitrosate and diazotize, thus acting as a nitrosonium salt

$$(NO^+)(F_3B\text{-}NO_3)^-$$

Raman spectroscopic investigations of the solid complex, carried out by Evans and Olah[128] showed only a relatively weak band at 1400 cm^{-1}. It was therefore suggested that the addition compound $N_2O_4 \cdot BF_3$, which is not necessarily completely ionized in the solid state, can form an equilibrium mixture of the nitronium and nitrosonium ions

Table 22. Nitration of Aromatics with N_2O_4 + H_2SO_4[77]

Aromatic	Concentration of H_2SO_4 (%)	Yield (%)
Benzene	80.2	95
Toluene	77.6	94
o-Xylene	78.8	77
m-Xylene	77.6	89
p-Xylene	78.8	95
Chlorobenzene	84.5	95
Nitrobenzene	100.0	94
Anthraquinone	oleum (5% SO_3)	84.5

in agreement with its observed reactivity and the fact that N_2O_4 itself is the mixed anhydride of HNO_3 and HNO_2.

The major difficulty in Friedel–Crafts nitrations with N_2O_4 was the fact that the N_2O_4-catalyzed complexes were insoluble in the reaction media. This resulted not only in slow reactions and low yields, but also in many cases in undesirable side reactions.

Schaarschmidt reported[126] that when $AlBr_3$ was tried instead of $AlCl_3$ as catalyst only ring bromination took place and no nitro product was formed. The use of a fluoride catalyst, such as BF_3 in the work of Bachman,[127] eliminated halogenation as a side reaction but still had to deal with the heterogeneous reaction medium.

Olah et al. found[89] that homogeneous Friedel–Crafts type reactions with N_2O_4 and Lewis acid catalyst such as $TiCl_4$, BF_3, BCl_3, PF_3, and AsF_5 can be carried out in sulfolane solution. It is not necessary to isolate the catalyst–N_2O_4 complex. Instead, a solution of N_2O_4 and the catalyst is prepared and is added to a sulfolane solution of the aromatic to be nitrated.

Nitrobenzene was obtained from the nitration of benzene in yields of 32–67%, fluoronitrobenzenes from fluorobenzene in 28–76% yields, the relative order of activity of the catalysts used being $AsF_5 > PF_5 > BF_3 > TiCl_4 > BCl_3$. With the chloride catalysts, a considerable amount of chlorobenzene also was formed in the reaction, as was the case with $AlCl_3$.

Bromide Lewis acids such as $AlBr_3$, BBr_3, and $TiBr_4$, in agreement with previous observations of Schaarschmidt about $AlBr_3$, gave a high amount of ring bromination but about 10% of nitroaromatics were also formed.

Subsequent investigations have proved that aluminum, titanium, and boron halides tend to react with N_2O_4 in the following way

$$2N_2O_4 + TiCl_4 \longrightarrow 2NO_2Cl + 2\,NOCl + TiO_2$$

$$3N_2O_4 + Al_2Br_6 \longrightarrow 3NO_2Br + 2NOBr + Al_2O_3$$

$$3N_2O_4 + 2BBr_3 \longrightarrow 3NO_2Br + 3NOBr + B_2O_3$$

NO_2Br, being unstable, decomposes to $N_2O_4 + Br_2$ and the bromine, formed in the presence of the catalyst, brominates the aromatic. In a similar manner, but to a less extent, chlorination takes

Table 23. Nitration of Aromatics with
Dinitrogen Tetroxide over Nafion–H Catalyst
at 0°C in CCl_4 Solution[89]

Substrate	Percent isomer distribution
Toluene	2-nitro (49)
	3-nitro (6)
	4-nitro (45)
o-Xylene	3-nitro (41)
	4-nitro (59)
m-Xylene	2-nitro (16)
	4-nitro (84)

place with chloride catalysts.

The use of solid–acid catalysts in aromatic nitration frequently represents substantial advantages over liquid–acid catalyst systems.

When alkylbenzenes in the liquid phase were nitrated with N_2O_4 in the presence of Nafion–H at 0°C, nitration was slow, rendering it unsuitable for preparative purposes.[89] In general, solid–acid catalysts are rather ineffective at such low temperatures. The isomer ratios of nitroarene products (Table 23), however, show that the products were obtained via a typical electrophilic aromatic substitution.

In addition to nitroarenes, products of side–chain substitution were also obtained. Thus, for example, nitration of toluene with N_2O_4 provided, in addition to three isomeric nitrotoluenes, significant amounts of phenylnitromethane and benzaldehyde. Phenylnitromethane could arise either via a free–radical pathway[129] or through ipso substitution in an electrophilic mechanism as illustrated in the following Scheme:

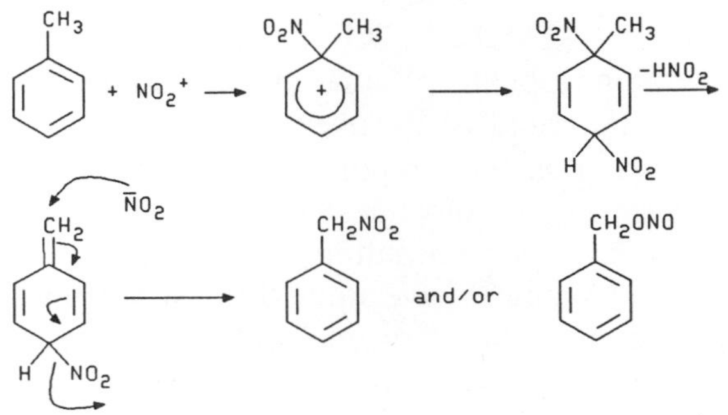

Similar formation of quinonoidal-type intermediates has been postulated before to explain the side chain substitution in nitrations with acetyl nitrate.[93,94] The formation of a quinonoid intermediate is quite likely in this system because NO_2^- is a good nucleophile that can capture the ipso Wheland intermediate, thus leading ultimately to phenylnitromethane (see also Chapter 3).

Since N_2O_4 is always in equilibrium with NO_2

$$N_2O_4 \rightleftharpoons 2NO_2$$

(thus we use both rather interchangeably in our discussion) it is understandable that in gas-phase reactions nitration by the latter dominates increasingly.

The nitration of toluene in the presence of perfluorinated resinsulfonic acid (Nafion-H) for two hr gave nitrotoluenes (20%) and benzaldehyde (30%).[89] The isomer distribution of nitrotoluenes (ortho, 48.5%; meta, 7.0%; para, 44.5%) showed that the products are obtained essentially by electrophilic substitution, but the somewhat increased amount of meta isomer indicated some competing reaction by a free-radical process or just a temperature effect. Under the same conditions, but without catalyst, nitrotoluenes (5%), benzaldehyde (25%), and phenylnitromethane (30%) were obtained.[130] The isomer distribution (ortho, 24.5%; meta, 53.5%; para, 26%) showed that the reaction proceeded via free-radical (i.e., NO_2) nitration of toluene (see Section 2.27 and Chapter 3, part II). The absence of any phenylnitromethane in the case of the Nafion-H catalyzed nitration can be explained by its acid catalyzed reaction which, transforms phenylnitromethane into benzaldehyde.

Several supported solid catalysts have been reported for vapor-phase nitration of benzene with NO_2, mostly in patent literature. Examples include WO_3-MoO_3,[131] $SiO_2-Al_2O_3$,[132] $SiO_2-Al_2O_3$ containing SO_4^{2-} ion,[133] metal phosphates,[134] and metal sulfates.[135] With these catalysts, however, the yield of nitrobenzene (based on benzene) was only 4-50%. Suzuki *et al.* have reported on the vapor-phase nitration of benzene over polyorganosiloxanes with a yield of 80%.[136] More recently, silica supported benzenesulfonic acid gave a 70 - 93% yield.[137]

Schumacher and Wang[138] have studied the vapor phase nitration of aromatic compounds including benzene and chlorobenzene with NO_2 over zeolites (molecular sieves) at 80-190°. Aromatic hydrocarbons, including halo- and alkoxybenzenes, were nitrated in the gas-phase over zeolite catalysts with NO_2.[138a,b] For example the nitration of chlorobenzene over Mordenite at 175°C gave 10% *o*-, 0.3% *m*-, and 33% *p*-chloronitrobenzene in 43% conversion.[138a] With N_2O_4 as nitrating agent using H-ZSM-5, benzene was nitrated to nitrobenzene in 98% conversion.[138c] Shape selectivity of the zeolite should greatly prefer para-nitration of toluene, although data on isomer distribution are unavailable.

2.23.3 Dinitrogen Pentoxide

N_2O_5, the anhydride of HNO_3, can be prepared by the dehydration of HNO_3 with P_2O_5 or by ozonolysis of N_2O_4. Solid N_2O_5 as shown by Ingold *et al.* is a colorless, crystalline ionic salt, i.e., nitronium nitrate $NO_2^+NO_3^-$. This structure was later confirmed by Raman spectroscopy and X-ray crystallography.

A study of nitrations with nitrogen pentoxide gave evidence, however, for the fact that other carriers of the NO_2^+ ion such as molecular N_2O_5 itself may also play a role. Decomposition to N_2O_4 and oxygen should also be considered if the N_2O_5 used is not entirely pure.

Aromatics are nitrated by N_2O_5 in carbon tetrachloride solution.[140] Benzotrichloride[141] and benzoyl chloride[142] can be nitrated without hydrolysis. However, decomposition of N_2O_5 to N_2O_4 and O_2 complicate the reaction. Klemenc and Scholler [143] have studied the nitration of aromatics with N_2O_5 in HNO_3 and H_2SO_4. Raman spectroscopic studies show that solutions of N_2O_5 in HNO_3 are completely ionized to NO_2^+, as shown by the absorptions at 1050 and 1400 cm^{-1}.[32] Gillespie's cryoscopic investigations also confirm the equilibrium

$$N_2O_5 + 3H_2SO_4 \rightleftharpoons 2NO_2^+ + 3HSO_4^- + H_3O^+$$

Nitration of aromatics with N_2O_5/P_2O_5 also allows obtaining nitro compounds with complete utilization of nitrogen.[144]

Lewis acid catalyzed nitration of aromatic compounds with the complex $N_2O_5 \cdot BF_3$, first prepared by Schmeiser[146] was reported by Bachman.[145] The active nitrating agent was originally postulated to be $NO_2^+BF_3ONO_2^-$. Raman and infrared studies of the complex have shown it to be $NO_2^+BF_4^-$,[128] probably formed via the reaction

$$3N_2O_5 + 8BF_3 \longrightarrow 6NO_2^+BF_4^- + B_2O_3$$

Mixing of N_2O_5 and BF_3 produces a white solid complex slightly soluble in sulfolane and nitromethane, and gives a colorless, stable liquid layer in CCl_4, probably of the polarized complex.[144] The complex nitrates even deactivated aromatics in high yield, thus indicating high activity (Table 24).

Subsequent investigations by Olah et al.[77,147] showed that Friedel–Crafts nitrations can be carried out efficiently by using N_2O_5 and Lewis acid catalysts such as BF_3, $TiCl_4$, $SnCl_4$, and PF_5 in sulfolane solution, without isolation of the intermediate complexes. Alkylbenzenes (using excess aromatics) and halobenzenes can be nitrated in 79–95% yield.

Lewis acid catalyzed nitration with N_2O_5 can also be carried out in anhydrous HF. However, because of the poor (~2%) solubility of aromatics in HF, the heterogeneous reaction mixture has to be stirred vigorously. The use of pyridinium polyhydrogen fluoride, a convenient HF-like solvent, overcomes much of these difficulties.[147] Although N_2O_5 does not seem to react with HF under the reaction conditions, addition of Lewis acids, such as BF_3, SbF_5, PF_5, AsF_3, SiF_4, NbF_5, TaF_5, and WF_6, results in the quantitative formation of the corresponding nitronium salts. Such solutions are extremely reactive and will nitrate even nitrobenzene and benzotrifluoride in ~90% yield.

Table 24. Nitration of Aromatics with $N_2O_5 \cdot BF_3$ in CCl_4[77,145]

Aromatic	Product	Yield (%)
Nitrobenzene	m-Dinitrobenzene	86
Benzoic Acid	3,5-Dinitrobenzoic acid	70
Benzoic Acid	m-Nitrobenzotrifluoride	84
Chlorobenzene	o,m,p-Chloronitrobenzenes	78
Benzonitrile	m-Nitrobenzoic acid	81

2.24 Nitronium Salts

Hydrocarbons are efficiently nitrated by nitronium salts under anhydrous conditions, as shown by Ingold *et al.* and subsequently developed by Olah *et al.*[148] as a general preparative nitration method

$$RH + NO_2^+MX_n^- \longrightarrow RNO_2 + HX + XN_{n-1}$$

As already mentioned (see Section 2.6) Hantzsch first reported[29] the reaction of HNO_3 with $HClO_4$. He claimed that the product formed was a mixture of nitracidium perchlorate and hydronitracidium perchlorate.

It was left to Goddard, Hughes, and Ingold[30] to show that Hantzsch's preparation gave a mixture of nitronium perchlorate and hydronium perchlorate, from which the nitronium salt could be isolated only with difficulty. They themselves developed a preparation of pure nitronium perchlorate, and a number of nitronium sulfates were also reported

$$N_2O_5 + HClO_4 \longrightarrow NO_2^+ClO_4^- + HNO_3$$

$$HNO_3 + 2SO_3 \longrightarrow NO_2^+HS_2O_7^-$$

$$N_2O_5 + 2SO_3 \longrightarrow (NO_2^+)_2S_2O_7^{2-}$$

$$N_2O_5 + HSO_3F \longrightarrow NO_2^+SO_3F^- + HNO_3$$

These and related salts were characterized by Raman spectroscopy, and other physical measurements.

Reaction of NO_2Cl with Lewis acid halides was also examined as a route to nitronium salts.[77] Reactions of NO_2Cl with $SnBr_4$ and SO_3 indicated the formation of $(NO_2)_2SnCl_6$ and $(NO_2)(ClS_2O_6)$ respectively.

The best indication for the formation of a nitronium salt from NO_2Cl came from its reaction with $SbCl_5$. Anion exchange reaction of the resulting product with $(CH_3)_4N^+ClO_4^-$ and $(CH_3)_4N^+BF_4^-$ allowed the isolation of $NO_2^+ClO_4^-$ and $NO_2BF_4^-$, respectively.[149] Reaction of NO_2F with a large number of Lewis acid fluorides gave nitronium salts containing complex fluoride anions[150]

$$NO_2F + MF_n \rightleftharpoons NO_2^+MF_{n+1}^-$$

$$2NO_2 + MF_m \rightleftharpoons (NO_2^+)_2MF_{m+2}^{2-}$$

$$MF_n = BF_3, PF_5, AsF_5, SbF_5, SeF_4, IF_5; \quad MF_m = SiF_4, SnF_4, GeF_4$$

Nitronium salts can also be prepared by reaction of strongly oxidizing NO_2F with elements such as Si, Ge, Sn, P, As, Sb, Te, Br_2, and I_2.[77,150]

The reaction of metal oxides with excess NO_2F also gave nitronium salts.

Woolf and Emeleus and later Clark and Emeleus synthesized nitronium salts by reaction with N_2O_5 and BrF_3 and the appropriate oxides (such as B_2O_3)[149,150]. Schmeisser and Elisher[146] and subsequently Olah, Kuhn, and Mlinko[148] prepared $NO_2^+BF_4^-$ and $(NO_2^+)_2SiF_6^{2-}$ respectively by the reaction of the appropriate Lewis acid fluoride with N_2O_5 and HF in nitromethane. This reaction can also be used to prepare other nitronium salts. A simple and efficient preparation of nitronium tetrafluoroborate was achieved by Olah et al. by letting a 2 mole excess of BF_3 react with an equimolar mixture of HNO_3 and anhydrous HF.[148]

$$HNO_3 + HF + nBF_3 \longrightarrow NO_2^+BF_4^- + H_2O \cdot (n-1)BF_3$$

Water formed as byproduct in the reaction is bound by BF_3 as a stable hydrate from which BF_3 can be easily regenerated by distilling with H_2SO_4 or oleum. Other nitronium salts (PF_6^-, AsF_6^- can also be prepared in a similar fashion. The reaction requires large amounts of PF_5 and AsF_5 because of their hydrolytic instability. The method thus can be used to prepare $NO_2^+BF_4^-$, $NO_2^+PF_6^-$, $NO_2^+AsF_6^-$, $NO_2^+SbF_6^-$, and $(NO_2^+)_2SiF_6^{2-}$. Since HNO_3, if not carefully purified, always contains nitrous acid (nitrogen oxides) the nitronium salts obtained can contain nitrosonium ion (NO^+) salts (vide infra).

Kuhn has found that HNO_3 can be replaced by alkyl nitrates (free of nitrites) in the preparation of nitronium salts.[153]

$$R{-}O{-}NO_2 + HF + 2BF_3 \longrightarrow NO_2^+BF_4^- + R{-}O{-}H\ BF_3$$

SbF_5 and AsF_5 react explosively with alkyl nitrates. Therefore, the reaction is limited to the preparation of $NO_2^+BF_4^-$, $NO_2^+PF_6^-$, and $(NO_2^+)SiF_6^{2-}$. This method provides extremely pure nitronium salts free of NO^+.

Coon et al.[154] have reported the preparation of nitronium trifluoromethanesulfonate based on the analogy of the related preparation of the perchlorate or fluoroborate. Hydronium trifluoromethanesulfonate is, however, difficult to separate from the nitronium salt

$$HNO_3 + 2CF_3SO_3H \longrightarrow NO_2^+CF_3SO_3^- + H_3O^+CF_3SO_3^-$$

Nitronium trifluoromethanesulfonate can also be readily prepared by the reaction of N_2O_5 with either trifluoromethanesulfonic anhydride (Yagupolskii)[155] or trifluoromethanesulfonic acid (Effenberger).[156] The former gives pure salt free of HNO_3 or hydronium triflate

$$(CF_3SO_2)_2O + N_2O_5 \longrightarrow 2NO_2^+CF_3SO_3^-$$

$$CF_3SO_3H + N_2O_5 \longrightarrow NO_2^+CF_3SO_3^- + HNO_3$$

Nitronium salts are colorless, crystalline, hygroscopic compounds. Nitronium perchlorate, sulfate and hexafluoroiodate are unstable. The spontaneous decomposition (explosive nature) of the perchlorate was experienced by Ingold.[157] It is in all probability due to the equilibrium with the covalent nitrate

$$NO_2^+ClO_4^- \rightleftharpoons O_2NClO_3$$

Consequently, its use is not generally recommended and extreme caution is called for. In contrast, complex fluoride salts such as the tetrafluoroborate and hexafluorophosphate are very stable. Only on heating to higher temperatures (>180–200°C) do they decompose into

NO_2F and the corresponding Lewis acid fluoride.

Specific conductivity of nitronium tetrafluoroborate varies linearly with concentration. Cryoscopic measurements in sulfolane solution indicated that the nitronium salt is present as ion pairs and that the conductance must be due to ion triplets and not separated ions.[148]

The nitronium salts have been well characterized by infrared and Raman spectroscopy, by ^{15}N NMR, and X-ray crystallography. All spectroscopic and crystallographic evidence indicates that the nitronium ion has a linear structure. These studies are well reviewed[77] and no further discussion is necessary.

Some nitronium salts with complex fluoride anions are commercially available or can be prepared readily in the laboratory by the procedure of Olah et al. from HNO_3.[148] Since HNO_3 always contains some HNO_2, the commercial salt generally needs to be purified from nitrosonium ion impurities. This can, according to Ridd, be readily achieved in the case of $NO_2^+PF_6^-$ by recrystallization using nitromethane.[158] In the case of $NO_2^+BF_4^-$ because of its limited solubility, such purification is difficult. To obtain pure $NO_2^+BF_4^-$ free of $NO^+BF_4^-$ starting from HNO_3, it is necessary to purify it from HNO_2 by treatment with urea and immediately convert it into NO_2^+ ion.[159] Alternatively, alkyl nitrates, which can be prepared in pure form, can be converted into nitronium salts free of NO^+.[153]

Sulfolane is a relatively good solvent for nitronium salts. $NO_2^+BF_4^-$ has an about 7% solubility.[160] Acetonitrile is also applicable for nitrations with nitronium salts but the nitrile group strongly interacts with NO_2^+ and causes acetonitrile to slowly oligomerize even at room temperature. Thus only freshly prepared solutions should be used at low temperature. Nitromethane and nitropropane can also be used as solvents for nitrations particularly with remarkably soluble $NO_2^+PF_6^-$ (~25%). On the other hand, nitronium tetrafluoroborate has little solubility in nitromethane (<0.5%). All solvents should be thoroughly dried and purified. Although sulfolane has a relatively high melting point for a solvent (+28.9°), its large molar freezing point depression (66.2°) allows nitrations to be carried out in a wide temperature range. Methanesulfonic acid, trifluoromethanesulfonic acid, fluorosulfuric acid, and even H_2SO_4 have also been used as solvents especially for deactivated substrates. It is important to choose a solvent that by itself does not react with the nitronium salt and preferably provides homogeneous solutions.

2.24.1 Nitronium Tetrafluoroborate

Nitronium tetrafluoroborate is the most frequently used nitronium salt for nitrating aromatics:

$$ArH + NO_2^+BF_4^- \longrightarrow ArNO_2 + HF + BF_3$$

The obtained byproducts HF and BF_3 can be readily recycled on an industrial scale and thus the nitration made catalytic.

Results of preparative nitration of arenes, haloarenes, and nitroarenes are summarized in Tables 25–27. Since HF and BF_3 are the only byproducts of the reaction, nitration with nitronium salts can be carried out under anhydrous conditions. This is advantageous in the nitration of aromatics containing functional groups sensitive to hydrolysis. Thus, aromatic nitriles, acid halides, and esters can be nitrated in high yield without any difficulty (Tables 28–29).

Table 25. Nitration of Arenes with $NO_2^+BF_4^{-77}$

Substrate	Product	Percent yield of mononitro product
Benzene	Nitrobenzene	93
Toluene	Nitrotoluenes	95
o-Xylene	Nitroxylenes	91
m-Xylene	Nitroxylenes	90
p-Xylene	Nitro-p-xylene	93
Mesitylene	Nitromesitylene	89
Ethylbenzene	Nitroethylbenzenes	93
n-Propylbenzene	Nitro-n-propylbenzenes	91
Isopropylbenzene	Nitro-isopropylbenzenes	93
n-Butylbenzene	Nitro-n-butylbenzenes	90
s-Butylbenzene	Nitro-s-butylbenzenes	92
t-Butylbenzene	Nitro-t-butylbenzenes	88
Biphenyl	Nitrobiphenyls	94
Naphthalene	Nitronaphthalenes	79
Phenanthrene	Nitrophenanthrene	89
Anthracene	9-Nitroanthracene	85
Fluorene	2-Nitrofluorene	79
Chrysene	6-Nitrochrysene	73
Benzo[a]pyrene	6-Nitrobenzo[a]pyrene	79
Anthrathrene	Nitroanthanthrenes	81

Table 25. Nitration of Arenes with $NO_2^+BF_4^-$[77] (cont.)

Substrate	Product	Percent yield of mononitro product
Pyrene	1-Nitropyrene	85
Triphenylene	Nitrotriphenylenes	77
Perylene	3-Nitroperylene	85

Table 26. Nitration of Haloarenes and Haloalkanes with $NO_2^+BF_4^-$[77]

Substrate	Product	Percent yield of mononitro product
Fluorobenzene	o,p-Fluoronitrobenzenes	90
Chlorobenzene	o,p-Chloronitrobenzenes	92
Bromobenzene	o,p-Bromonitrobenzenes	87
Iodobenzene	o,p-Iodonitrobenzenes	90
Benzotrifluoride	m-Nitrobenzotrifluoride	20
p-Fluorobenzotrifluoride	3-Nitro-4-fluorobenzotrifluoride	85
o-Dichlorobenzene	Nitro-o-dichlorobenzenes	70
m-Dichlorobenzene	Nitro-m-dichlorobenzenes	74
p-Dichlorobenzene	Nitro-p-dichlorobenzenes	80
o-Difluorobenzene	Nitro-o-difluorobenzenes	82
m-Difluorobenzene	Nitro-m-difluorobenzenes	79
p-Difluorobenzene	Nitro-p-difluorobenzenes	85
α-Fluoronaphthalene	Nitro-α-fluoronaphthalenes	75
β-Fluoronaphthalene	Nitro-β-fluoronaphthalenes	79
Benzyl chloride	Nitrobenzyl chlorides	52
β-Fluoroethylbenzene	Nitro-β-fluoroethylbenzenes	69
β-Chloroethylbenzene	Nitro-β-chloroethylbenzenes	82
β-Bromoethylbenzene	Nitro-β-bromoethylbenzenes	78

Table 27. Nitration of Nitroarenes with Nitrohaloarenes with $NO_2^+BF_4^-$[77]

Substrate	Product	Percent yield of mononitro product
Nitrobenzene	m-Dinitrobenzene	81
α-Nitronaphthalene	Dinitronaphthalenes	85
p-Fluoronitrobenzene	2,4-Dinitrofluorobenzene	78
o-Fluoronitrobenzene	2,4-Dinitrofluorobenzene	84
2,4-Dinitrofluorobenzene	Picryl fluoride	40

Table 27. Nitration of Nitroarenes with Nitrohaloarenes with $NO_2^+BF_4^{-77}$ (cont.)

Substrate	Product	Percent yield of mononitro product
p-Nitrochlorobenzene	2,4-Dinitrochlorobenzene	75
o-Nitrochlorobenzene	2,4-Dinitrochlorobenzene	77
2,4-Dinitrochlorobenzene	Picryl chloride	80

Table 28. Nitration of Arylcarboxylic Acid Esters and Halides with $NO_2^+BF_4^{-77}$

Substrate	Product	Percent yield of mononitro product
Methyl benzoate	Methyl m-nitrobenzoate	88
Ethyl benzoate	Methyl m-nitrobenzoate	79
Propyl benzoate	Methyl m-nitrobenzoate	82
Ethyl m-nitrobenzoate	Ethyl 3,5-dinitrobenzoate	60
Benzoyl fluoride	m-Nitrobenzoyl fluoride	69
Benzoyl chloride	m-Nitrobenzoyl chloride	70

Table 29. Nitration of Aryl and Aralkyl Nitriles with $NO_2^+BF_4^{-77}$

Substrate	Product	Percent yield of mononitro product
Benzonitrile	3-Nitrobenzonitrile	85
o-Toluonitrile	2-Methyl-5-nitrobenzonitrile	90
m-Toluonitrile	Nitrotoluonitriles	85
p-Toluonitrile	4-Methyl-3-nitrotoluonitrile	92
Nitro-o-toluonitrile	3,5-Dinitro-o-toluonitrile	93
Nitro-m-toluonitrile	Dinitro-m-toluonitriles	84
Nitro-p-toluonitrile	3,5-Dinitro-o-toluonitrile	89
p-Fluorobenzonitrile	4-Fluoro-3-nitrobenzonitrile	90
p-Chlorobenzonitrile	4-Chloro-3-nitrobenzonitrile	92
1-Naphthonitrile	Nitronaphthonitrile	91
Benzyl cyanide	Nitrobenzyl cyanides	84

Nitration of aromatics with nitronium tetrafluoroborate is usually carried out in sulfolane or with the more soluble nitronium hexafluorophosphate in nitromethane solution. Reactions of reactive aromatics can be carried out from −20°C to room temperature and short reaction times (5–10 min). Deactivated aromatics need higher temperatures and longer reaction times. They are preferentially carried out in strongly acidic solutions (CF_3SO_3H, FSO_3H, HF, H_2SO_4). The nitro products are generally formed in very high yield. Mononitrations, with limited (<3%) polynitration are achieved using an excess of the aromatic substrate. This is different form mixed acid nitrations, where the high solubility of mononitro compounds in the acid layer frequently results in the formation of increased amounts of dinitro byproducts.

The reactivity of nitronium salts is further enhanced in strong acids such as FSO_3H. Such solutions can be used to even trinitrate benzene to 1,3,5-trinitrobenzene, a reaction which was previously reported only in low yield.[161-163] 1,3,5-Trinitrobenzene is usually obtained only indirectly,[164] but can be prepared in good yield by the nitration of meta-dinitrobenzene with NO_2BF_4 in FSO_3H.[165] Optimum reaction conditions require a reaction time of ~3 hr at 150°C, to yield 100% pure 1,3,5-trinitrobenzene in 50% yield.[166,167] The data in Table 30 show that higher yields can be obtained at shorter reaction times, with mixtures of di- and trinitro products necessitating purification by HPLC. Longer reaction times give pure 1,3,5-trinitrobenzene but also result in oxidative losses and hence lower the yield.

Nitronium salts clearly are the most effective electrophilic nitrating agents for the nitration of aromatic compounds under very mild conditions. They are also widely applied in the nitration of heterocyclic aromatic compounds. The nitration of heterocyclic compounds by nitronium salts was first studied in the case of pyridine.[76,79,168] N-nitration giving N-nitropyridinium ion is followed by ring opening, if excess pyridine is present, yielding glutaconic aldehyde

$$C_6H_5N + NO_2^+BF_4^- \longrightarrow C_6H_5NO_2^+BF_4^-$$

$$[C_6H_5NNO_2]^+BF_4^- \xrightarrow{C_5H_5N} [C_6H_5^+NCH=CH-CH=CH-CH=N-NO_2]^+BF_4^-$$

$$\xrightarrow{H_2O} [C_6H_5NCH=CH-CH=CH-CHO]BF_4^-$$

Table 30. Nitration of m-Dinitrobenzene to 1,3,5-Trinitrobenzene with Nitronium Tetrafluoroborate ($NO_2^+BF_4^-$) in Fluorosulfuric Acid (FSO_3H) Solution at 150°C[165]

Reaction time (h)	Recovery of nitro compounds (%)	1,3,5-Trinitrobenzene in total nitro products (%)	Yield of 1,3,5-trinitrobenzene (%)
0	100	0	0
0.5	95.2	38.0	36.2
1.0	90.3	60.4	54.5
1.7	82.5	80.0	66.2
2.2	77.7	85.0	66.0
3.0	64.8	95.0	61.6
3.4	56.7	98.2	55.7
3.6	52.3	99.4	52.0
3.8	49.3	100	49.3
4.0	44.8	100	44.8
4.2	39.4	100	39.4

Reverse addition of pyridine to excess nitronium salt gives stable N-nitropyridinium salts.[169] No N-C migration of the nitro-group is, however, observed even on heating. On the other hand, alkylnitropyridinium salts are good transfer nitration agents[169 172] (see Section 2.25.2).

The preferential N-nitration of pyridine would seem to indicate that direct electrophilic C-nitration is difficult to achieve except when the non-bonded nitrogen electron pair is occupied such as in the case of pyridine-N-oxides which are readily nitrated in the 4 position. Pyridinium salts have deactivated rings and are nitrated only with difficulty in the 3 position.

Olah and Kuhn reported that thiophene forms nitrothiophene in 91% yield on nitration with $NO_2^+BF_4^-$,[77] while the nitration of furan results in a 14% yield of nitrofuran. 3-Nitro-6-phenyl-2-pyridone has been obtained in 40% yield by the nitration of 6-phenyl-2-pyridone with $NO_2^+BF_4^-$.[173]

C-Nitrazoles are known to be difficult to obtain by direct nitration with HNO_3.[174a] The use of $NO_2^+BF_4^-$ in acetonitrile made it possible to obtain a series of 3-substituted 1,2,4-triazoles.[174b] The reaction proceeds via a stage involving the formation of the N-nitro-derivatives, which isomerize to the C-nitro-compounds. Isopropylimidazole is nitrated analogously.[175]

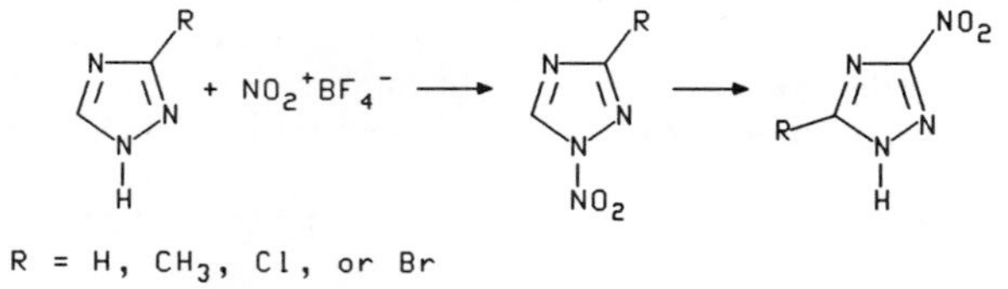

R = H, CH₃, Cl, or Br

2-Isopropyl-1-trimethylsilylimidazole when nitrated with $NO_2^+BF_4^-$ gave 2-isopropyl-1,4-dinitroimidazole. De-silylative N-nitration is involved in the reaction (see Section 2.26.5).

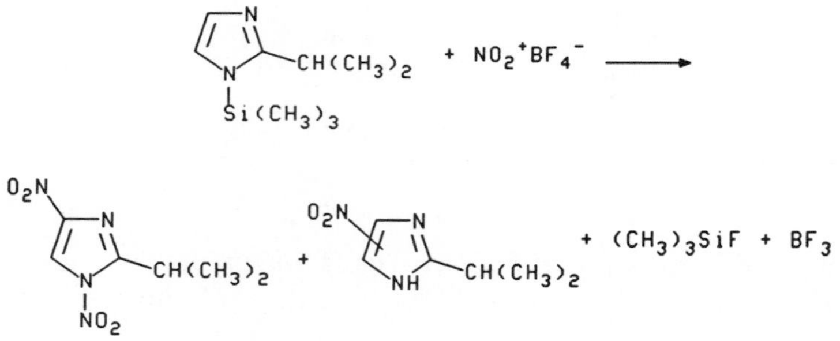

Similar nitration of N-trimethylsilyl-1,2,4-triazoles makes it possible to obtain N-nitrotriazoles, which can be converted to the C-nitro derivatives in high yield.[174]

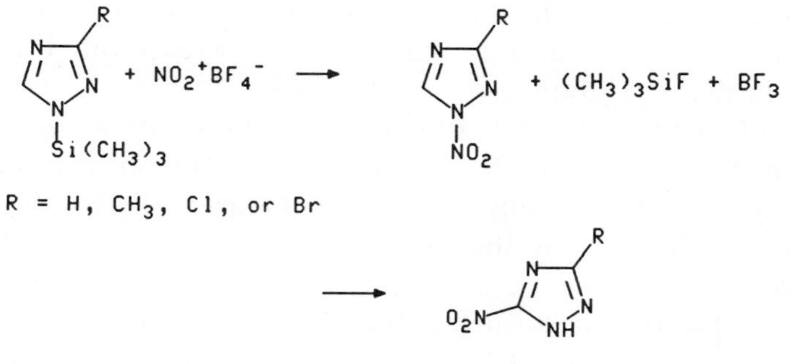

R = H, CH₃, Cl, or Br

Uracil and its 1-methyl and 1,3-dimethyl derivatives are nitrated to 5-nitro derivatives in high yield by $NO_2^+BF_4^-$ in sulfolane:

R^1 and R^2 = H or CH_3

Attempts to nitrate a pyrimidine nucleoside (a 2'-deoxyglycoside) by $NO_2^+BF_4^-$, as a rule lead to the degradation of the molecule or the formation of 5-nitrouracil, but the corresponding nucleotide (nucleoside 5'-monophosphate) is nitrated in the 5 position in the pyrimidine ring.[176]

2.24.2 Nitronium Hexafluorophosphate

The most serious limitation of the use of NO_2BF_4 is its low solubility in many solvents. The most convenient solvent is sulfolane, in which the tetrafluoroborate is soluble in about 7%. In nitromethane its solubility is only ~ 0.2%. Therefore, there is a significant need for more soluble nitronium salts that at the same time are stable.

Nitronium hexafluorophosphate ($NO_2^+PF_6^-$) in contrast was found to be much more soluble in many solvents. Its solubility in nitromethane for example is >30%. Consequently it was recognized as a very convenient nitrating agent for aromatics, (as well as aliphatics, see Section 4.3.9). Like the tetrafluoroborate, it can be prepared by using HF and PF_6. The availability of PF_5 may be, however, a limitation.

2.24.3 Nitronium Trifluoromethanesulfonate (Triflate)

Nitration of aromatics with the nitronium trifluoromethanesulfonate (containing hydronium trifluoromethanesulfonate) formed in the HNO_3-CF_3SO_3H system have been studied by Coon et al.[154] (see Section 2.12). Selective mono- and dinitration of toluene in 98% yield can be carried out under heterogeneous conditions by varying the reaction temperature. Low reaction temperature (−60 to −110°C) result in the formation of ortho- and para-nitrotoluenes, with meta-nitrotoluene limited to 0.2 to 0.5%. The very limited meta-nitration is, however, primarily due to the low reaction temperatures. The heterogeneous nature of these nitrations precludes comparison of data with homogeneous nitrations with $NO_2^+BF_4^-$.

Nitric acid-triflic anhydride (trifluoromethanesulfonic anhydride) was found by Olah et al. to be a very effective nitrating agent.[12] The system can be used in sulfolane or nitromethane solution. HNO_3-$(CF_3SO_2)_2O$ acts as nitronium triflate according to

$$HNO_3 + (CF_3SO_2)_2O \longrightarrow CF_3SO_3H + NO_2^+CF_3SO_3^-$$

2.25 Transfer Nitrating Agents

According to Ingold, the reactivity of a nitrating agent NO_2-X is highly dependent upon the electron affinity of X.[1] It is therefore possible to alter the reactivity of the NO_2^+ ion by using different carriers of varying electron affinity. If a prepared nitro (or nitrito) onium ion is used as the nitrating agent, transfer of the nitro group to the substrate occurs. These reactions are called *transfer nitrations* utilizing nitro and nitrito onium salts generally derived from suitable O, S, and N containing heteroorganic compounds.

The term *transfer nitration* is thus defined as a nitration carried out by reacting the incipient nitronium ion bound to a suitable career (the delivery system), in order to modify the reactivity and reaction conditions of nitrations. The term is, however, arbitrary. There is frequently only a fine dividing line between solvated nitronium ions and nitro onium ions, although in other cases the transfer nitrating agents are stable and well defined.

2.25.1 Nitronium Tetrafluoroborate-Crown Ether Complexes

The regioselectivity of nitration of toluene with nitronium salts has been successfully altered by their prior complexation with crown ethers. Complexation of $NO_2^+BF_4^-$ by 18C6 crown ether substantially altered the selectivity in nitration of toluene and benzene as reported by Elsenbaumer and Wasserman.[177] Similar effect was observed with polyethylene oxides. Savoie *et al.* reported isolation of the 18C6-$NO_2^+BF_4^-$ complex and its characterization.[178] Masci has carried out the most detailed study yet of the effect of crown ethers on the selectivity of electrophilic aromatic nitration.[179]

Equimolar amounts of $NO_2^+BF_4^-$ and 21-crown-7 or 18-crown-6 ethers yielded homogeneous solutions in nitromethane and dichloromethane. In using these systems in nitrating benzene and toluene, both substrate and positional selectivities were altered and were dependent upon the nature of the crown ether and crown ether/$NO_2^+BF_4^-$ ratio. The linear nitronium ion seems to form a guest-host complex with the crown ether. The ortho-para ratio in nitration of toluene can be varied from 1.5 to 0.3 on changing the 21-crown-6-ether/$NO_2^+BF_4^-$ ratio from 1 to 6. The isomer distribution of the nitration of toluene with $NO_2^+BF_4^-$ in CH_2Cl_2 with crown 18C6 was 53% ortho, 4% meta, and 43% para (ortho/para ratio, 1.2), and with crown 21C7, 19% ortho, 3% meta, 78% para (ortho/para ratio 0.25), showing the much better complexing ability of the latter, giving very high preference for para nitration. Since crown-ether complexations affect the NO_2^+ cation, the high preference for para-nitration reflects not only a bulkier reagent, but also a much more selective nitration with a weaker electrophile.

2.25.2 N-Nitropyridinium and N-Nitroquinolinium Salts

Olah *et al.* in 1965 reported the preparation of N-nitropyridinium tetrafluoroborate from pyridine and nitronium tetrafluoroborate.[169] The salt showed only limited reactivity in carrying out transfer C-nitration of aromatic hydrocarbons, probably because of the insolubility of N-nitropyridinium tetrafluoroborate in the reaction medium. Transfer nitration of *n*-donor heteroorganic substrates (alcohols, etc.) was, however, readily accomplished. Cupas and Pearson then extended the scope of transfer C-nitration by the preparation and use of a variety of N-nitropyridinium and N-nitroquinolinium salts.[170] Comprehensive studies by Olah *et al.*[171,172] allowed the design of reagents of varying reactivity by appropriate choice of the heterocyclic base and also the counter ions (PF_6^- vs. BF_4^-). Nitration with these reagents occurs under basically neutral conditions because the proton eliminated in the aromatic nitration reaction is bound by the heterocyclic base.

$$ArH + \underset{\underset{NO_2}{|}}{\underset{R}{\bigcirc_N^+}} PF_6^-(BF_4^-) \longrightarrow ArNO_2 + \underset{\underset{H}{|}}{\underset{R}{\bigcirc_N^+}} PF_6^-(BF_4^-)$$

The N-nitropyridinium and N-nitroquinolinium salts are stable, but moisture-sensitive, crystalline reagents, well characterized by spectroscopic methods (NMR, IR and Raman). They are prepared in essentially quantitative yield by the slow addition of the corresponding pyridine to an equivalent amount of the nitronium salt in acetonitrile, nitromethane, or sulfolane solution. It is important to add the pyridine to the solution of the nitronium ion, because excess pyridine present during the reaction can lead to opening of the pyridinium ring. The N-nitropyridinium salts can be used as isolated compounds or they can be generated *in situ*.

N-Nitropyridinium hexafluorophosphate does not react with benzene and toluene at room temperature, whereas N-nitro-2-picolinium tetrafluoroborate nitrates well under similar reaction conditions. This ease of nitration is due to the methyl group causing steric hindrance to resonance with concomitant weakening of the N–N bond. It seems that non-bonded interaction with one α-methyl group is sufficient to impede completely the resonance interaction, since N-nitro-2,6-lutidinium salt did not further change the selectivity of the reagent. The steric hindrance to resonance can also be achieved by utilizing the peri interaction in N-nitroquinolinium salts.

There is generally no significant change of positional selectivity in the nitration of aromatics with N-nitropyridium ions. In case of toluene, the isomer distribution is 62–64% ortho, 2–4% meta, and 33–36% para-nitrotoluene.

2.25.3 Pyridine-N-Oxide- and Dimethylsulfoxide-Nitrosonium Tetrafluoroborate

N-Nitrito-4-nitropyridinium salts are isomeric with the previously discussed N-nitropyridinium ions. Similarly, dimethylnitritosulfonium salts are isomeric with N-nitrosulfonium ions formed from nitronium salts and dimethyl sulfide. The nitro–onium salts are prepared from nitrosonium hexafluorophosphate with 4–nitropyridine–N–oxide and dimethyl sulfoxide, respectively.[111]

The nitrito–onium salts act as nitrating agents for aromatics that do not undergo nitrosation. Their nitrating ability is, however, considerably less compared to that of the corresponding nitro–onium salts; e.g., toluene is nitrated only at $\geq 60°C$ whereas the nitro–onium salts nitrate at $\leq 25°C$.

S–Nitrosulfonium salts isomerize to S–nitritosulfonium salts at $-20°C$. As a consequence dialkyl sulfides are readily oxidized to their sulfoxides with nitronium salts. When triarylphosphines are reacted with nitronium salts, only the nitrophosphonium ions are observed spectroscopically (by NMR). They subsequently give the corresponding phosphine oxides and nitrosonium ion. These observations can also be rationalized as a consequence of the ambident reactivity of the nitronium ion reacting not on nitrogen, but on oxygen and thus acting as an oxidizing agent (single electron transfer can also be involved).

2.25.4 N-Nitropyrazole

N-Nitropyridinium salts are good transfer nitrating agents but they require careful handling because of their sensitivity to moisture. Consequently, to overcome this difficulty Olah *et al.* have developed[180] other N-nitramine-based transfer nitrating agents such as N-nitrapyrazole.

$$ArH + \underset{NO_2}{\text{pyrazole}} \xrightarrow{H^+} ArNO_2 + \underset{H}{\text{pyrazole}}$$

N-nitrapyrazole is a very convenient reagent because of its weak N–NO$_2$ bond. The N–NO$_2$ bond (1.399Å) is considerably longer than that in dimethylnitramine (1.372 Å). The N–NO$_2$ bond becomes more labile for cleavage upon protonation or complexation with a Lewis acid. Reactions are conveniently catalyzed with boron trifluoride etherate or protic acids (including methanesulfonic acid and triflic acid). Table 31 shows that aromatic nitro products are obtained in high yield on nitration with N-nitropyrazole/boron trifluoride etherate. The isomer distributions show lower ortho/para ratios than in nitronium salt nitrations.

The reactive transfer nitrating agent is considered to develop from the complex of boron trifluoride with N-nitropyrazole.

Transfer nitrations with N-nitropyrazole can also be carried out in the presence of Bronsted acids.

Table 31. Boron Trifluoride Etherate Catalyzed Nitration of Alkylbenzenes with N–Nitropyrazole[180]

Substrate	Percent yield	Percent isomer distribution			o/p
		o	m	p	
Benzene	89				
Toluene	92	58	3	39	1.49
Ethylbenzene	87	44	3	53	0.83
n-Propylbenzene		41	3	56	0.73
Isopropylbenzene	90	22	4	74	0.30
tert-Butylbenzene	86	10	7	83	0.12
o-Xylene		52% 3-nitro, 48% 4-nitro			
m-Xylene		15% 2-nitro, 85% 4-nitro			
p-Xylene	96				
Mesitylene	41				
1,2,4-Trimethylbenzene		11% 3-nitro, 61% 5-nitro, 28% 6-nitro			
1,2,3,4-Tetramethylbenzene	–				

2.25.5 9-Nitroanthracenium Ion

Electrophilic aromatic nitration is generally irreversible, although cases of intermolecular isomerization in nitroarenium ion intermediates have been reported,[181] particularly in highly activated systems such as anthracene.[182] This promised to allow transfer nitration with some of these systems. 9-Nitroanthracene was indeed found to act as a transfer nitrating agent[183a] under superacid catalysis (HF–TaF$_5$ or Nafion–H) because in its *ipso* protonated arenium ion the strong peri–interaction of the nitro group with two neighboring hydrogens tilts the nitro group out of plane of the aromatic ring, thereby weakening the carbon–nitrogen bond. It is therefore possible to carry out transfer nitration of benzene, toluene, and mesitylene with 9-nitroanthracene in the presence of superacid catalysts.

Transfer nitration of toluene yielded predominantly p-nitrotoluene (>95%) indicating the bulky nature of the transfer nitrating agent, but perhaps also due to the heterogeneous reaction conditions and an obviously late arenium ion-like transition state of the reaction. To our knowledge this is the highest para-selectivity reported in nitrations of toluene, exceeding even those effected by crown-ethers (see Section 2.25.1).

2.25.6 Nitrohexamethylbenzenium Ion

The nitrohexamethylbenzenium ion, prepared from hexamethylbenzene and nitronium salts, was studied by low-temperature NMR spectroscopy,[183b] showing intramolecular nitro group migration

Olah *et al.* also found that the ion is capable of transfer nitrating benzene and mesitylene.[183a] The transfer nitrating ability of the nitrohexamethylbenzenium ion is interesting since addition of hexamethylbenzene as a complexing agent to nitronium salt nitrations of aromatics can affect regioselectivity.

2.26 Oxidative Nitration with Nitrosonium Salts

Kim and Kochi[184] found that benzene, methylbenzenes, halobenzenes, naphthalene, and anthracene are nitrated by nitrosonium salts, such as $NO^+PF_6^-$, in acetonitrile, nitromethane, or methylene chloride solution upon introduction of dioxygen, which oxidizes the initially formed colored charge-transfer complexes

$$ArH + NO^+PF_6^- \rightleftharpoons [ArH \cdot NO^+PF_6^-] \xrightarrow{\frac{1}{2}O_2} ArNO_2 + HPF_6$$

Addition of hindered bases, such as 2,6-di-*tert*-butyl-4-methylpyridine, which binds the acid formed, helps to minimize side reactions. Photolysis was also found to promote oxidative aromatic nitration via NO^+ arene charge-transfer complexes.

2.27 Nitration via Metallation

When toluene is nitrated with conventional electrophilic nitrating agents, the product distribution usually shows 60–65% ortho, 3–4% meta, and 27–30% para nitrotoluene. Only in nitrosative nitrations, crown ether complexed or transfer nitrations, and heterogeneous solid-acid catalyzed nitrations is there a significant change in isomer distribution, reflecting increased steric hindrance to ortho substitution and late arenium ion-like transition states of highest energy.

One of the most successful approaches for altering the regioselectivity of aromatic nitration involves nitration via metallation.[84] This method was discovered as a catalytic nitration, which at the same time also provides unusual isomer distribution. The most important metallative nitration reactions involve metallation with mercury, palladium, and thallium salts.

2.27.1 Nitration via Mercuration

The first report of catalytic nitration via mercuration was a patent issued to Wolffenstein and Boeters[185] at the beginning of the century. They reported a procedure for the synthesis of dinitrophenol and picric acid via oxynitration of benzene with mercuric nitrate and 50–55% HNO_3. The mechanism of oxynitration was delineated by Westheimer et al.[186] and other research groups, working under the National Defense Research Committee sponsorship during the second World War. The established reaction Scheme is as shown

$$C_6H_6 + Hg(NO_3)_2 \rightleftharpoons C_6H_5-HgNO_3 + HNO_3$$

$$C_6H_5-HgNO_3 + N_2O_4 \longrightarrow C_6H_5-NO + Hg(NO_3)_2$$

$$C_6H_5-NO + 2\,NO \longrightarrow C_6H_5-N\equiv N^+ + NO_3^-$$

$$C_6H_5-N\equiv N^+ + H_2O \longrightarrow C_6H_5-OH + N_2 + H^+$$

$$C_6H_5-OH + HNO_3 \xrightarrow{NO_2} HO-C_6H_3(NO_2)_2\ (o,p) + H_2O$$

or

$$C_6H_5-NO \xrightarrow[HNO_3]{\text{oxidation and rearrangement}} O_2N-C_6H_4-OH$$

Nitrophenols subsequently can undergo further nitration.

Davis et al. showed that mercury was the only active catalyst among many examined by them.[187] Lead, copper, silver, arsenic, aluminum, manganese, vanadium, cerium, and uranium salts were found to be inactive. These studies firmly established the intermediacy of phenylmercuric nitrate in the reaction, its conversion to nitrosobenzene, and finally to nitrophenols via benzenediazonium nitrate. The formation of phenolic products can be suppressed by using concentrated HNO_3 instead of 50–55% acid. Under such conditions, nitrobenzene is obtained in good yield. Davis et al.,[187] Tsutsumi et al.,[188] and Yoshida et al.[189] discovered that the ortho–para ratio of nitrotoluenes can be significantly altered via mercurative-nitration generally increasing para-substitution. The ortho–para ratio could be changed from 2 : 1 to 1 : 2 via mercuration nitration. Stock et al. later confirmed and extended these experimental findings.[190] The reaction can be catalyzed by mercuric oxide, mercuric acetate, mercuric nitrate, and to a lesser extent by mercuric sulfate. The effect of reaction conditions on isomeric distribution was examined. The ortho–para ratio decreases as the reaction progresses. It was established that nitrotoluenes are formed via initial nitroso-demercuration followed by oxidation of nitrosotoluenes.

It was left to Komoto and co-workers[191] to demonstrate the preparative usefulness of mercurative-nitration in the synthesis of 1,2-bis(4-aminocyclohexyl)ethane, an important intermediate in the synthesis of polyamides, polyimides, and polyurethanes.

Because of the advantages of using solid superacidic catalysts in electrophilic aromatic nitration and in acid catalyzed reactions in general, Olah et al. have examined the mercury(II)-promoted azeotropic nitration of aromatics using Nafion-H solid superacidic catalyst.[192] Azeotropic removal of water accelerates the rate of reaction by mitigating the dilution of HNO_3 in a static reaction system. The yield of nitroaromatics varied from 48–77% (Table 32).

As the water formed is removed azeotropically, the mercury-impregnated Nafion-H catalyst can be recovered by filtration without any loss of activity and can be recycled. Comparison of data with nitration in the absence of mercury catalyst shows that formation of less-hindered isomeric nitroarenes are favored. It is interesting to note that attempted azeotropic nitration of ethylbenzene with nitric acid-Nafion-H yielded only acetophenone via side-chain oxidation, whereas in the presence of mercury salt under similar reaction conditions, nitroethylbenzenes were obtained in good yield with only 13% of product due to side-chain oxidation.

Table 32. Hg^{2+}-Promoted Nitration of Aromatics over Nafion–H Catalyst[192]

Substrate	Yield[a] (%)	Isomer (% distribution)
Benzene	71	
Toluene	67	2-nitro (33), 3-nitro (7), 4-nitro (60)
Ethylbenzene	66	2-nitro (38), 3-nitro (5), 4-nitro (44), acetophenone (13)
tert-Butylbenzene	72	2-nitro (11), 3-nitro (17), 4-nitro (72)
o-Xylene	56	3-nitro (33), 4-nitro (67)
m-Xylene	48	2-nitro (11), 4-nitro (89)
Chlorobenzene	59	2-nitro (37), 3-nitro (2), 4-nitro (61)
Bromobenzene	76	2-nitro (44), 4-nitro (56)
Naphthalene	77	1-nitro (97), 2-nitro (3)

[a]Yields are based on the amount of nitric acid.

2.27.2 Nitration via Palladation

Palladium salts are known to react with aromatic compounds yielding products of oxidation and dimerization. Schramm et al. reported[193] the synthesis of palladium complexes by the reaction of palladium with $NO^+BF_4^-$, $NO_2^+BF_4^-$, and N_2O_4. Tisue and Downs showed[194] that the same Pd(II) complexes could be obtained by the reaction of $NaNO_2$, N_2O_4 or $NO-O_2$ with suspensions of colloidal palladium in acetic acid. These complexes catalyzed the oxidative substitution of benzene to yield mixtures of phenylacetate and nitrobenzene in the ratio of 1 : 2 and 2 : 1 depending upon the nature of the reagents used. The yield of phenyl acetate and nitrobenzene was estimated to correspond to a turnover of approximately 60 for the palladium catalyst. Ichikawa et al. showed that small amounts of nitro compounds are obtained on reaction of aromatic hydrocarbons with palladium nitrate in acetic acid.[195] Henry's work on Pd(II)catalyzed substitution of aromatics and mercurated aromatics in the presence of nucleophiles and oxidants ($NaNO_3$, $NaNO_2$) showed the formation of nitrobenzene as a byproduct. He proposed a palladation/oxidative nitration pathway.[196]

$$\text{ArH} + PdX_2 \longrightarrow \text{Ar-PdX} \xrightarrow[X^-]{\text{oxidant}} \text{Ar-X} + PdX_2$$

Based on foregoing observations, Norman et al.[197] developed a semicatalytic process for the nitration of aromatics with palladium acetate–sodium nitrite in chloroacetic acid. The isomer distribution for nitrotoluenes at 100°C was 9% ortho, 42% meta, and 49% para. The product was contaminated with arylchloroacetates. The reaction can be made catalytic by replacing sodium nitrate with N_2O_4. With this system, nitrobenzene was obtained in 18–30% yield (based on palladium(II)) indicating good turnover. The authors have tentatively suggested the following mechanism:

$$\text{Ph-Pd}^{II} \xrightarrow{NO_2^-} \left[\text{(H,NO}_2\text{)-Pd}^{2+} \longleftrightarrow \text{(H,NO}_2\text{)-Pd}^0 \right] \longrightarrow PhNO_2 + H^+ + Pd^0$$

The results are in agreement with the earlier work of Ichikawa et al.[195]

2.27.3 Nitration via Thallation

The work of McKillop and Taylor on regioselective thallation and nitrosodethallation of aromatic compounds[198] has led to the discovery of nitro–dethallation by Davies et al.[199] The reaction of aromatics with $Tl(OCOCF_3)_3$ in trifluoroacetic acid was carried out while passing NO_2 through the reaction mixture. High yields of nitroarenes were obtained. The nitro compounds are predominantly ortho and para substituted in the case of alkylbenzenes, and aromatics containing o,p-directing substituents. The para–substituted products predominate. The isomer ratios change if the intermediate arylthallium species are allowed to equilibrate by heating under reflux.

Uemura et al. have shown[200] that nitroarenes are also obtained by the reaction of arylthallium(III) compounds with $NaNO_2$, KNO_2, and $AgNO_2$ in trifluoroacetic acid solution. They showed 100% specific nitro–dethallation under the reaction conditions. The reaction can also be carried out via *in situ* thallation. Nitrosoarenes were shown to be intermediates in the reaction. It was shown that the nitro compounds are formed via nitroso-dethallation followed by *in situ* oxidation of the nitrosoarenes.

$$R-C_6H_4-TlXY \xrightarrow[-TlXY^+]{+NO^+} R-C_6H_4-NO \xrightarrow{[O]} R-C_6H_4-NO_2$$

The source of NO^+ is HNO_2 and N_2O_3 generated under the reaction conditions:

$$NaNO_2 + CF_3CO_2H \rightleftharpoons HNO_2 + CF_3CO_2Na$$

$$2HNO_2 \rightleftharpoons N_2O_3 + H_2O$$

$$HNO_2 + H^+ \rightleftharpoons NO^+ + H_2O$$

$$N_2O_3 + H^+ \rightleftharpoons NO^+ + HNO_2$$

Even though there are claims that $NaNO_2$ in trifluoroacetic acid is itself the nitrating agent, it is clear from the nitro-dethallation experiments on isolated arylthallium trifluoroacetates that nitro-dethallation actually proceeds via nitrosodethallation-oxidation sequence.

2.27.4 Nitro-Demetallation with Other Metal Compounds

Uemura *et al.* showed that nitro-demetallation reactions can be carried out with arylmetal compounds of Si, Sn, Pb, and Bi.[201]

The nitro-demetallation of organomercury, palladium, and thallium compounds, discussed previously, can be also conveniently carried out.

Nitration via metallation/nitrosodemetallation and oxidation sequences allows the use of different metallations for obtaining specific nitro products with increased selectivity.

2.27.5 Desilylative Nitration

Desilylative nitration of arylsilanes involves the needed *ipso* nitroarenium ion intermediate. In the reactions the major products are nitrated arylsilanes formed from the reaction at the ortho, meta and para positions.

[Reaction scheme: ArSiR$_3$ + NO$_2^+$ → Wheland intermediate (R$_3$Si, NO$_2$ on same carbon) ⇌ (with X$^-$) → PhNO$_2$ + R$_3$SiX; alternatively, intermediate with SiR$_3$ and NO$_2$ on different carbons → nitrated arylsilane (o, m, p-)]

Nitro–desilylation is usually faster than nitro–deprotonation because the nitrosilylbenzenium ion is stabilized by the silyl substituent. Deans and Eaborn showed that 1,4-*bis*(trimethylsilyl)benzene undergoes nitro–desilylation on nitration with acetyl nitrate.[202] Acetyl nitrate is the reagent of choice, because nitration with HNO$_3$–H$_2$SO$_4$ will lead primarily to proto–desilylation.

[Reaction: 1,4-bis(trimethylsilyl)benzene + HNO$_3$ + (CH$_3$CO)$_2$O → 2-nitro-1,4-bis(trimethylsilyl)benzene]

Nitro–deprotonation is, however, favored in the nitration of 4-tolyltriethylsilane.

[Reaction: 4-tolyltriethylsilane + Cu(NO$_3$)$_2$ / (CH$_3$CO)$_2$O → 3-nitro-4-methylphenyltriethylsilane]

This is perhaps due to the bulkier nature of the triethylsilyl substituent, resulting in steric hindrance to *ipso* attack.[203]

Benkeser,[204] Speier,[205] and Eaborn[206] have shown that intact ring nitration generally predominates and proto-desilylation plays a significant role under the reaction conditions.

As mentioned earlier, Uemura *et al.* have observed nitro-demetallation with arylmetal compounds of Sn, Pb, and Bi as well as Si.[201] Under harsh conditions indirect nitration via nitroso-desilylation/oxidation complicates the reactions.[207]

Olah and Narang[208] found that the reaction of nitronium tetrafluoroborate with phenyltrimethylsilane gives only limited (10%) nitro-desilylation in addition to intact (predominantly para) ring nitration.

Acid-sensitive systems represent a particularly useful application for desilylative nitration with nitronium salts. Mononitration of imidazoles and triazoles is difficult because acid formed in the reactions even with $NO_2^+BF_4^-$ tend to catalyze denitration. However, the nitration of trimethylsilyl derivatives with $NO_2^+BF_4^-$ overcomes this difficulty. It has been shown for 2-isopropyl-1-trimethyl-silylimidazole that it is possible to obtain nitro compounds that could not be obtained previously, for example, 2-isopropyl-1,4-dinitroimidazole.[209]

The nitration of N-trimethylsilyl-1,2,4-triazoles makes it possible to obtain N-nitrotriazoles, which can be used to synthesize C-nitro-derivatives in yields up to 90%.[210]

R = H, CH₃, Cl, or Br

The method is also most convenient for the preparation of 3(5)-nitro-1,2,4-triazoles.

II. Homolytic (Radical) Nitration

Electrophilic acid catalyzed nitration allows preparation of aromatic nitro compounds under a great variety of conditions. Regardless, there is also substantial need for carrying out nitrations without acid catalysis. Subsequently, such nitrations are discussed which are generally of homolytic nature, (mechanisms of nitration are discussed in Chapter 3). It should be pointed out here, however, that aromatic nitration reactions usually produce acids as by-products, by replacing ring hydrogens by the nitro group, and consequently autocatalysis by *in situ* formed acids must be considered.

When discussing nitrations taking place via homolytic free-radical processes we differentiate two types. In reactions that take place by oxidation of the aromatics of sufficiently low oxidation potential compared to the oxidizing agent (which can be N_2O_4, metal complexes, bidentate metal nitrates, or in some instances the NO_2^+ ion itself), aromatic radical cations are first formed, which then react with $\cdot NO_2$ giving nitroarene products. These nitrations can be categorized as radical ion nitrations. It is alteratively possible to achieve radical nitration of aromatics directly by $NO_2\cdot$ generated via thermal, photolytic, or radiolytic homolysis of N_2O_4 or other NO_2-carriers. These nitrations are characterized as free-radical nitrations.

2.28 Nitrogen Dioxide (Dinitrogen Tetroxide)

2.28.1 Solution Nitration

By itself N_2O_4 (we are using interchangeably NO_2 and N_2O_4 in our discussion as they are generally in equilibrium) like other lower oxides of nitrogen in solution, is generally not effective for nitrating simple aromatic hydrocarbons such as benzene and toluene; they are nitrated, but only very slowly. On the other hand, activated aromatics such as phenol, anisole, polymethylated benzenes, and polycyclic aromatic hydrocarbons can be nitrated with N_2O_4 alone. Phenol, anisole, and N,N-dimethylaniline give high proportions of para-nitro derivatives and are possibly nitrated via nitrosation.[211] Ando and Nakaoka have reported[212] the nitration of acetanilide to give 4-nitroacetanilide in 87.6% yield by the reaction with N_2O_4 at 15°C in 2 hr.

Underwood and co-workers reported that mesitylene and other polymethylated benzenes are nitrated at reasonable rates with N_2O_4 in solvents such as nitromethane and methylene chloride.[213] Nitration was accompanied by color change to brown or red. In pentane, a nonpolar solvent, nitration was not observed, nor were there color changes. One mole of nitroaromatic was formed for every mole of N_2O_4, the other half of the nitrogen presumably ending up as HNO_2 although its formation was not confirmed.

Yoshida et al. have examined the nitration of aromatics with N_2O_4 in carbon tetrachloride solution.[214] Nitration of benzene, fluorobenzene, chlorobenzene, bromobenzene, benzonitrile, naphthalene, and acenaphthalene yielded products of predominant ring nitration. However, hydrogen abstraction rather than ring nitration, was the major reaction with toluene, anisole, p-nitrotoluene, and triphenylmethane.

Eberson and Radner reported that unlike simple monocyclic aromatics, polycyclics can be cleanly nitrated with N_2O_4 in methylene chloride at room temperature.[215] High yields of nitrated products, generally in excess of 90%, were obtained form substrates such as perylene, pyrene, anthracene, and chrysene. In most cases the reaction was complete in several hours to a day (Table 33). The work-up procedure consists of little more than evaporating the solvent and excess N_2O_4, followed by column chromatography for purification of the nitroaromatics. For less activated substrates like fluorene and fluoroanthene, addition of a catalytic amount of CH_3SO_3H reduced the reaction time from 24 hr to about 1 hr. More importantly, these nitrations exhibited a high degree of positional selectivity. Thus, nitration of perylene by N_2O_4 in methylene chloride gave 3-nitroperylene almost exclusively. The amount of 1-nitroperylene was 100 times less than that of the 3-nitro isomer. In contrast, nitration with HNO_3 in Ac_2O, the method of choice for acid nitration of perylene, gave the two isomers in a 23 to 1 ratio.

Table 33. Nitration of Polycyclic Aromatic Hydrocarbons with Dinitrogen Tetraoxide[215]

Aromatic	Catalytic amount of CH_3SO_3H added	Reaction time (hr)	Yield (%)	Isomer distribution (%)	
Perylene	No	0.2	95	3-nitro	99.2
				1-nitro	0.8
Pyrene	No	0.5	97	1-nitro	100
Anthracene	No	1	>90	9-nitro	100
Chrysene	No	24	>90	6-nitro	97
				other mononitro	3
Naphthalene	No	48	59	1-nitro	96
				2-nitro	4
Fluorene	No	24	>90	2-nitro	90
	Yes	2	92	3-nitro	1
				4-nitro	9
Fluoranthrene	No	24	75	3-nitro	63
	Yes	0.4	90	8-nitro	27
				other mononitro	10
Binaphthyl	No	24	>90	4-nitro	100
	Yes	1	89		
Triphenylene	No	120	50	1-nitro	22
	Yes	2	92	2-nitro	78

4-Methyl-2,6-di*tert*-butylphenol reacts with NO_2 in the liquid phase to yield the 4-nitrocyclohexadiene derivative.[216]

If the reaction is allowed to proceed for another 4 hr, a dinitrocyclohexanone derivative is obtained as the predominant product.

[Structure: cyclohexenone with (CH₃)₃C, C(CH₃)₃, OH, NO₂, H₃C, NO₂ substituents]

A free-radical mechanism was proposed for the formation of these two compounds. In a related study, Plate *et al.* found that NO_2 reacts with alkaline solutions of guaiacol to yield nitroguauacols.[217]

Heterocyclic compounds are nitrated readily with a solution of NO_2/N_2O_4. Quinoline yields 7-nitroquinoline with NO_2/N_2O_4, whereas nitration occurs in the 5 and 8 positions with mixed acid.[218] Furylpropenones undergo ring nitration as well as nitration of the double bond. The ring products are obtained by subsequent nitration of the exocyclic nitro product.[219]

$$\text{furyl}-CH=CH-\overset{O}{\underset{\|}{C}}-C_6H_4R + N_2O_4 \longrightarrow$$

$$\text{furyl}-CH=CH-\overset{O}{\underset{\|}{C}}-C_6H_4R \longrightarrow \text{furyl}-CH=CH-\overset{O}{\underset{\|}{C}}-C_6H_4R$$

Octaethylporphyrin and mesoporphyrin diethyl ester complexes of various metals react readily with NO_2 in dichloromethane solution.[220] The nature of the nitrated product is dependent upon the oxidation state of the metal. The +3 metalloporphyrins gave $-CH_2NO_2$ groups in the four meso positions. The +2 metalloporphyrins gave substitution of the nitro group in each meso position. Reaction of thebaine with N_2O_4 provided $^{14}\beta$-nitrocodeinone and 8-nitrothebaine in low yield.[221] Thus products of both aromatic substitution and nitro-dealkylation were observed. The Wheland intermediate from attack at the 14 position is highly susceptible to demethylation.

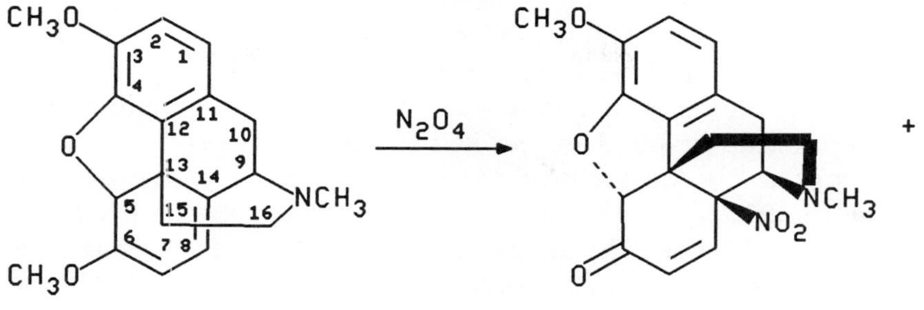

Homolytic (Radical) Nitration

[Structures shown: reaction of a methoxy-substituted morphinan-type alkaloid with N_2O_4 giving a 23% nitro-dienone product and a 7% nitro product, plus another nitro-substituted structure.]

23%

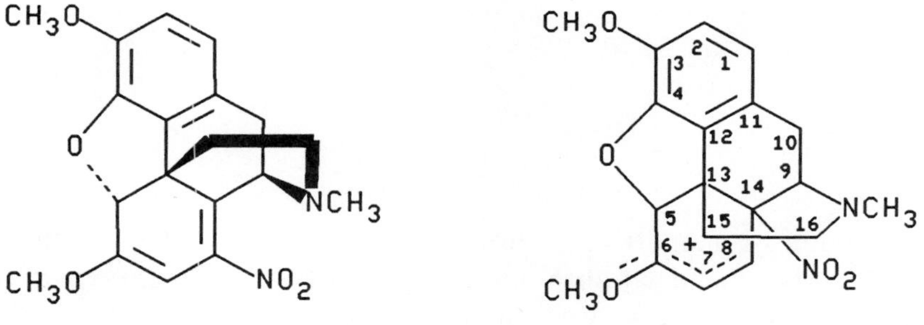

7%

1,2-Dithioacenaphthalene is nitrated readily with N_2O_4. Surprisingly, only products of nitration in the naphthalene ring are obtained, without any oxidation to yield S-oxides.[222] It should be noted that N_2O_4 (as well as nitronium salts) generally oxidize sulfides or disulfides.

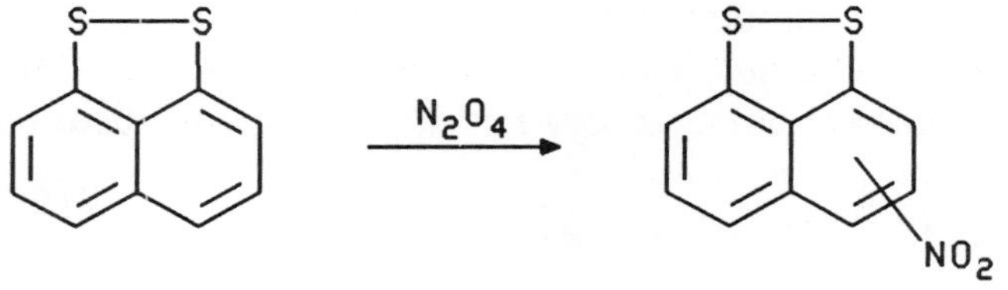

[Reaction scheme: 1,2-dithioacenaphthalene + N_2O_4 → nitro-substituted 1,2-dithioacenaphthalene (NO_2).]

Phenoxides have long been known to be nitrated with tetranitromethane.[223] Both water and pyridine are useful solvents. Azulene has been nitrated with tetranitromethane in pyridine in high yield.[224] Nitration of amino acids has shown that tetranitromethane does not nitrate tryptophan, but it is specific for tyrosine.[225]

Tyrosine is quantitatively converted to 3-nitrotyrosine with tetranitromethane. The optimum conditions are between pH 8 and 9. At higher pH, hydroxide causes breakdown of the tetranitromethane, and below pH 7 no nitration occurs.

$$\text{HOC}_6\text{H}_4\text{CH}_2\text{CH}(\overset{+}{\text{NH}}_3)\text{CO}_2^- + \text{C(NO}_2)_4 \longrightarrow$$

$$\text{HOC}_6\text{H}_3(\text{NO}_2)\text{CH}_2\text{CH}(\overset{+}{\text{NH}}_3)\text{CO}_2^- + \text{HC(NO}_2)_3$$

The nitration of phenols with tetranitromethane was studied by Bruice and was shown to be a radical ion reaction[226] (see Chapter 3, part II).

Nitrations with NO_2 are generally mild reactions but are not yet fully explored for synthetic purposes. They are especially useful for easily oxidizable aromatics, which tend to give extensive side reactions with acid-nitration systems. The reactions are generally carried out in solvents such as dichloromethane, leading to very simple work-up procedures. The reactions are more selective than NO_2^+ ion-effected ones, and yields are high, frequently 90-95%. By addition of nitrosonium ion, nitration of less-reactive aromatics becomes possible, the practical limit being mesitylene and compounds of similar reactivity. Conversely, for very-reactive substrates the activity of NO_2 can be decreased by the addition of a hindered pyridine base, like 2,5-di*tert*-butyl-4-methylpyridine. Thus N_2O_4 in dichloromethane solution is frequently the reagent of choice for the nitration of reactive aromatics.[227]

2.28.2 Gas-Phase Nitration

Titov carried out[228] nitration of simple aromatic hydrocarbons, such as benzene, with low concentrations of NO_2 or with dilute HNO_3 (containing nitrogen oxides). The reaction proceeds at 135–150°C giving nitrobenzene, dinitrobenzene, nitrophenols and oxidation products. The reactions were considered to be of a radical nature involving the stable ·NO_2 radical (see Section 2.27). Similar reaction of toluene (and other methylbenzenes) at 100°C with NO_2 or dilute HNO_3 gave extensive side-chain reaction (oxidation) with the produced mixture containing phenylnitromethane, phenyldinitromethane, and benzoic acid.[229] Ring-nitration products were not further discussed.

Olah and Overchuck studied the ring nitration of benzene, toluene, and fluorobenzene with N_2O_4 under UV irradiation or with irradiation using a Van der Graff generator, as well as by the thermal nitration (>300°C) with tetranitromethane.[230] Isomer distributions obtained were close to statistical. Similar nitration of naphthalene also gave about equal amounts of 1- and 2-nitronaphthalene.[231] Photochemical nitration of benzene had been reported earlier.[232]

Olah and Piteau found that when N_2O_4/NO_2 was reacted with toluene at 190°C while passing through a glass tube reactor about 5% nitrotoluenes were obtained in addition to 30% phenylnitromethane and 25% benzaldehyde. The isomer distribution of nitrotoluenes is 25% ortho, 42% meta, and 33% para, characteristic of radical substitution. If the same reaction was carried out over solid superacidic Nafion-H catalyst, but otherwise under the same conditions, the isomer distribution changed to 49% ortho, 7% meta, and 44% para, indicating predominantly ionic nitration. Similar radical nitration of naphthalene gave a 1-nitro-/2-nitronaphthalene isomer ratio of ~1, whereas ionic nitration gave a ratio of 10.[223]

The cited examples of free-radical nitration indicate a potentially useful way for obtaining otherwise difficult to prepare isomers. However, when statistical mixtures of isomers are formed, they must be separated.

2.29 Photochemical Nitration with Nitrogen Oxides

Gas phase photochemical reactions of nitrogen oxides are of considerable interest to environmental chemists. Watanabe *et al.* have studied the photochemical nitration of benzene with NO_2 in the presence and absence of oxygen. It was found that oxygen greatly accelerates the rate of photochemical formation of nitrobenzene from benzene and NO_2.[234a] Atkinson studied extensively the role of environmental nitrogen oxides, especially NO_3[235b].

Nojima et al. have also studied the environmental fate of nitrogen oxides and halogenated insecticides in air.[235] They found that chlorobenzene was converted to various chloronitrophenols, m-chloronitrobenzene, and p-nitrophenol when exposed to UV radiation in the presence of nitrogen oxides in air. Similarly, p-dichlorobenzene yielded 2,5-dichloronitrobenzene, 2,5-dichlorophenol, 2,5-dichloro-6- nitrophenol, and 2,5-dichloro-4-nitrophenol. The initial product is 2,5-dichloronitrobenzene, which undergoes nitro-nitrite rearrangement on O–N cleavage to yield 2,5-dichlorophenol, which subsequently is nitrated.

Photochemical vapor-phase nitration of bromobenzene, a well- known hepatotoxic compound yields phenol, 4-nitrophenol, 2,4-dinitrophenol, 4-bromophenol, 3-bromonitrobenzene, 3-bromo-2-nitrophenol, 3-bromo-4-nitrophenol, 3-bromo-6-nitrophenol, 2-bromo-4-nitrophenol, and 2,6-dibromo-4-nitrophenol.[236]

Photochemical nitration of phenanthrene with N_2O_4 in CCl_4 solution results in the formation of products of nitration, oxidation, halogenation, and rearrangement.[237]

Homolytic (Radical) Nitration

Photochemical Vapor-phase Nitration of Bromobenzene

2.30 Nitration with $NO/O_2/N_2O_4$

While it is well known that NO and O_2 react to give NO_2 and also that NO_2 is not very effective for nitrating benzene and toluene in solution at modest temperature, it is interesting that simultaneous passage of NO and O_2 through a solution of benzene in N_2O_4 at 0°C results in the formation of nitrobenzene as well as some dinitrobenzene, nitro- and nitrophenols, and traces of trinitrobenzene. Ross came upon this surprising result while attempting to prepare N_2O_4.[238a] The dinitrobenzene isomer ratio was observed to be a $o:m:p$ = 11:62:27. The substantially higher amounts of $o-$ and $p-$dinitro products are in contrast to the ratio commonly found in mixed-acid nitrations, 7:91:2. Since nitrobenzene itself is not nitrated under the same conditions, dinitrobenzene is formed by a pathway that does not involve prior formation of nitrobenzene.

Nitration of toluene also gave nitro- and dinitrotoluenes along with varying amounts of nitrocresols.[238b] The isomer ratio of nitrotoluenes is $o:m:p$ = 51:5:44. This ratio is only slightly different form the one found in mixed acids, in that the $m-$ isomer is somewhat higher. Like nitrobenzene, nitrotoluene also does not undergo nitration under these conditions, nevertheless, the dinitrotoluenes are substantially different from those obtained with mixed acids. Under conventional conditions, most of the product is composed of an 80 : 20 mixture of 2,4- and 2,6-dinitrotoluenes. With $NO/O_2/N_2O_4$, only trace amounts of the 2,6-isomer are obtained with about equal quantities of the remaining five isomers.

The product yield and composition varies with the ratio of the feed gases O_2 and NO.[238b] The efficiency with which the added NO results in nitration changes with the feed-gas ratio. At O_2/NO ratios between 1.0 and 1.8, some of the added NO is merely oxidized to NO_2; above the value of 1.8, all of the added NO is used for nitration of the substrate. Production of dinitrobenzenes is favored at low O_2/NO ratios, which results in an inefficient use of NO. The production of phenolic materials does not seem to correlate with the ratio of feed gases. The ultimate value of benzene-converted/NO ratio reflects the stoichiometry between the two species: each nitroaromatic results from one NO.

$$ArH + NO + O_2 \longrightarrow ArNO_2 + HNO_3$$

Like other nitrations with lower oxides of nitrogen, this reaction produces HNO_3 as a by-product. This interesting reaction needs further study. On possibility is that the initial product of the reaction of NO and O_2, NO_3, being a powerful oxidant, oxidizes the aromatic to its radical cation. Subsequent reaction of the radical cation with NO_2 leads to the reaction products. The polynitrated product is possibly formed by the addition of N_2O_4 to the aromatic radical cation.

2.31 Metal Complex Promoted Nitration with N_2O_4

Following the possibility of one-electron oxidation in nitrations with $NO/O_2/N_2O_4$, Ross, Malhotra, and Johnson studied related nitrations promoted by metal oxidants.[239a] Most salts are too ionic to dissolve in liquid N_2O_4, which has a dielectric constant of only 2.42. However, inner complexes of transition metals, such as acetylacetonates, that have their ionic as well as their coordination valencies simultaneously satisfied, are soluble in media of low dielectric constant. Accordingly, acetylacetonate complexes of a variety of metals were investigated. Nitroaromatics are formed slowly over a period of several hours when the aromatic is added to a solution of Fe(III) or Co(III) acetylacetonate in $NO_2^+BF_4^-$. If on the other hand, the acetylacetonate is added to a solution of the aromatic in N_2O_4 the nitration is rapid and complete in about fifteen minutes. Addition of N_2O_4 to even small quantities of metal acetylacetonates (100 mg) results in a brilliant flame. Experiments were conducted by careful addition of small amounts of the metal complex to liquid N_2O_4 with stirring. Other metal acetylacetonates, such as Mn(III), Cu(II), Ge(IV), Fe(II), and Li(I) were also found effective in nitrating benzene and toluene. The fact that Fe(II) and Li(I) acetylacetonates also promoted nitrations clearly demonstrates that this nitration is not dependent upon the oxidizing power of the metal and the operative mechanism is as yet not known. The amount of nitroaromatics formed is about equimolar to the acetylacetonate complex used.

Under anhydrous conditions, N_2O_4 converts many metal salts into the corresponding nitrates and nitrato complexes of the type $NO^+[M(NO_3)_n]$[240a,241] These complexes are very hygroscopic and need to be handled in a dry-box. Addison and co-workers have shown that the complexes of Fe and U, but not those of Cu and Zn, can nitrate benzene and toluene.[240b] They performed these reactions in excess hydrocarbon and observed stoichiometric nitrations.

Suspecting the formation of such nitrato complexes in the reaction of metal acetylacetonates and N_2O_4, Ross, Malhotra, and Johnson studied the reaction of $NO^+[Fe(NO_3)_4]^-$ with benzene and toluene in a mixture of nitromethane and N_2O_4 at 0°C.[238b] They observed a catalytic nitration of the aromatics. In the case of toluene some dinitro products were also formed with the 2,4- and 2,6- isomers predominating. The $o{:}m{:}p$ isomer ratio of the nitrotoluene products was 53 : 3: 44. The catalytic behavior may be a result of regeneration of the nitrato complex by N_2O_4. Further, the presence of excess N_2O_4 appears to make the system more tolerant towards moisture.

2.32 Bidentate Metal Nitrates

Whereas monodentate metal nitrates are inert towards aromatic hydrocarbons and nitrate only when activated by acids, metal nitrates with bidentate ligands exhibit extremely high reactivity. In fact $Ti(NO_3)_4$ and $Sn(NO_3)_4$ react extremely violently with hydrocarbons giving nitro as well as varied oxidation products.[242]

In reactions with aromatics, isomer distribution data of nitro compounds, obtained with $Ti(NO_3)_4$ and HNO_3/H_2SO_4, show that nitrations are electrophilic in nature, each molecule of $Ti(NO_3)_4$ giving two-mole equivalents of nitro product with 82% efficiency.[243] Spectroscopic evidence indicates the absence of detectable quantities of NO_2^+ in solution of $Ti(NO_3)_4$. Therefore, the nitration was proposed by Amos et al.[244] to proceed as

Similar behavior has been observed for $VO(NO_3)_3$ and $CrO_2(NO_3)_2$, which also have bidentate nitrate ligands.

Because of the extreme speed of reaction, Coombes et al.[245] decided to study the nitration in dilute solution in carbon tetrachloride, a solvent inert to $Ti(NO_3)_4$. Although a homogeneous solution was obtained, addition of the aromatic resulted in the immediate formation of a precipitate. Furthermore, it was shown [246] that the product distribution in the nitration of toluene is significantly dependent upon the relative concentration of the substrate and the nitrating agent.

$Ti(NO_3)_4$, along with $Zr(NO_3)_4$ and $Fe(NO_3)_4NO$ has found good use in the nitration of pyridine and quinoline. Nitration of pyridine with $Ti(NO_3)_4$ gives 3-nitropyridine albeit in modest yield. The reaction of quinoline with $Zr(NO_3)_4$ provides 7-nitroquinoline in 90% yield. These reagents thus have great potential for nitration of various heterocyclic substrates.[247]

Ceric ammonium nitrate also nitrates aromatic hydrocarbons without the presence of any catalysts.[248] For reactive aromatics, the nitration is thought to proceed via oxidation of the aromatic hydrocarbon to the radical cation followed by further reaction (see Section 3.6.5). The radical cation can readily yield products of side-chain substitution. In order to obtain products of ring substitution, the aromatic hydrocarbon is treated with ceric ammonium nitrate in the presence of NO_2. The intermediate radical cation is assumed to be captured by NO_2 to yield ring-substituted nitro compounds via arenium ions.[247]

$$ArCH_3 \xrightarrow{Ce\ IV} ArCH_3^{+\bullet} \xrightarrow{-H^+} ArCH_2^\bullet \xrightarrow{NO_3^\bullet} ArCH_2ONO_2$$

$$ArCH_3^{+\bullet} + NO_2^\bullet \longrightarrow \left[Ar\begin{smallmatrix}CH_3\\NO_2\end{smallmatrix} \right]^+ \xrightarrow{-H^+} Ar\begin{smallmatrix}CH_3\\NO_2\end{smallmatrix}$$

In cases where the aromatic hydrocarbons have low reactivity or where there is no side-chain substitution possible, ring nitration has been observed, e.g., t-butylbenzene, anisole, and naphthalene provide products of aromatic substitution, and toluene gives 66% nitrotoluenes and 34% benzyl nitrate. The reactions appear to be electrophilic in nature, as judged from isomer distribution of the products. The reacting species appears to be polarized ceric ammonium nitrate serving as a source of coordinated incipient NO_2^+ ion.[250]

$$ArH + Ce(NO_3)_6^{2-} \rightleftharpoons ArHCe(NO_3)_6^{2-}$$

$$ArHCe(NO_3)_6^{2-} \rightleftharpoons ArHNO_2^+ + CeO(NO_3)_5^{3-}$$

$$ArHNO_2^+ CeO(NO_3)_5^{3-} \longrightarrow ArNO_2 + Ce(OH)(NO_3)_5^{2-}$$

Ceric ammonium nitrate nitrations, so far, had little preparative use, but have served as valuable probes in mechanistic studies.

The use of copper(II) nitrate in Ac_2O is a useful reagent for nitrating phenylsilanes that otherwise undergo facile proto-disilylation.

Arylsilanes, such as trimethylphenyl- and trimethylbenzylsilane, undergo nitration with cupric nitrate-Ac_2O.[251]

III. Nucleophilic Nitration

Aromatic nitration can also be carried out by nucleophilic substitution reactions. Nucleophilic aromatic nitration is, however, considerably less studied than electrophilic nitration or homolytic radical nitration.

2.33 Nitro-Dehalogenation

Nucleophilic replacement of activated (by electron withdrawing ortho/para groups) halogens in aromatics generally shows the reactivity sequence

$$I > Br >> Cl$$

Lütgert obtained 1,2,4-trinitrobenzene by reacting 4-iodo-1,3-dinitrobenzene with aqueous sodium nitrite at room temperature[252]

The reaction clearly has the characteristics of an S_N2 displacement reaction, but was not further investigated. In aqueous solution, hydrolysis is a strong competing reaction and limits the utility of this method. Even in nonaqueous solutions, the nitrite ion being an ambient nucleophile tends to give aryl nitrites and, through them, phenols. Attempted reactions with silver nitrite gave some improvement, but still generally no satisfactory yields for preparative nitration.[253]

Nucleophilic displacement of halogen in nitration of aromatics should be clearly differentiated from electrophilic nitrodehalogenations observed in some electrophilic, acid catalyzed nitrations.[254]

2.34 Nitro-Dediazoniation

Nitroarenes can be obtained in good yields by treatment of arenediazonium ion salts with sodium nitrite preferentially in the presence of cuprous ion in neutral or alkaline solution (nitro-dediazoniation). The reaction is similar to the Sandmeyer reaction.

$$ArN_2^+ + NaNO_2 \xrightarrow{Cu^{2+}-Cu^+} ArNO_2$$

Diazotization of aromatic amines followed by treatment with sodium nitrite, generally in the presence of a copper sulfate catalyst converts arylamines into nitroarenes. The reaction has been extensively used to prepare nitro derivatives of naphthalene inaccessible by direct nitration.[255]

In contrast, 2- and 4-aminopyridines can be diazotized only under special conditions.[256]

The replacement of the diazonium group by nitrite ion can generally only be effected in neutral or basic media. To achieve neutrality or slight alkalinity various methods are used: addition of calcium carbonate[257] or sodium bicarbonate,[258] or precipitation (and washing free of acid) of the diazonium salts as the sulfates,[259] fluoroborates,[260] or colbaltinitrites.[261] Diazonium fluoroborates give fluoroarenes in a competing Schieman reaction.

Although the use of diazonium fluoroborates is described in Organic Synthesis for the preparation of 2- and 4-dinitrobenzenes,[260] improved yields can be obtained by the method of Ward and co-workers by adding the solution of diazonium sulfate to a solution of excess sodium nitrite and sodium bicarbonate.[258] If electron withdrawing groups are present in the aromatic ring, no catalyst is needed and $NaNO_2$ alone gives high yields.

To isolate the diazonium sulfates as solid compounds, excess ether is added to the diazonium ion prepared in H_2SO_4-AcOH mixture. Solid cobaltinitrites are precipitated form aqueous medium by adding a small excess of sodium cobaltinitrite to the solution of diazonium sulfate or diazonium chloride, previously neutralized with calcium carbonate and filtered.

Results of the nitro-dediazonation of substituted amines by various methods are shown in Table 34.[257,261]

2.35 Nitrolysis of Diarylhalonium Ions

Nesmeyanov first reported[262] that the reaction of phenyl-p-tolyliodonium tetrafluoroborate with sodium nitrite yielded a mixture of nitrobenzene and p-nitrotoluene in a ratio of 2.5 : 1. In a similar reaction McEwen found[263] that diphenyliodonium tetrafluoroborate when allowed to react with sodium nitrite in aqueous dioxane gave nitrobenzene in 70% yield. Olah, Sakakibara, and Asensio[264] carried out a detailed study of the nucleophilic nitrolysis of diarylhalonium ions (both symmetrical and unsymmetrically substituted diarylchloronium, bromonium, and iodonium tetraphenylborates and hexafluorophosphates) with sodium nitrite in aqueous acetone solution.

Table 34. Yields (%) of Nitro Compounds Obtained by
Reaction of Neutralized Solutions and
Isolated-Solid Diazonium Salts with Sodium Nitrite

Diazotized amine	Solution		Solid	
	$Ca(CO_3)^{257}$	$NaHCO_3^{258}$	$(ArN_2)SO_4^{259}$	$(ArN_2)_3Co(NO_2)_6^{261}$
Aniline	35	–	–	75.5
2-Nitroaniline	70	97	–	67.4
4-Nitroaniline	76	97	–	75
4-Chloroaniline	35	–	–	82.5
4-Anisidine	16	–	–	68
2-Anisidine	–	–	–	63
4-Toluidine	–	–	–	69
2-Toluidine	–	–	–	61
2-Naphthylamine	15	–	57	60
4-Nitro-1-naphthylamine	25	50	65	–
5-Nitro-2-naphthylamine	15	–	55	–
Benzidine	10	–	16	–

The reaction of unsymmetrically substitute diarylhalonium tetraphenylborates with nitrite ion gave, in a typical nucleophilic nitrolysis reaction, a mixture of the corresponding nitroarenes. The nitroarene product composition reflects the effect of the substituents on the course of the nucleophilic nitrolysis (Table 35).

$$Ar-X^+-Ar^1\ B(C_6H_5)_4^- \xrightarrow{NaNO_2} ArNO_2 + Ar^1NO_2 + ArX + Ar^1X$$

$$X = Cl, Br, I$$

In the studied reactions, nitroarenes were obtained in 70–75% yield, and only relatively small amounts of phenols (<8%) and biphenyls (<10%) were detected. The presence of water in the reaction media as well as the ambident nature of the nitrite ion can account for the formation of phenols. Since biphenyl was detected even from ditolylchloronium tetraphenylborate, but in substantially decreased yield (~2%) in the case of diphenylchlorinium hexafluorophosphate, it must be mostly formed from the tetraphenylborate anion and not as a reaction by-product from the diarylhalonium ion. In the nitration of phenyl-4-tolylchloronium and 3,3'-ditolylchloronium tetraphenylborate, a small amount of toluene and trace amounts of 4-tolylbenzene and 3,3'-dimethylbiphenyl, respectively, were detected. These results indicate that radical side reactions take place only to a minor degree.

The effect of the halonium center is mostly reflected in its inductive effect on the aryl rings. As expected from the greater electronegativity of chlorine and bromine relative to iodine, diarylchloronium and -bromonium ions were found to be more reactive than the corresponding diaryliodonium ions. In fact, when diphenylchloronium or diphenylbromonium salts were allowed to react with sodium nitrite under similar conditions, the reactions were completed within 2 hr to give nitrobenzene in 75% yield, whereas 65% of the starting diphenyliodonium salt was recovered under the same conditions.

Nitration of 2,2'-3,3'- and 4,4'-ditolylchloronium tetraphenylborate gave 2-, 3-, and 4-nitrotoluene, respectively. Not even trace amounts of other isomers were detected, showing that only *ipso* attack occurred. This rules out the possible formation of a benzyne intermediate or attack at other ring positions, which would be the case if the diarylhalonium ions were to show ambident character.

In the nucleophilic substitution reactions of 4-substituted diaryliodonium salts, nucleophiles generally attack the phenyl ring carrying electron-withdrawing groups. This trend is also observed in the nucleophilic nitrolysis of phenyl-4-tolychloronium hexafluorophosphate, where the obtained nitrobenzene-to-4-nitrotoluene ratio is 3.0 : 1.0. Nitration of phenyl-3-tolylchloronium hexafluorophosphate gave a mixture of nitrobenzene and nitrotoluene in a ratio of 1.0 : 1.2, indicating the weak effect of the 3-methyl relative to the 4-methyl substitution on the course of the reaction. On the other hand, 2-methyl substitution of one of the aryl rings was shown to exert the opposite effect on the relative reactivity of the rings. 2-Nitrotoluene was formed 4.7 times faster than nitrobenzene in the nitration of phenyl-2-tolylchloronium hexafluorophosphate

Nearly identical product distributions were obtained in the nitrolysis of phenyl-4-tolychloronium, -bromonium, and -iodonium salts. However, the *o*-methyl substituent effect increased in the sequence phenyl-2-tolylchloronium < -bromonium < -iodonium ions. The amount of 2-nitrotoluene obtained from the iodonium ion was approximately twice as much as that obtained from the chloronium ion. Nitration of phenyl-2,6-xylylchloronium, -bromonium, and -iodonium ions gave 2-nitro-m-xylene and the corresponding haloarene almost exclusively, indicating the reinforced ortho effect. The effect of the

Table 35. Ratio of Nitroarenes in the Nitrolysis

$$Ar-X^+-Ar' \; B(C_6H_5)_4^- \xrightarrow{NaNO_2} ArNO_2 + Ar'NO_2 + ArX + Ar'X^{264}$$

Halonium ion (Ar–X–Ar¹)	Products ArNO₂	Ar'NO₂	ArNO₂/Ar'NO₂ ratio Observed	Calculated
2-CH₃C₆H₄–Cl⁺–C₆H₅	2-Nitrotoluene	Nitrobenzene	4.7 : 1.0	
3-CH₃C₆H₄–Cl⁺–C₆H₅	3-Nitrotoluene	Nitrobenzene	1.2 : 1.0	
4-CH₃C₆H₄–Cl⁺–C₆H₅	4-Nitrotoluene	Nitrobenzene	1.0 : 3.0	
2-CH₃C₆H₄–Cl⁺–3'-CH₃C₆H₄	2-Nitrotoluene	3-Nitrotoluene	3.9 : 1.0	3.9 : 1.0
2-CH₃C₆H₄–Cl⁺–4'-CH₃C₆H₄	2-Nitrotoluene	4-Nitrotoluene	13.5 : 1.0	14.1 : 1.0
3-CH₃C₆H₄–Cl⁺–4'-CH₃C₆H₄	3-Nitrotoluene	4-Nitrotoluene	3.4 : 1.0	3.6 : 1.0
2,4-(CH₃)₂C₆H₃–Cl⁺–C₆H₅	4-Nitro-m-xylene	Nitrobenzene	1.8 : 1.0	1.6 : 1.0
2,4-(CH₃)₂C₆H₃–Cl⁺–4'-CH₃C₆H₄	4-Nitro-m-xylene	4-Nitrotoluene	3.7 : 1.0	4.7 : 1.0
2,4-(CH₃)₂C₆H₃–Cl⁺–2'-CH₃C₆H₄	4-Nitro-m-xylene	2-Nitrotoluene	1.0 : 1.7	1.0 : 3.0
2,3-(CH₃)₂C₆H₃–Cl⁺–C₆H₅	3-Nitro-o-xylene	Nitrobenzene	11.0 : 1.0	5.6 : 1.0
2,3-(CH₃)₂C₆H₃–Cl⁺–2'-CH₃C₆H₄	3-Nitro-o-xylene	2-Nitrotoluene	1.9 : 1.0	1.2 : 1.0
2-FC₆H₄–Cl⁺–C₆H₅	2-Nitrofluorobenzene	Nitrobenzene	1.0 : 4.5	
3-FC₆H₄–Cl⁺–C₆H₅	3-Nitrofluorobenzene	Nitrobenzene	1.3 : 1.0	
2-ClC₆H₄–Cl⁺–C₆H₅	2-Nitrochlorobenzene	Nitrobenzene	1.8 : 1.0	
4-ClC₆H₄–Cl⁺–C₆H₅	4-Nitrochlorobenzene	Nitrobenzene	1.0 : 2.0	
2-CH₃C₆H₄–Br⁺–C₆H₅	2-Nitrotoluene	Nitrobenzene	7.6 : 1.0	
4-CH₃C₆H₄–Br⁺–C₆H₅	4-Nitrotoluene	Nitrobenzene	1.0 : 3.1	
2-CH₃C₆H₄–I⁺–C₆H₅	4-Nitrotoluene	Nitrobenzene	9.6 : 1.0	
4-CH₃C₆H₄–I⁺–C₆H₅	4-Nitrotoluene	Nitrobenzene	1.0 : 2.7	

chloronium center on the relative reactivity of the phenyl and the chlorophenyl rings in the nitration of phenyl-4-chlorophenyl- and phenyl-2-chlorophenylchloronium ions was shown to be smaller than that observed in the case of methyl substitution, although attack by the nitrite ion took place in the same direction. However, in the case of the phenyl-2-fluorophenylchloronium ion, a reversed ortho effect was observed; the 2-nitrofluorobenzene to nitrobenzene ratio was 1.0 : 4.5.

Substitution of tetraphenylborate for hexafluorophosphate as counterion did not affect the product distributions and suggests that the counterion has no particular effect on the reactions.

IV. Conclusions

Nitric-sulfuric acid (mixed acid) continues to be the most widely used practical nitrating agent. Due to the strongly oxidizing nature of the system coupled with serious environmental problems caused by spent-acid disposal both academic and industrial nitrations increasingly apply more selective reagents and methods. The use of more suitable acid (superacid) catalysts and HNO_3 derivatives, (such as alkyl or acyl nitrates and nitryl halides) allows nitrations to be carried out under anhydrous conditions with recyclable acid systems. Hydrogen fluoride-boron trifluoride (or phosphorous pentafluoride) are readily recoverable, volatile superacid catalysts. They also form stable nitronium salts with nitric acid that can be isolated and used as highly efficient nitrating agents in organic solvents. When using N-nitropyridinium salts as a transfer nitrating agent, the nitrations can be carried out under completely acid-free conditions, since the pyridine base immediately neutralizes the acid formed in the proton elimination step of the nitration. Nitrogen oxides are of interest not only in acid-catalyzed oxidative nitrations, but also in homolytic nitrations. Nitrations via metallation (such as mercurative and thallative nitrations) and homolytic nitrations allow to obtain isomer distributions substantially different form those of acid-catalyzed HNO_3 nitrations. Use of solid superacid catalysts in heterogeneous reactions eliminates altogether the use of corrosive acids and their disposal problems.

References

1. Ingold, C. K. *"Structure and Mechanism in Organic Chemistry"*, Bell and Sons: London, 1953, 2nd ed., Cornell University Press, Ithaca, N. Y. and London, 1969 and references therein.

2. Seidenfaden, W., Pawellek, D. In: Houben-Weyl. *"Methoden der Organischen Chemie"*, Vol. XII, Thieme: Stuttgart, 1971, pp. 477, 488.

3. Houben-Weyl. *"Methoden der Organischen Chemie"*, Vol. XII, Thieme: Stuttgart, 1971, p. 472.

4. Euler, H. *Ann. Chem.* **1903**, *330*, 280.

5. Schofield, K. *"Aromatic Nitration"*, Cambridge University Press: Cambridge, 1980 and references therein.

6. Ciaccio, L. L., Marcus, R. A. *J. Am. Chem. Soc.* **1962**, *84*, 1838; Ciaccio, L. L., Ph. D. Thesis, Polytechnic Institute, Brooklyn, 1962.

7. Ridd, J. H., Draper, M. R. *J. Chem. Soc., Perkin II*, **1981**, 94.

8. Carr, R. V. C., Zoseland, B. A., Eur. Pat. Appl. EP 169, 441; U. S. Appl. 630, 788; *Chem. Abstr.* **1986**, *104*, 151269m.

9. Harrar, J. E., Pearson, R. K. *J. Electrochem. Soc.* **1983**, *130*(1), 108-112.

10. Carr, R. V. S., Ross, D. S., Toseland, B. A., U. S. Pat. Appl. 638,436; Eur. Pat. Appl. EP 173,131; *Chem. Abstr.* **1986**, *105*, P24055j.

11. Tedder, J. M. *Chem. Rev.* **1955**, *55*, 287.

12. Olah, G. A., *et al.*, unpublished results.

13. Alazard, J. P., Kagan, H. B., Setton, R. *Bull. Soc. Chim. Fr.* **1977**, 499.

14. (a) Ingold, C. K. *"Structure and Mechanism in Organic Chemistry"*, Bell and Sons: London, 1953, 2nd ed., Cornell University Press, Ithaca, N. Y. and London, 1969 and references therein; (b) Seidenfaden, W., Pawellek, D. In: *"Methoden der Organischen Chemie"*, Vol. XII, Houben-Weyl. Thieme: Stuttgart, 1971, pp. 477, 488; (c) Houben-Weyl. *"Methoden der Organischen Chemie"*, Vol. XII, Thieme: Stuttgart, 1971, p. 472; (d) Schofield, K. *"Aromatic Nitration"*, Cambridge University Press: Cambridge, 1980 and references therein.

15. Olah, G. A., Kuhn, S. J., Flood, S. H., Evans, J. C. *J. Am. Chem. Soc.* **1962**, *84*, 368.

16. Schofield, K. *"Aromatic Nitration"*, Cambridge University Press: Cambridge, 1980, pp. 240-245.

17. Albright, L. F. *Chem. Eng.* **1966**, *April 25,* 169.

18. Rys, P. *Pure & Appl. Chem.* **1981**, *53*, 209.

19. Chadwick, D. H., Cleveland, T. H. *"Kirk-Othmer Encylcopedia of Chemical Technology"*, Wiley-Interscience:New York, 1981, p. 789.

20. Desvergnes, L. *Chim. Ind. (Paris)* **1931**, *23*, 291.

21. Rochester, C. H. *"Acidity Functions,"* Academic Press: London, 1970.

22. Olah, G. A. *"Friedel-Crafts Chemistry,"* Wiley-Interscience: New York, 1973, pp. 337-339.

23. Moodie, R. B., Schofield, K. Wait, A. R. *J. Chem. Soc., Perkin II*, **1984**, p. 921.

24. Tsang, S. M., Paul, A. P., DiGiaimo, M. P. *J. Org. Chem.* **1964**, *29*, 3387.

25. Harris, G. F. P. In: *"Industrial and Laboratory Nitrations"*, Albright, L. F., Hanson, C., ed., American Chemistry Society Symposium Series Vol. **22**, Washington, 1976, pp. 300-313.

26. Pearson, D. E., Frazer, M. G., Frazer, V. S., Washburn, L. C. *Synthesis*, **1976**, p. 621.

27. Moodie, R. B., Schofield, K., Thomas, P. N. *J. Chem. Soc., Perkin II*, **1978**, p. 318.

28. Robinson, R. *J. Chem. Soc.* **1941**, p. 238.

29. Hantzsch, A. *Ber.* **1925**, *58*, 941; *Z. Phys. Chem.* **1930**, *149*, 161.

30. Goddard, D. R., Hughes, E. D., Ingold, C. J. *J. Chem. Soc.* **1950**, p. 2559.

31. Cox, E. G., Jeffrey, G. A., Truter, M. R. *Nature*, **1948**, *162*, 259.

32. Millen, D. J. *J. Chem. Soc.* **1950**, p. 2606.

33. Gillsepie, R. J., Millen, D. J. *Quart. Revs.* **1948**, *2*, 277.

34. Simons, J. H., Passino, M. J., Archer, S. *J. Am. Chem. Soc.* **1941**, *63*, 608.

35. Houben-Weyl. *"Methoden der Organischen Chemie"*, Thieme: Stuttgart, Vol. X, 1971, Part 1, p. 774.

36. Houben-Weyl. *"Methoden der Organischen Chemie"*, Thieme: Stuttgart, Vol. X, 1971, Part 1, p. 744.

37. *Methodium Chemicum*. Vol. 6, 1.1, ref. 171.

38. Olah, G. A., Kuhn, S. J. *Chem. Ind.* **1956**, p. 98; Olah, G. A., Kuhn, S. J., Mlinko, A. *J. Chem. Soc.* **1956**, p. 4257.

39. Thomas, R. J., Anzilotti, W. F., Hennion, G. F. *Ind. Eng. Chem.* **1940**, *32*, 408; Hennion, G. F. U.S. Patent, 2,314,212 (1943).

40. Csürös, Z., Deák, Gy, Fenichel, L., Török, L., Kalmár, A. *Acta Chim. Acad. Sci. Hung.* **1969**, *59*, 401.

41. Redlich, O., Hood, G. C. *Discuss Faraday Soc.* **1957**, *24*, 87.

42. Bourne, E. J., Stacey, M., Tatlow, J. C., Tedder, J. M. *J. Chem. Soc.* **1952**, p. 1695.

43. Brown, H. C., Wirkkala, R. A. *J. Am. Chem. Soc.* **1966**, *88*, 1447.

44. Spitzer, U. A., Stewart, J. *J. Org. Chem.* **1974**, *39*, 3936.

45. Moodie, R. B., Schofield, K., Tobin, G. D. *J. Chem. Soc., Perkin II*, **1977**, p. 1688.

46. Barnett, J. W., Moodie, R. B., Schofield, K., Taylor, P. G., Weston, J. B. *J. Chem. Soc., Perkin II*, **1979**, p. 747.

47. Olah, G. A., Prakash, G. K. S., Sommer, J. *"Superacids"*, Wiley-Interscience: New York, 1985, pp. 36-37.

48. Coon, C. L., Blucher, W. G., Hill, M. E. *J. Org. Chem.* **1973**, *38*, 4243.

49. Olah, G. A., Lin, H. C. *J. Am. Chem. Soc.* **1974**, *96*, 549.

50. (a) Kameo, T., Nishimura, S., Manabe, O. *Nippon Kagaku Kaishi*, **1974**, *1*, 122.
    (b) Wright, O. L., Teipel, J., Thoennes, D. *J. Org. Chem.* **1965**, *30*, 1301.

51. Tapia, R., Torres, G., Valderrama, J. A. *Syn. Comm.* **1986**, *16*, 681.

52. Olah, G. A., Malhotra, R., Narang, S. C. *J. Org. Chem.* **1978**, *43*, 4628.

53. Milligan, B., Miller, D. G. U.S. Patent 3,957,889 (1976).

54. Topchiev, A. V. *Nitration of Hydrocarbons and Other Organic Compounds"*, Pergamon Press: New York, 1959.

55. Sprague, R. W., Garrett, A. B., Sisler, H. H. *J. Am. Chem. Soc.* **1960**, *82*, 1059.

56. Narang, S. C. Ph.D. Thesis, Flinders University, Adelaide, South Australia, 1975.

57. Olah, G. A., Fung, A. P., Narang, S. C., Olah, J. A. *J. Org. Chem.* **1981**, *46*, 3533.

58. Uemura, S., Toshimitsu, A., Okano, M. *J. Chem. Soc., Perkin I*, **1978**, p. 1077.

59. Crivello, J. V. *J. Org. Chem.* **1981**, *46*, 3056; Crivello, J. V. U.S. Pat. 3,634,520 (1972), U.S. Pat. 3,715,323 (1973).

60. Sastry, S., Kudav, N. A. *Ind. J. Chem. Sect. B.* **1979**, *18B*, 198.

61. Majumdar, M. P., Kudav, N. A. *Ind. J. Chem. Sec. B.* **1976**, *14B*, 1012.

62. Even, C., Fauquenoit, C., Claes, P. *Bull. Soc. Chim. Belg.* **1980**, *89*, 559.

63. Riham, T. I., Mustafa, H. T., Habib, R. M. *Egypt. J. Chem.* **1977** (pub. 1979) *20*, 215.

64. Menke, J. B. *Rec. Trav. Chim.* **1925**, *44*, 141, 270.

65. Menke, J. B. *Rec. Trav. Chim.* **1928**, *48*, 618.

66. Sondheimer, F., Shani, A. *J. Am. Chem. Soc.* **1964**, *86*, 3168.

67. Horita, H., Sakata, Y., Misumi, S. *Tetrahedron Lett.* **1976**, p. 1509.

68. (a) Fukunaga, K., Kimura, M. *Nippon Kagaku Zasshi*, **1973**, p. 1306; (b) Fukunaga, K. *Nippon Kagaku Zasshi*, **1974**, P. 2231.

69. (a) Cornélis, A., Laszlo, P., Pennetreau, P. *J. Org. Chem.* **1983**, *48*, 4771.
(b) Cornélis, A., Laszlo, P., Pennetreau, P. *Bull. Soc. Chim. Belges.* **1984**, *93*, 961.

70. Cornélis, A., Laszlo, P. *Synthesis* **1985**, p. 909.

71. Laszlo, P., Pennetreau, P. *J. Org. Chem.* **1987**, *52*, 2407.

72. Cornélis, A., Delaude, L., Gerstmans, A., Laszlo, P. *Tetrahedron Lett.*, **1988**, p. 5657.

73. Dewar, M. J. S., Mole, T., Urch, D. S., Warford, E. W. T. *J. Chem. Soc.* **1956**, p. 3576.

74. Macsi, B. *Tetrahedron Lett.* **1989**, *45*, 2719.

75. Houben-Weyl. *"Methoden der Organischen Chemie"*, Thieme: Stuttgart, Vol. X, 1971, Part 2, p. 789.

76. Olah, G. A., Olah, J. A., Overchuck, N. A. *J. Org. Chem.* **1965**, *30*, 3373.

77. Olah, G. A., Kuhn, S. J. *"Friedel-Crafts and Related Reactions"*, Vol. III, Part II, Olah, G. A., ed., Wiley: New York, 1964, p. 1393-1491.

78. Raudnitz, H. *Chem. Ber.* **1927**, *60*, 738.

79. Wright, H. R., Donalson, W. J. U.S. Pat. 2,416,974 (1947); *Chem. Abstr.* **1947**, *41*, 3485.

80. Colonna, H. *Pubbl. Inst. Chim. Univ. Bologna*, **1943**, *2*, 3; *Chem. Abstr.* **1945**, *41*, 754.

81. Colonna, H., Andrisano, R. *Pubbl. Inst. Chim. Univ. Bologna*, **1944**, *3*, 3; **1945**, *4*, 3; *Chem. Abstr.* **1945**, *41*, 754.

82. Plazak, E., Roupuszynski, S. *Rocs. Chem.* **1958**, *32*, 681; *Chem. Abstr.* **1959**, *53*, 3111.

83. Tsang, S. M., Paul, A. P., Di Giaims, M. P. *J. Org. Chem.* **1964**, *29*, 3387.

84. Topchiev, A. V. *"Nitration of Hydrocarbons"*, translated by C. Mattews, Pergamon Press: New York, 1959.

85. Bodtker, E. *Bull. Soc. Chim. Fr.* **1908**, *3*, 726.

86. Titov, A. I. *J. Gen. Chem. USSR* **1948**, *18*, 2190.

87. Slavinskaya, R. A. *Zs. Obsc. Chim.* **1957**, *27*, 844; *Chem. Abstr.* **1958**, *52*, 2734a.

88. Wilkinson, J. H., Finar, I. L. *J. Chem. Soc.* **1948**, p. 288.

89. Olah, G. A., Malhotra, R., Narang, S. C. *J. Org. Chem.* **1978**, *43*, 4628.

90. Tronow, B. W., Ssigba-Aullin, N. C. *Zs. Obsc. Chim.* **1930**, *62*, 2267.

91. Olah, G. A., Lin, H. C. *J. Am. Chem. Soc.* **1974**, *96*, 2892.

92. Emmons, W. D., Freeman, J. P. *J. Am. Chem. Soc.* **1955**, *77*, 4391.

93. Narang, S. C. Ph.D. Thesis, Flinders University, Adelaide, South Australia, 1975.

94. (a) Narang, S. C., Thompson, M. J. *Austr. Chem.* **1975**, *28*, 385; (b) Schmidt, M., Schmidbauer, H. *Angew. Chem.* **1959**, *71*, 220.

95. Olah, G. A., Malhotra, R., Narang, S. C. *J. Org. Chem.* **1978**, *43*, 4628.

96. König, W. *Angew. Chem.* **1955**, *67*, 157.

97. Houben-Weyl. "*Methoden der Organischen Chemie*", Vol. XII, Thieme: Stuttgart, 1971, p. 757.

98. Oxford, A. E. *J. Chem. Soc.* **1926**, p. 2004.

99. Hayashi, E., Inana, K., Ishikawa, T. *J. Pharm. Soc. Japan*, **1959**, *79*, 972.

100. Scheinbaum, M. L. *Chem. Comm.* **1969**, p. 1235.

101. Halvarson, K., Melander, L. *Ark. Kemi.* **1957**, *11*, 77.

102. Doering, W. v. E., DePuy, C. H. *J. Am. Chem. Soc.* **1953**, *75*, 5955.

103. Hoggett, J. G., Moodie, R. B., Penton, J. R., Schofield, K. In: "*Nitration and Aromatic Reactivity*", Cambridge University Press: London, 1971.

104. Bordwell, F. G., Garbisch, E. W. *J. Am. Chem. Soc.* **1960**, *82*, 3588.

105. Fischer, A., Read, A. J., Vaughan, J. *J. Chem. Soc.* **1964**, p. 3691.

106. Bodor, N., Dewar, M. J. S. *Tetrahedron*, **1969**, *25*, 5777.

107. Ridd, J. H. In: "*Studies on Chemical Structure and Reactivity*", Ridd, J. H. ed., John Wiley and Sons: New York, 1966, pp. 133.

108. Gold, V., Hughes, E. D., Ingold, C. K. *J. Chem. Soc.* **1950**, p. 2467.

109. Norman, R. O. C., Radda, G. K. *J. Chem. Soc.* **1961**, p. 3030.

110. Hartshorn, S. R., Hoggett, J. G., Moodie, R. B., Schofield, K., Thompson, M. J. *J. Chem. Soc.* **1971**, *B*, 2461.

111. Olah, G. A., Lin, H. C., Olah, J. A., Narang, S. C. *Proc. Natl. Acad. Sci. USA* **1978**, *75*, 1045.

112. Collis, M. J., Gintz, F. P., Goddard, D. R., Hebdon, E. A. *Chem. and Ind.* **1955**, p. 1742.

113. Price, C. C., Sears, C. A. *J. Am. Chem. Soc.* **1953**, *75*, 3276.

114. Paul, R. C., Sigh, D., Malhotra, K. C. *J. Chem. Soc.* **1969**, *A*, 1396.

115. Kuhn, S. J., Olah, G. A. *J. Am. Chem. Soc.* **1961**, *83*, 4564.

116. Dachlauer, K. Germ. Pat. 509,405, 1929; *Chem. Abstr.* **1931**, *25*, 781.

117. Lin, H. C. Ph.D. Thesis, Case Western Reserve University, Cleveland, Ohio, 1972.

118. Hetherington, O., Robinson, F. L. *J. Chem. Soc.* **1954**, p. 3512.

119. Millen, D. J. *J. Chem. Soc.* **1950**, p. 2600.

120. Bachman, G. B., Hokama, T. *J. Am. Chem. Soc.* **1957**, *79*, 4370.

121. Pinck, J. A. *J. Am. Chem. Soc.* **1927**, *49*, 2536.

122. Titov, A. I., Banyshnikova, A. N. *J. Gen. Chem.* **1936**, *6*, 1800; Titov, A. I. *J. Gen. Chem.* **1937**, *7*, 667.

123. Goulden, J. D. S., Millen, D. J. *J. Chem. Soc.* **1950**, p. 2620; Millen, D. J. *J. Chem. Soc.* **1950**, p. 2600.

124. Gillespie, R. J., Graham, J., Hughes, E. D., Ingold, C. K., Peeling, E. R. A. *J. Chem. Soc.* **1940**, p. 2504.

125. U.S. Pat. 4,123,466 (Upjohn); *Chem. Abstr. 90*, 96994v.

126. Schaarschmidt, A. *Ber.* **1924**, *57*, 2065; *Angew. Chem.* **1926**, *39*, 1457.

127. Bachman, G. B., Feuer, H., Bluestein, B. R., Vogt, C. M. *J. Am. Chem. Soc.* **1955**, *77*, 6188; Bachman, G. B., Vogt, C. B. *J. Am. Chem. Soc.* **1958**, *80*, 2987.

128. Evans, J. C., Rinn, H. W., Kuhn, S. J., Olah, G. A. *Inorg. Chem.* **1964**, *3*, 857.

129. Kurz, M. E., Yang, L. t. A., Zahora, E. P., Adams, R. C. *J. Org. Chem.* **1973**, *38*, 2271.

130. Olah, G. A., Overchuck, N. A. *Can. J. Chem.* **1965**, *43*, 3279.

131. Jpn. Kokai Tokkyo Koho, 58-162557 and Eur. Patent Appl., 92372.

132. Kokai Tokkyo Koho, 58-183644.

133. Kokai Tokkyo Koho, 58-180459, Sumitomo; *Chem. Abstr.* **1984**, *100*, 85380e; U.S. 4,426,643 (1984); Monsanto; *Chem. Abstr.* **1984**, *100*, 102918.

134. Jpn. Kokai Tokkyo Koho, 50-126626; 50-126627 and Brit. Patent 586732.

135. Jpn. Kokai Tokkyo Koho, 58-185543.

136. Suzuki, E., Tohmori, K., Ono, Y. *Chem. Lett.* **1986**, p. 747.

137. Suzuki, E., Tohmori, K., Ono, Y. *Chem. Lett.* **1987**, p. 2273.

138. (a) Schumacher, I., Wang, K. B. Europ. Pat. 53031 (1982), Europ. Pat. 78247 (1983), U.S. Pat. 4,107,220 (1978); (b) Jpn. Pat. 59-216851 (1984); (c) Jpn. Pat. 58-15748 (1983).

139. Jpn. Kokai Tokkyo Koho, 58-157748.

140. Haines, L. B., Adkins, H. *J. Am. Chem. Soc.* **1925**, *47*, 1419.

141. Speckles, E. *Ber.* **1919**, *52*, 315.

142. Cooper, K. E., Ingold, C. K. *J. Chem. Soc.* **1927**, p. 836.

143. Klemenc, A., Scholler, A. *Z. Anorg. Allgem. Chem.* **1924**, *141*, 231.

144. Caesar, G. V., U. S. Pat. 2,400,287 (1946); *Chem. Abstr.* **1946**, *40*, 6687.

145. Bachman, G. B., Dever, J. L. *J. Am. Chem. Soc.* **1958**, *80*, 5871.

146. Schmeisser, H. *Angew. Chem.* **1952**, *64*, 616.

147. Olah, G. A. In: *"Industrial and Laboratory Nitrations"*, Albright, L. F., Hanson, C., eds., ACS Symposium Series, Vol. **22**, Washington, D.C. 1976, p. 1.

148. Olah, G. A., Kuhn, S. J. *Chem. Ind.* **1956**, *98*; Olah, G. A., Kuhn, S. J., Mlinko, A. *J. Chem. Soc.* **1956**, p. 4257; Kuhn, S. J., Olah, G. A. *J. Am. Chem. Soc.* **1961**, *83*, 4564.

149. Seel, F., Nogradi, J. *Z. Anorg. Allgem. Chem.* **1952**, *269*, 188.

150. Aynsley, E. E., Hetherington, G., Robinson, P. L. *J. Chem. Soc.* **1954**, p. 1119.

151. Woolf, A. A., Eme1éus, H. J. *J. Chem. Soc.* **1950**, p. 1050.

152. Clark, H. C., Eme1éus, H. J. *J. Chem. Soc.* **1958**, p. 190.

153. Kuhn, S. J. *Can. J. Chem.* **1962**, *40*, 1660.

154. Coon, C. L., Bucher, W. G., Hill, M. E. *J. Org. Chem.* **1973**, *38*, 4243.

References

155. Yagupolskii, L. M., Maletina, I. I., Orda, V. V. *Zh. Org. Chim, USSR* **1974**, *10*, 2226, Eng. Transl. p. 2240.

156. Effenberger, F., Goke, J. *Synthesis* **1975**, p. 40.

157. Ingold, C. K. et al., *J. Chem. Soc.* **1950**, p. 2559 and personal communication to Olah, G. A. in December, 1956.

158. Yoshida, T., Ridd, H. J. In: *"Industrial and Laboratory Nitrations"*, Albright, L. F., Hanson, C., eds., American Chemical Society Symposium Series, Vol. 22, 1976, pp. 110-111.

159. Elsenbaumer, R. J. *J. Org. Chem.* **1988**, *53*, 437.

160. Olah, G. A., Kuhn, S. J., Flood, S. H. *J. Am. Chem. Soc.* **1961**, *83*, 4571.

161. Radcliffe, L. G., Pollitt, A. A. *J. Soc. Chem. Ind.* **1921**, *40*, 45T, 90T.

162. Drummond, A. A. *J. Soc. Chem. Ind.* **1922**, *41*, 338T.

163. Desvergnes, L. *Chim. Ind.* **1931**, *25*, 291.

164. Clarke, H. T., Hartman, W. W. *"Organic Syntheses"*, Coll. Vol. II, John Wiley and Sons, Inc.: New York, 1943, p. 526.

165. Olah, G. A., Lin, H. C. *J. Am. Chem. Soc.* **1974**, *96*, 549.

166. Olah, G. A., Lin, H. C. *Synthesis*, **1974**, p. 494.

167. Lin, H. C., Ph.D. Thesis, Case Western Reserve University, Cleveland, Ohio, 1972.

168. Jones J. *Tetrahedron Lett.* **1964**, p. 2177.

169. Olah, G. A., Olah, J. A., Overchuck, N. A. *J. Org. Chem.* **1965**, *30*, 3373.

170. Cupas, C. A., Pearson, R. L. *J. Am. Chem. Soc.* **1968**, *90*, 4742.

171. Olah, G. A., Narang, S. C., Pearson, R. L., Cupas, C. A. *Synthesis*, **1978**, p. 452.

172. Olah, G. A., Narang, S. C., Olah, J. A., Pearson, R. L., Cupas, C. A. *J. Am. Chem. Soc.* **1980**, *102*, 3507.

173. (a) Shusherina, N. P., Likhomanova, G. I. *Khim. Geterotsikl. Soed.* **1972**, p. 1374; (b) quoted by Guk, Yu. V., Ilyushin, M. A., Golad, E. L., Gidaspov, B. V. *Uspekhi Khimii*, **1983**, *52*, 499, Eng. Transl. **1983**, *52*(3), 284.

174. (a) Boyer, J. H., "Nitrazoles," VCH Pub.: Deerfield Beach, Florida, 1986; (b) Pevzner, M. S., Gidaspov, B. V., Tartakovskii, V. A. *Khim. Geterotsikl. Soed.* **1979**, p. 550; (c) quoted by Guk, Yu. V., Ilyushin, M. A., Golad, E. L., Gidaspov, B. V. *Uspekhi Khimii*, **1983**, *52*, 499, Eng. Transl. **1983**, *52*(3), 284.

175. Glass, R. S., Blount, J. F., Batler, D. *Can. J. Chem.* **1972**, *50*, 3472.

176. Huang, G. F., Torrenu, P. F. *J. Org. Chem.* **1977**, *42*, 3821.

177. (a) Elsenbaumer, R. L., Wasserman, E. *Abstr. Pap-Chem. Congr. North Am. Cont. 2nd* **1980**, *Abstr. 77*; (b) Elsenbaumer, R. L., Wasserman, E., U.S. Patent 4,392978, 1983.

178. Savoie, R., Pigeon-Gosselin, M., Rodrigue, A., Chénevert, R. *Can. J. Chem.* **1983**, *61*, 1248.

179. Masci, B. *J. Chem. Soc. Chem. Commun.* **1982**, p. 1262; *J. Org. Chem.* **1985**, *50*, 4081.

180. Olah, G. A., Narang, S. C., Fung, A. P. *J. Org. Chem.* **1981**, *46*, 2706.

181. Myhre, P. C. *J. Am. Chem. Soc.* **1972**, *94*, 7921.

182. Cerfontain, H., Telder, A. *Recl. Trav. Chim. Pays-Bas.* **1967**, *86*, 371.

183. (a) Olah, G. A., Narang, S. C., Malhotra, R., Olah, J. A. *J. Am. Chem. Soc.* **1979**, *101*, 1805; (b) Olah, G. A., Lin, H. C., Mo, Y. K. *J. Am. Chem. Soc.* **1972**, *94*, 3667.

184. Kim, E. K., Kochi, J. K. *J. Org. Chem.* **1989**, *54*, 1692.

185. Woeffenstein, R., Boeters, O. Germ. Pat. 194,883; *Chem. Abstr.* **1980**, *2*, 1861.

186. Westheimer, F. H., Segel, E., Schramm, R. *J. Am. Chem. Soc.* **1947**, *69*, 773.

187. Davis, T. L., Worrall, D. E., Drake, N. L., Heimkanys, R. W., Young, A. M. *J. Am. Chem. Soc.* **1921**, *43*, 594.

188. Tsutsumi, S., Iwata, E. *J. Chem. Soc. Jpn. Pure Chem. Sect.* **1951**, *72*, 741; *Chem. Abstr.* **1952**, *46*, 6604a.

189. Osawa, T., Yoshida, T., Namba, K. *Kogyo Kayaku Kyokashi* **1966**, *27*, 162.

190. Stock, L. M., Wright, T. L. *J. Org. Chem.* **1977**, *42*, 2875; *J. Org. Chem.* **1979**, *44*, 3467.

191. Komoto, H., Hoyano, F., Takami, T., Yamato, S. *J. Polym. Sci. Part A-1*, **1971**, *9*, 2983.

References

192. Olah, G. A., Krishnamurthy, V. V., Narang, S. C. *J. Org. Chem.* **1982**, *47*, 596.

193. Schramm, R. F., Wayland, B. B. *Chem. Comm.* **1968**, p. 898.

194. Tisue, T., Downs, W. *J. Chem. Comm.* **1969**, p. 410.

195. Ichikawa, K., Uemura, S., Okada, T. *Nippon Kagaku Zasshi* **1969**, *90*, 212.

196. Henry, P. M. *J. Org. Chem.* **1971**, *36*, 1886.

197. Norman, R. O. C., Parr, W. J. E., Thomas, C. B. *J. Chem. Soc., Perkin I*, **1974**, p. 369.

198. Taylor, E. C., Danforth, R. H., Mckillop, A. *J. Org. Chem.* **1973**, *38*, 2088.

199. Davies, B., Thomas, C. B. *J. Chem. Soc., Perkin I* **1975**, p. 65.

200. Uemura, S., Toshimitsu, A., Okano, M. *Bull. Chem. Soc. Japan* **1976**, 49, 2582.

201. Uemura, S., Toshimitsu, A., Okano, M. *J. Chem. Soc., Perkin I* **1978**, p. 1076.

202. Deans, F. B., Eaborn, C. *J. Chem. Soc.* **1957**, p. 498.

203. Benkeser, R. A., Landesman, H. *J. Am. Chem. Soc.* **1954**, *76*, 904.

204. Benkeser, R. A., Brumfield, P. E. *J. Am. Chem. Soc.* **1951**, *73*, 4770.

205. Speier, J. L. *J. Am. Chem. Soc.* **1953**, *75*, 2930.

206. Eaborn, C., Salih, Z. S., Walton, D. R. M. *J. Chem. Soc., Perkin II* **1972**, p. 172.

207. Chvalovsky, V., Bazant, V. *Collect. Czech. Chem. Comm.* **1951**, *16*, 580.

208. Olah, G. A., Narang, S. C. unpublished results.

209. Glass, R. S., Blount, J. F., Butler, D. *Can. J. Chem.* **1972**, *50*, 3472.

210. Pevzner, M. S., Kulibabina, T. N., Ioffe, S. L., Maslina, I. A., Gidaspov, B. V., Tartakovskii, V. A. *Khim. Geterotsikl. Soed.* **1979**, p. 550; quoted by Guk, Y. V., Ihyushin, M. A., Golod, E. L., Gidaspov, B. V. In: *"Nitronium Salts in Organic Chemistry", Russ. Chem. Revs.*: (Eng. Ed) **1983**, *52*, 284.

211. Milligan, B. *J. Org. Chem.* **1983**, *48*, 1495; Uemura, S., Toshimitsu, A., Okano, M. *J. Chem. Soc., Perkin I* **1978**, p. 1076; Davies, B., Thomas, C. B. *J. Chem. Soc., Perkin I* **1975**, p. 65; Effenberger, F., Kurtz, W., Fischer, P. *Chem. Ber.* **1974**, *107*, 1285; Bonner, T. G., Hancock, R. A., Yousif, G., Rolle, F. R. *J. Chem. Soc. B* **1969**, p. 1237.

212. Ando, W., Nakaoka, I. *Jpn. Kokai Tokkyo Koho* **1978**, *121*, 717.

213. Underwood, G. R., Silverman, R. S., Vanderwalde, A. *J. Chem. Soc., Perkin II* **1973**, p. 1170.

214. Yoshida, T., Saheki, K., Takahashi, K., Wakabayashi, T., Namba, K. *Kogyo Kayaku Kyokashi* **1974**, *35*(1), 7-11.

215. Radner, F. *Acta Chem. Scand., Ser. B* **1983**, *37*, 65; Eberson, L., Radner, F. *Acta Chem. Scand., Ser. B.* **1985**, *39*, 343.

216. Brunton, G., Cruse, H. W., Riches, K. M., Whittle, A. *Tetrahedron Lett.* **1979**, p. 1093.

217. Plato, A., Kesten, E. M., Lotz, F., Molinari, E. *An. Asoc. Quim. Argent.* **1977**, *65*(2), 121.

218. Shorygin, P. P., Topchiev, A. *Chem. Ber.* **1936** *69B*, 1974.

219. Sitkin, A. I., Klimenko, V. I., Fridman, A. L. *Zh. Org. Chim.* **1977**, *13*(3), 648.

220. Goug, L. C., Dolphin, D. *Can. J. Chem.* **1985**, *63*, 401.

221. Archer, S., Osei-Gymah, P. *J. Heterocycl. Chem.* **1979**, *16*(2), 389.

222. Oae, S., Nabeshima, T., Takata, T. *Heterocycles* **1982**, *18*, 41.

223. Schmidt, E., Fischer, H. *Chem. Ber.* **1920**, *53*, 1529.

224. Reid, D. H., Stafford, W. H., Stafford, W. L. *J. Chem. Soc.* **1958**, p. 1118.

225. Riordan, J. F., Sokolovsky, M., Vallee, B. L. *J. Am. Chem. Soc.* **1966**, *88*, 4104.

226. Bruice, T. C., Gregory, M. J., Walters, S. L. *J. Am. Chem. Soc.* **1968**, *90*, 1612.

227. Eberson, L., Radner, F. *Acct. Chem. Res.* **1987**, *20*, 53.

228. Titov, A. I. *Zh. Obshch. Khim.* **1947**, *17*, 382; **1948**, *18*, 190; **1952**, *22*, 1379; Titov, A. I. *Tetrahedron*, **1963** *19*, 558.

229. Titov, A. I. *Zh. Obshch. Khim.* **1940**, *10*, 382; 1878.

230. Olah, G. A., Overchuck, N. A. *Can. J. Chem.* **1965**, *43*, 3279.

231. Olah, G. A., Narang, S. C., Olah, J. A. *Proc. Natl. Acad. Sci. USA* **1981**, *78*, 3298.

232. Avanesov, D., Vyatskin, I. *Khim. Referat. 2h.,* **1939**, *2*, 43.

233. Olah, G. A., Piteau, M., unpublished results.

234. (a) Watanabe, K., Ishikawa, H., Ando, W. *Bull. Chem. Soc. Jpn.* **1978**, *51*, 1253.
(b) Winer, A. M., Atkinson, R., Pitts, J. N. Jr. *Science* **1984**, *224*, 156.

235. Nojima, K., Kanno, S. *Chemosphere*, **1980**, *9*, 437; Kanno, S., Nojima, K. *Chemosphere* **1979**, *9*, 225.

236. Nojima, K., Ikarigawa, T., Kanno, S. *Chemosphere*, **1980**, *9*, 421.

237. Barlas, H., Parlar, H., Kotzias, D., Korte, F. *Z. Naturforsch., B. Anorg. Chem. Org. Chem.* **1982**, *37B*(4), 486.

238. (a) Ross, D. S., Blucher, W., Report 1980, ARO-13831.3-CX, order No. ADA085324, avail. NTIS from Gov. Rep. Announce Inder (U.S.), **1980**, *80*, 4255; *Chem. Abstr.* **1981**, *94*, 120541; (b) Ross, D.S., Malhotra, R., Johnson, R. M., unpublished results.

239. (a) Ross, D. S., Johnson, R. M., Malhotra, R. U.S. Pat. 4,417,080; *Chem. Abstr.* **1984**, *100*, P67984a; (b) Ross, D. S., Johnson, R. M., Malhotra, R. U.S. Pat. 4,447,662; *Chem. Abstr.* **1984**, *101*, P38212a.

240. (a) Addison, C. C. *Chem. Rev.* **1980**, *80*, 21; (b) Addison, C. C., Boorman, P. M., Logan, N. *J. Chem. Soc.* **1965**, p. 4978.

241. Boughriet, A., Wartel, M., Fischer, J-C. *Can. J. Chem.* **1986**, *64*, 5.

242. Addison, C. C., Garner, D. C., Simpson, W. B., Sutton, D., Wallwork, S. C. *Proc. Chem. Soc.* **1964**, p. 367.

243. Field, B. O., Hardy, C. J. *J. Chem. Soc.* **1963**, p. 5278.

244. Amos, D. W., Baines, D. A., Flewett, G. W. *Tetrahedron Lett.* **1973**, p. 3191.

245. Coombes, R. G., Russell, L. W. *J. Chem. Soc., Perkin II*, **1974**, p. 830.

246. Coombes, R. G., Russell, L. W. *J. Chem. Soc., Perkin I*, **1974**, p. 1751.

247. Nikishin, G. I., Kaplan, E. P., Kapustina, N. I. *Izv. Akad. Nauk USSR Ser. Khim.* **1976**, p. 1434.

248. Draper, M. R., Ridd, J. H. *J. Chem. Soc., Chem. Comm.* **1978**, p. 445.

249. Suzuki, H., Nagae, K., Maeda, H., Ozuka, A. *J. Chem. Soc., Chem. Comm.* **1980**, p. 1245.

250. Dinçtürk, S., Ridd, J. H. *J. Chem. Soc., Perkin II* **1982**, p. 961.

251. Houben-Weyl. *"Methoden der Organischen Chemie"*, Vol. XII, Thieme: Stuttgart, 1971, p. 772; Benkeser, R. H., Brumfield, P. E. *J. Am. Chem. Soc.* **1951**, *73*, 4770; Benkesser, R. H., Landesman, H. *J. Am. Chem. Soc.* **1954**, *76*, 904; Shih-Hucé, W., Ugi-ie, K., Tanaka, K. *Bull. Chem. Soc. Jpn.* **1966**, *39*, 2227.

252. Lütgert, H. *Ber.* **1937**, *70*, 151.

253. Olah, G. A., Lin, H. C., unpublished results.

254. Houben-Weyl. *"Methoden der Organischen Chemie"*, Vol. XII, Thieme: Stuttgart, 1971, p. 821.

255. Hodgson, H. H., Mahadevan, A. P., Ward, E. R. In: *"Organic Synthesis"*, Col. Vol. III (Horning, E. C. ed.), John Wiley and Sons:New York, 1960, p. 341.

256. Craig, L. C. *J. Am. Chem. Soc.* **1934**, *56*, 231; Kirpal, A., Bohm, W. *Chem. Ber.* **1931**, *64*, 767; **1932**, *65*, 680.

257. Hodgson, H. H., Heyworth, F., Ward, E. R. *J. Chem. Soc.* **1948**, p. 1512.

258. Ward, E. R., Johnson, G. D., Hawkins, J. G. *J. Chem. Soc.* **1960**, p. 894.

259. Hodgson, H. H., Mahadevan, A. P., Ward, E. R. *J. Chem. Soc.* **1947**, p. 1392.

260. Starkey, E. B. In: *"Organic Synthesis"*, Col. Vol. II (Blatt, A. H. ed.), John Wiley and Sons: New York, 1943, p. 225.

261. Hodgson, H. H., Marsden, E. J. *J. Chem. Soc.* **1944**, p. 22.

262. Nesmeyanov, A. N., Tolstaya, T. P., Isaeva, L. S. *Dokl. Akad. Nauk. USSR* **1957** p. 996.

263. Lubinknowski, J. K., McEwen, W. E. *Tetrahedron Lett.* **1972**, p. 4817.

264. Olah, G. A., Sakakibara, T., Asensio, G. *J. Org. Chem.* **1978**, *43*, 463.

CHAPTER 3.
Mechanisms of Aromatic Nitration

I. Electrophilic Nitration

3.1 The Ingold-Hughes Mechanism

3.1.1 Nitration by the Nitronium Ion

Aromatic compounds are nitrated by HNO_3 in a variety of media. HNO_3 may be used neat, in organic solutions, in aqueous media, or in other acids. The kinetics of the reaction in these different media have been studied extensively and for a large number of substrates. Nitration through the NO_2^+ ion was originally proposed by Euler[1] in 1903. Subsequently, the NO_2^+ ion was frequently invoked on indirect grounds[2] but the existence of the ion was not conclusively demonstrated until 1946.[3] The realization of the connection between activation of aromatics and orientation of nitration,[4a] and the introduction of the competitive method of rate determination,[4b] were significant in these studies. Ingold showed that the rate expression takes diverse forms under different reaction conditions and yet for all of the cases studied, the reaction proceeded through the intermediation of the NO_2^+ ion.

Isolation of a nitronium salt dates back to Hantzsch.[5] It was recognized later by Ingold *et al.* that Hantzsch's salt was nitronium perchlorate in a mixture of hydronium perchlorate ($H_3O^+ClO_4^-$) (see Sections 2.6; 2.24). The study of the exchange of ^{18}O label between H_2O and HNO_3 and its relation to the zero-order nitrations was studied by Bunton.[6] The seminal work of Westheimer and Kharasch[7] comparing the rates of nitration of nitrobenzene with the ionization of 4,4',4"-trinitrotriphenylmethyl alcohol in H_2SO_4 solution was extremely important to the elucidation of the mechanism of aromatic nitration by HNO_3.

Rather than reviewing all of the extensive experimental evidence[8] that led to the establishment of the NO_2^+ ion mechanism of aromatic nitration, we will start our discussion with the mechanism proposed by Ingold and Hughes[3,8,9] and examine some of the experimental observations in its light. The Ingold mechanism of electrophilic aromatic nitration consists of four steps.

$$HNO_3 + HA \rightleftharpoons H_2NO_3^+ + A^- \qquad (1)$$

$$H_2NO_3^+ \rightleftharpoons NO_2^+ + H_2O \qquad (2)$$

$$ArH + NO_2^+ \rightleftharpoons ArHNO_2^+ \qquad (3)$$

$$ArNO_2^+ + A^- \longrightarrow ArNO_2 + HA \qquad (4)$$

The first two steps involve the acid-catalyzed transformation of HNO_3 into the NO_2^+ ion, the reactive electrophile. The NO_2^+ ion then reacts with the aromatic to give the Wheland intermediate[10] (later called σ-complex by Brown[11]). It is interesting to note that Wheland never referred to it as an intermediate, rather, he called it a transition state. It was only in the mid-1950s that the arenium (i.e., cyclohexadienyl) ion nature of this intermediate was firmly established though isolation and study of long-lived arenium ions by Olah et al.[12] Deprotonation of the Wheland intermediate completes the nitration by regenerating the aromaticity of the system. The last step also regenerates the acid catalyst.

When considering the reaction of the NO_2^+ ion with the aromatic we must keep in mind that, whereas the linear NO_2^+ ion (isoelectronic with CO_2) does not contain an empty bonding molecular orbital, a gradual displacement of a π-electron pair to the more electronegative O atom takes place when the NO_2^+ ion approaches the aromatic substrate (facilitated by attraction of the cation by the nucleophilic nature of the aromatic). That is, the NO_2^+ ion is a polarizable electrophile. Concurrent bending of the developing nitro group allows eventual bonding with the aromatic ring.

The Ingold mechanism readily accounts for the observed substituent effects. Electron-donating substituents on the benzene nucleus increase its reactivity and conversely, electron-withdrawing ones decrease it. These observations are in accord with the electrophilic nature of the reaction. Furthermore, electron-donating groups generally direct the incoming nitro group to the ortho ($o-$) and para ($p-$) positions in relation to the substituent, while electron-withdrawing groups lead to meta ($m-$) nitration. Again, since electron-donating groups increase the electron density at $o-$ and $p-$positions (as well as the ipso position), the predominant formation of $o-$ and $p-$ nitro products in these cases is explained. In the case of electron-withdrawing substituents, electron density at the $o-$ and $p-$positions is decreased whereas the $m-$position is not much affected, hence the observation of $m-$nitration in these cases. These rationalizations are based on electronic considerations of the starting aromatics. It should also be pointed out that, in addition, ring carbons already attached to a substituent, the so-called *ipso* positions, can also be involved in bonding of the nitro group (ipso attack, see Section 3.5). The electronic effects of substituents have differing origins. Chlorobenzene, for example, is less reactive than benzene and yet nitration of chlorobenzene yields $o-$ and $p-$nitrochlorobenzenes. Consideration of opposing inductive and conjugative effects in stabilizing the intermediate arenium ions leading to the various isomers helps to resolve this anomaly. These arguments are commonly presented in textbooks of organic chemistry and need not be further elaborated here.

Isomer distributions in reactions involving product-determining transition states, which lie relatively early on the reaction coordinate, will reflect the charge distribution of the aromatic hydrocarbon and not that of the arenium ions. Consideration of the charge distribution of the aromatics, such as toluene, as well as of their corresponding arenium ions are in accord with this conclusion.[13] The latter give preference for the para-substituted isomer, while in the former charge distributions are highest around the ortho and para positions. Substitution at a meta position is not favored in any case.

In treating directing effects in electrophilic aromatic substitution, Fukui's frontier orbital approach was successfully applied by Klopman[14a] and subsequently by Elliott *et al.*[14b] (see Section 3.6.7).

Nitration in mineral acids, depending on the nature of the aromatic and the reaction condition, exhibits second-order kinetic behavior; first order in each HNO_3 and the aromatic. By invoking steps (3) or (4) of the Ingold mechanism to be rate limiting, this behavior can be easily explained. However, if the proton elimination, step (4), were to be rate limiting, nitration of deuterated aromatics would show a primary kinetic hydrogen isotope effect, which generally is not the case (*vide infra*). In mineral-acid media, steps (1) and (2) are rapid and reversible, and quickly establish an equilibrium concentration of the NO_2^+ ion. Furthermore, the limited solubility of the aromatics in these media and the presence of sufficient quantities of water generally result in step (3) being much slower than the reverse of step (2).

In organic media, nitration of reactive substrates is often zero-order in the aromatic. In these cases, there is generally a large excess of the aromatic substrate and a much reduced activity of water. These factors contribute to make step (3) faster than step (2). Step (2), i.e., formation of the NO_2^+ ion, is then rate limiting which results in the observed zero-order in aromatics.

In the absence of other strong acids, nitrations in solvents such as sulfolane and carbon tetrachloride are very strongly dependent upon the HNO_3 concentration. Fourth- to fifth-order behavior has been observed in carbon tetrachloride for very low concentrations of HNO_3.[15] It has been argued that up to five molecules of HNO_3 may be required to support the ionization process. At high concentrations of HNO_3, the medium is sufficiently polar and places less stringent requirements on the ionization with the concomitant drop in the observed order in HNO_3.

As mentioned earlier, zero-order in aromatic substrate results when reaction (2) is slower than reaction (3). Thus, for a given set of conditions a more reactive aromatic, such as mesitylene, may exhibit zero-order behavior while a less reactive one, such as benzene, can show first-order dependence. At somewhat higher acid concentrations, benzene will also show zero-order dependence. Furthermore, the zero-order rates of nitration of mesitylene and benzene are the same. From the mechanism proposed by Ingold *et al.* involving steps (1-4), we can easily see that the limiting rate can only be the rate of formation of the active nitrating agent, i.e., the NO_2^+ ion. However, the same argument would also hold if the nitrating agent were the nitracidium ion. Recognizing the implication of zero-order nitrations, (i.e., a rate determining step which involves HNO_3 alone) and that nitration is an electrophilic substitution, Ingold argued that the nitracidium ion cannot be the reactive electrophile since its formation involves only proton transfer, which is instantaneous in oxy-acids and could hardly account for the rate-limiting step of forming the *de facto* nitrating agent.[9] Ingold and his associates also succeeded in isolating and characterizing (by Raman spectroscopy) nitronium perchlorate from a mixture of HNO_3 and $HClO_4$. Subsequently, more stable nitronium salts (such as the fluoroborate) were prepared and characterized by IR, Raman, NMR spectroscopy, and X-ray crystallography (see Section 2.24). At the same time, the nitracidium ion $H_2NO_3^+$ remained elusive.

Bunton and co-workers[6] investigated the rate of ^{18}O exchange between HNO_3 and $^{18}OH_2$ for various concentrations of HNO_3 and compared it with the rates of nitration of a number of aromatic substrates at these acidities. The kinetic form of nitration varied from first to zero-order depending upon the reactivity and concentration of the aromatic. The rates of nitration were always less than the rate of exchange of the label, but as the reaction order approached zero, the nitration rate approached the exchange rate. Later, Bunton reported[6c] the zero-order nitration rates of 2-mesitylethanesulfonic acid to be only about 15% higher than the extrapolated exchange rates for given HNO_3 concentrations. Considering the uncertainties in the extrapolation and the effect of increased ionic strength on the various processes (2-mesitylethanesulfonic acid was introduced as the sodium salt), the agreement is remarkable.

Bunton's investigations provided strong kinetic evidence for the involvement of the NO_2^+ ion. This was significant, since the previous studies by Ingold on the kinetics of nitration were consistent with either protonated HNO_3, i.e., $H_2NO_3^+$ ion, or the NO_2^+ ion as the reactive species, although Ingold did present strong arguments in favor of the NO_2^+ ion.

Independent strong evidence for the involvement of the NO_2^+ ion was furnished by Westheimer and Kharasch.[7] Qualitatively it had been well recognized that nitration rates increase with increasing acidity. This observation is consistent with both NO_2^+ and $H_2NO_3^+$ ions as the nitrating agent. Following a suggestion by Hammett, Westheimer and Kharasch compared the dependence of the second-order rate constant for the nitration of nitrobenzene with the ionization of *tris*(*p*-nitrophenyl)methyl alcohol in 80%-100% H_2SO_4. Cryoscopic measurements by Hantzsch[5] had helped to establish the similarity of the ionization of triarylmethyl alcohols and HNO_3 in H_2SO_4, both processes producing four particles:

$$HNO_3 + 2H_2SO_4 \rightleftharpoons NO_2^+ + H_3O^+ + 2HSO_4^-$$

$$Ar_3COH + 2H_2SO_4 \rightleftharpoons Ar_3C^+ + H_3O^+ + 2HSO_4^-$$

Consequently, it was argued, if NO_2^+ ion was truly the reactive electrophile the acidity dependence of the rates of nitration ($d\log k_2/d\%H_2SO_4$) should equal that for the ionization of a suitable indicator triarylmethyl alcohol ($d\log K/d\log\%H_2SO_4$). Here, k_2 is the observed second-order rate constant of nitration and K is the equilibrium constant for the ionization of the triarylmethyl alcohol.

Acidity profiles of the nitration of nitrobenzene and the ionization of a suitable triarylmethyl alcohol, in this case *tris*(*p*-nitrophenyl)methyl alcohol, were found to be parallel up to 90% H_2SO_4. This parallel strongly suggests that NO_2^+ ion is the active electrophile.

Subsequently, Deno examined the acidity dependence of the rates of nitration and the ionization of triarylmethyl alcohols in greater detail.[16] He defined a scale of acidity (H_R, formerly C_O) based on the ionization of arylmethyl alcohols and showed a linear correlation between the nitration rates and the H_R acidity function. The range of acidities considered extended to below 50% H_2SO_4, where only vanishingly small quantities of NO_2^+ ion could exist. Spectroscopically, NO_2^+ ion has not been detected below 85% H_2SO_4. Deno and Stein reported that the correspondence of the observed second-order rate constant of nitration was best with $-H_R$ acidity function and not with $-H_O$ or other acidity functions. A correspondence with the $-H_O$ acidity function would have implied that the activation of HNO_3 to the electrophile required, by analogy to the definition of H_O, only a proton transfer:

$$HNO_3 + H^+ \rightleftharpoons H_2NO_3^+$$

$$B + H^+ \rightleftharpoons BH^+ \quad \text{(equation used for defining } H_O\text{)}$$

This line of reasoning was later shown not to be entirely accurate.[16] The acidity profiles of $\log k_2$ and $-H_R$ happen to be parallel at 25°C, but the two quantities have different slopes at other temperatures; in fact, while the slope of $\log k_2$ line increases with increasing temperature, that of $-H_R$ decreases. Thus the unit slope obtained by plotting $\log k_2$ and $-H_R$ was rather a coincidence.

It has long been noted that the acidity profile of the rates of nitration of a number of compounds showed a maximum around 90% H_2SO_4. At first this fact was taken as evidence for step (4) of the Ingold mechanism, the loss of proton from the Wheland intermediate, as becoming rate limiting in these strongly acidic solutions, where the concentration of the base HSO_4^- is exceedingly small. If proton loss were to be rate limiting, a primary kinetic isotope effect should be observed upon substituting deuterium, or tritium, for protium in the aromatic. Melander nitrated 2-T toluene in HNO_3-H_2SO_4 mixtures and found that the product 2,4-dinitrotoluene had 50% of the original tritium.[17] In other words, tritium and protium were substituted with equal facility, showing the absence of a primary kinetic isotope effect in the nitration.[18] Thus, the loss of proton from the Wheland intermediate could not be the rate-limiting step.

It was only by considering medium effects on the activity of the different species that the decline in the rates of nitration above 90% H_2SO_4 could be explained. Using the Bronsted rate law, the rate of nitration can be expressed as:

$$\text{rate} = k_r[\text{ArH}][\text{NO}_2^+] f_{\text{ArH}} f_{\text{NO}} / f^{\ddagger}$$

where k_r is the specific rate of the reaction of the NO_2^+ ion with the aromatic and fs denote the activity coefficients of the respective species.

Since the NO_2^+ ion and the nitration transition state are both positively charged species, it may be argued that their activity coefficients are likely to vary in the same manner, and to a first approximation the ratio of these coefficients will be constant. Thus, any acidity dependence of the second-order rate constant beyond the point where all of the HNO_3 has been converted into NO_2^+ ion is primarily a consequence of the changes in the activity coefficient of the aromatic. Vinnik and co-workers[19] determined the activity coefficient of several aromatics, and also the second-order rate constants for their nitrations over a range of acidities. Plots of the observed second-order rate constant against the acidity of the medium for p-nitrotoluene, o-nitrochlorobenzene, and p-nitrochlorobenzene showed usual maxima. However, if the observed rate constants are corrected for the variations in the activity coefficient of the substrate, then the plots do not show any maximum. This lack of maximum demonstrates that the decline in the observed second-order rate constant for nitration in H_2SO_4 of strengths greater than about 90% was primarily a result of the medium effects. It should be noted that, in 90%–100% H_2SO_4 media, the substrates considered by Vinnik exist largely as the free bases and not as their conjugate acids.

3.1.2 Acidity Dependence of the Nitric Acid/Nitronium Ion Equilibrium

A discussion of the mechanism of nitrations with HNO_3–H_2SO_4 mixtures must include consideration of the acidity dependence of the $HNO_3 + H^+ \rightleftharpoons NO_2^+ + H_2O$ equilibrium, and the kinetics of the forward and reverse reactions. A number of spectroscopic techniques such as UV, Raman, and NMR, have been used to determine the fractional conversion of HNO_3 into NO_2^+ ion as a function of the acidity of the medium. Chedin performed extensive Raman spectroscopic studies over a wide range of H_2SO_4 concentrations.[20a] Later, studies were also extended to some temperature–dependence measurements.[20b] The results from these studies of the ternary system, N(V), H_2SO_4, and H_2O are shown in the triangular plot of Fig. 1. N(V) denotes all nitrogen species in the +5 oxidation state. It is important to note that the fractional conversion also depends upon the total N(V) species added. Most studies on the kinetics of nitration are performed along the H_2SO_4–H_2O axis with very small quantities of N(V). On the other hand, preparative nitrations are generally performed with considerably larger amounts of N(V).

Marziano and co–workers[21] also conducted Raman spectroscopic studies and showed that the acidity dependence of the HNO_3–NO_2^+ ion equilibrium is satisfactorily described in terms of the M_c acidity function. Furthermore, they showed excellent linear correlation between the rates of nitration of a large variety of aromatics and of the M_c function over 50%–98% H_2SO_4 between closely related structures.

Ross, Kuhlmann, and Malhorta[22] have used ^{14}N-NMR studies of HNO_3–NO_2^+ equilibrium studies to determine the fractional conversions and also to directly obtain the rate constants for the formation, k_1, and hydration, k_{-1}, of the NO_2^+ ion. NMR spectra were recorded for 0.5 M solutions of 100% HNO_3 in 80–98% aqueous H_2SO_4. A single resonance was observed below 85% and above 93% H_2SO_4. These resonances were assigned to HNO_3 and to the NO_2^+ ion, respectively. In the intermediate range, broad singlets of varying intensities were

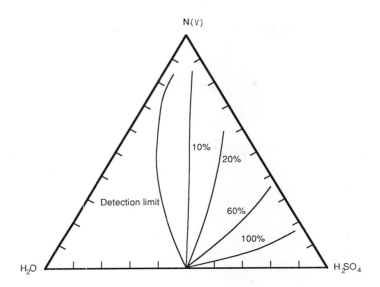

Figure 1. Equilibrium conversion of N(V) species into nitronium ion in $N(V)/H_2SO_4/H_2O$ systems.

observed. Line-shape analysis of these resonances provided the fractional conversions and the rate constants k_1 and k_{-1} (Table 36).

Table 36. Rate and Equilibrium Data for the HNO_3/NO_2^+ System in Aqueous Sulfuric Acid[22]

Percent H_2SO_4	$-H_o$[a]	$-\log a_w$	Percent $[NO_2^+]/[N(V)]$	$\log[k_1/s^{-1}]$	$\log[k_{-1}/s^{-1}]$
81.0	–	–	0	–	–
86.2	8.46	2.96	12	2.26	3.11
87.7	8.68	3.15	34	2.45	2.74
88.6	8.82	3.26	54	2.48	2.41
89.5	8.95	3.36	66	2.59	2.30
91.2	9.21	3.55	92	3.44	2.18
92.6	9.41	3.72	98	3.51	1.91
96.7	10.03	–	100	–	–

[a]Hammett acidity function.

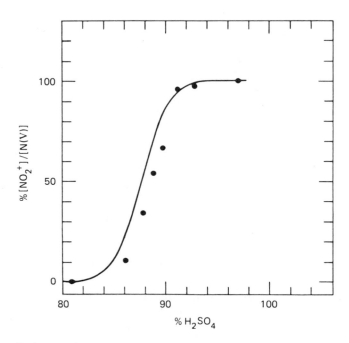

Figure 2. Fractional conversion of HNO_3 into NO_2^+ in H_2SO_4 of varying strength. ____, Raman Spectroscopy Study[21]; •, ^{14}N NMR Study using HNO_3.[22]

The data obtained from these NMR studies are generally in good agreement with those from previous studies. For example, as shown in Fig. 2, the fractional conversion of N(V) into NO_2^+ are fully in accord with previous Raman studies.

Olah and Prakash carried out similar studies using ^{15}N–NMR spectroscopy.[23]

The acidity dependence of rate constants for the formation and hydration of NO_2^+ ion (k_1 and k_{-1}) were previously estimated by Moodie, Schofield, and Taylor[24] from the kinetics of nitration of aromatics in 60–80% H_2SO_4. The NMR data provide these rates directly, albeit at a different (higher) range of acidities. Nonetheless, extrapolation shows that the two sets of data are generally in reasonable agreement. Figure 3 shows the variation of the pseudo first-order rate constant of hydration of the NO_2^+ ion (k_{-1}) as a function of the acidity of the medium as determined by the two studies. The decrease in k_{-1}

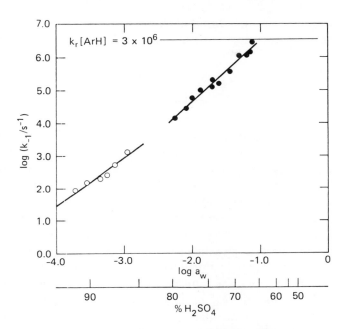

Figure 3. Acidity dependence of the pseudo first-order rate constant for hydration of nitronium ion (k_{-1}). $d(\log k_{-1})/d(\log a_w) = 1.95$[24]; $d(\log k_{-1})/d(\log a_w) = 1.51$.[22]

with increasing acidity is expected because this pseudo first-order rate constant is a product of a rate constant for hydration (k'_{-1}) and the activity of water (a_w) raised to some power: $k_{-1} = k'_{-1} a_w^n$. The value of n determined by the NMR and nitration studies is 1.5 and 1.95, respectively, although a single line could well describe both sets of data.

Temperature dependence of the HNO_3/NO_2^+ ion equilibrium was studied in 88.6% H_2SO_4 solution from 14.2 to 43.0°C. Line-shape analysis was used to obtain k_1 and k_{-1} values, and the enthalpies of activation were calculated from the Arrhenius plot:

$$k_1: \Delta H^\ddagger = 14.3 \pm 2.6 \text{ kcal/mol}$$

$$k_{-1}: \Delta H^\ddagger = 16.8 \pm 2.4 \text{ kcal/mol}$$

Previously, Strachan and co-workers[25] determined the enthalpy of activation for the formation and hydration of NO_2^+ ion in 77.3% and 78.5% H_2SO_4 using temperature dependence of the nitration of toluene. Since the nitration is extremely rapid at these acidities, the kinetics were studied with the help of a stop-flow apparatus. The values determined for 78.5% H_2SO_4 were:

$$k_1: \Delta H^{\ddagger} = 18.3 \text{ kcal/mol}$$

$$k_{-1}: \Delta H^{\ddagger} = 12.0 \text{ kcal/mol}$$

Despite these agreements, the NMR study pointed out some inconsistencies between the observed rates and orders of nitrations and the data from the HNO_3/NO_2^+ ion equilibrium. For example, it was noted that whereas the NO_2^+ ion concentration increases by only 20% in going from 90% to 95% H_2SO_4, the corrected (according to Vinnik) rate constant for nitration of p-nitrotoluene increases by a factor of 2.3 over the same acidity range. Furthermore, the lack of zero-order nitration of substrates such as anisole in 72% H_2SO_4 is also not consistent with the values for k_{-1} as projected by the NMR study, not with the value determined at that acidity in the nitration study by Moodie, Schofield, and Taylor.[24]

The rate expression for the nitration of a substrate proceeding according to the Ingold-Hughes scheme is given by:

$$\text{rate} = \frac{k_1 k_r [HNO_3][ArH]}{k_{-1} + k_r[ArH]}$$

Here k_1 and k_{-1} refer to the first-order rate constants for the formation and hydration of the NO_2^+ ion and k_r is the specific second-order rate constant for the reaction of the aromatic substrate and the NO_2^+ ion.

The usually limited solubility of the aromatic in H_2SO_4 generally results in k_{-1} being considerably greater than $k_r[ArH]$. Consequently, the second term in the denominator can be neglected and nitration is observed to be first-order in the aromatic. The value for k_{-1} at 72% H_2SO_4 was determined to be about $3.0 \times 10^5 s^{-1}$ with [anisole] concentration around 10^{-2}, and a k_r for anisole 3×10^8 M^{-1} s^{-1}. We see that k_{-1} is substantially larger than $k_r[ArH]$, and a zero-order behavior is expected. Moodie and co-workers report only a departure from

Electrophilic Nitration 129

strict first-order kinetics.

These discrepancies probably point to the need for a more rigorous treatment of the rate data with careful consideration of the activity coefficients. Availability of rate constants for nitration under a fixed set of conditions should create a better basis for comparison of substrates with widely differing reactants. In this regard, use of Mariziano's M_c function[21] for extrapolating thermodynamic rate constants for nitration to a standard state of nitration in pure water may prove to be particularly helpful. Prior attempts at placing all the substrates on a single scale have always involved choosing a standard set of conditions for nitrations. Exemplary as such an attempt is the work of Katritzky and co-workers.[26] As standard conditions, they chose nitrations in 75% H_2SO_4 (H_o = −6.6) at 25°C. For most substrates, nitration data were available under conditions close to the "standard" conditions and only minor interpolation was involved. However, for many cases, extrapolations from other temperatures and acidities were necessary, and corrections had to be made to accommodate the fact that the substrate was reacting as a conjugate acid rather than as the free base. This treatment did not result in any general relation between structure and activity of the substrates, although partial correlations were found.

3.1.3 Effect of Nitrous Acid

The effect of HNO_2, (all N(III) species in solution that, upon dilution with water can be estimated as HNO_2, are included in the term "nitrous acid") upon HNO_3 nitration is manifested by its anticatalytic effect and in the way this is superceded in nitration of very reactive aromatics by an indirect mechanism of nitration via nitrosation or oxidative radical-cation formation.[27]

In aqueous H_2SO_4 or $HClO_4$ solutions of less than 50% concentration, HNO_2 is predominantly present in its molecular form, with some N_2O_3 also present. In acid solutions above 60%–65% concentration it is essentially present as the NO^+ ion.[28a,b] (as shown by Raman spectroscopic studies).[29]

In an excess of HNO_3, HNO_2 exists essentially as N_2O_4, which is nearly completely ionized[30]

$$N_2O_4 \rightleftharpoons NO^+ + NO_3^-$$

with some NO_2^+ and NO_2^- also present. In organic solvent–HNO_3 systems, however, N_2O_4 is little ionized.

Ingold et al. observed first the anticatalytic effect of HNO_2 on zero-order nitrations in nitromethane and AcOH.[9,15] For nitrations in HNO_3 the anticatalytic effect of HNO_2 was also demonstrated, but it is smaller and the kinetic form was not established.

Ingold interpreted the retarding effect of HNO_2 on NO_2^+ ion nitrations by considering its behavior in the involved systems. In strong aqueous HNO_3 systems added HNO_2, which gets ionized to NO^+ and NO_3^-, serves to increase the NO_3^- ion concentration at the expense of molecular HNO_3. In nitration, the nitrate ions formed can deprotonate the nitroacidium ion, $H_2NO_3^+$, the precursor of the NO_2^+ ion, thus leading to retardation of the reaction. With higher concentrations of HNO_2 and some water present the observed stronger anticatalysis is explained by HNO_2 also forming N_2O_3.

$$2N_2O_4 + H_2O \rightleftharpoons N_2O_3 + 2HNO_3$$

which then ionizes to give nitrite ion:

$$N_2O_3 \rightleftharpoons NO^+ + NO_2^-$$

The nitrite ion, being a stronger base than the nitrate ion, deprotonates the $H_2NO_3^+$ ion much more efficiently and thus retards nitration by NO_2^+ ion. In the presence of H_2SO_4 there is no retardation of nitration with N_2O_4. In H_2SO_4 and with other strong acids, N_2O_4 acts as a nitrating agent[31] (see Section 2.23).

Whereas HNO_2 retards nitration of aromatic hydrocarbons with HNO_3, nitration of phenols and anilines were long found to be catalyzed by it. Studies by Ingold et al. have shown[32] that, in the nitration of these and related compounds in AcOH solution, two reactions take place: the zero-order NO_2^+ ion nitration of the reactive aromatics and a new reaction that has the form

$$\text{rate} = k_2[\text{ArH}][\text{HNO}_2]$$

(where HNO_2 denotes all N(III) species)

Evidently it is the latter reaction that causes positive catalysis of the reaction. It can be nearly completely eliminated by removing the

HNO_2 (with urea or other traps) and increasing the HNO_3 concentration sufficient to make NO_2^+ ion freely available. The kinetics of the catalyzed process strongly indicate initial C-nitrosation followed by rapid oxidation of the formed C-nitroso compound by the HNO_3, reforming the HNO_2 consumed in the initial slow stage:

$$ArH + HNO_2 \xrightarrow{slow} ArNO + H_2O$$

$$ArNO + HNO_3 \xrightarrow{fast} ArNO_2 + HNO_2$$

The reactive nitrosating agent was identified as the NO^+ ion, which can be formed from HNO_2 through $H_2NO_2^+$ or by ionization of N_2O_4 or N_2O_3.

The question why only reactive aromatics such as phenols and anilines show catalysis of their nitration by HNO_2, can be answered by first remembering that these aromatics are highly reactive towards electrophilic reagents in general. Thus to carry out mononitration relatively free of dinitration, it is necessary to choose conditions that render the NO_2^+ ion not easily available, allowing the much less reactive, but much more plentiful carriers of the NO^+ ion to react with the aromatics.

The initial nitrosation of phenol during nitration is supported by isolation of some p-nitrosophenol from interrupted nitration of phenol[33] and by the observation that the ortho : para ratio (9 : 91%) of the catalyzed reaction in aqueous media is similar to that of the formation of nitrosophenols in the absence of HNO_3.

Ingold referred to nitrations via nitrosation as "special" nitrations. Special nitrations have generally been reported for substituted phenols and anilines. The use of this mode for purely hydrocarbon systems is not well documented, although the literature is abundant with reports of nitrations of activated hydrocarbons in mixed acids displaying spurious and complex kinetics. These aberrations have been attributed to build-up of HNO_2 resulting in "special" nitration.[17] Consequently, urea or hydrazine is generally added to scavenge the HNO_2 when determining the kinetics of nitration with HNO_3.

Recently, some studies have shown catalytic action of lower nitrogen oxides that does not involve prior nitrosation. As with "special" nitrations, it is again the activated aromatic substrates that undergo lower nitrogen oxide-catalyzed nitrations not involving nitrosation. The phenomenon has been observed for heteroatom-substituted, as well as for purely hydrocarbon aromatic substrates: N,N-dimethylaniline, phenol, 1,2,3-trimethoxy-5-nitrobenzene, naphthalene, and polyalkylated benzenes. Addition of trace quantities of HNO_2 to mixtures of these substrates and HNO_3 in H_2SO_4 results in a dramatic increase in the rate of nitration. However, using HNO_2 alone, even at much higher concentrations, does not result in nitrosation.

Ridd studied the nitration of N,N-dimethylaniline in 82%-87% aqueous H_2SO_4 solutions.[34] At these acidities the substrate is almost entirely protonated and, accordingly, formation of p-nitro product is not expected to dominate. However, as the acidity was decreased from 87% to 82% the amount of p-nitration increased. Concurrently, autocatalytic behavior was noted for the nitration reaction, and a number of other products, particularly benzidine derivatives, were formed. The slow initial rate was consistent with the expected straight NO_2^+ ion nitration, and in order to study the kinetics of the catalyzed reaction, the component due to straight nitration was algebraically subtracted from the observed rates.

Addition of N(III) eliminated the induction period, however, in the absence of N(V), no nitrosation of the substrate was detected. At lower acidities, C-nitrosation of N,N-dimethylaniline to give the p-nitroso derivatives is well known. Clearly, here was a case of catalysis of nitration by N(III) which did not proceed by the "special" mechanism. The catalyzed nitration was found to be first order in the substrate and N(III). The reaction became zero order in N(V) at N(V) concentrations greater than 0.15 M. The rate law for the limiting catalyzed reaction can be expressed as:

$$\text{rate} = k_2[C_6H_5NH(CH_3)_2^+][N(III)]$$

While studying N(III)-catalyzed nitration of phenols in 56% H_2SO_4, Ross and co-workers observed the curious phenomenon that the o/p ratio of the product nitrophenols was sensitive to the order of addition of N(V) and N(III).[35] If N(III) was added first, the product was almost exclusively p-nitrophenol. However, if N(III) was added to a solution already containing N(V), a significant quantity of o-nitrophenol was formed. The two nitrogen acids were added 30 seconds apart. Previous studies by Challis and Lawson had established that nitrosation would have been complete in this period giving almost pure p-nitrosophenol.[36] The formation of p-nitrophenol when N(III) is added first is therefore expected. In 56% H_2SO_4, direct nitration of phenol gives an *o/p* ratio of 0.78; however, it is very slow and in the 30-second period less than 2% of the phenol could have reacted by this pathway. Therefore, the altered *o/p* ratio is an indication of a pathway for N(III) catalysis not involving prior nitrosation.

Almost identical behavior was reported by Schofield[37a] and by Ross[37b] in the kinetics of 1,2,3-trimethoxy-5-nitrobenzene and naphthalene, respectively. The catalyzed reaction is first order in N(III). With respect to N(V), the reaction is first order at low concentrations of N(V) changing to zero order at high concentrations. With respect to the aromatic substrate, the reaction displays orders varying between one and two. In other work, phenol, mesitylene, and *p*-xylene were also found to give similar rate laws, although the first two substrates do undergo C-nitrosation.[37c] Under similar conditions, benzene and toluene were shown not to display such catalysis.

Milligan has reported on the nitration of benzene and toluene in trifluoroacetic acid by lower oxides of nitrogen.[38] The reactions were slow, yet the formation of nitroaromatics is remarkable. Initial formation of a complex between the aromatic and the NO^+ is indicated by the intense red color of the solution. Upon standing, the color faded to yellow if the ratio of the nitrite to substrate was greater than three, otherwise the red color persisted until aqueous workup. Nitroaromatics were also formed if NO_2 was used instead of sodium nitrite. The product nitrotoluenes had an *o/p* ratio ranging from 1.2 - 1.4 with *m*-nitration always less than 2%.

Analysis of the reaction mixture at intermediate stages did not show any nitrosoarenes. Thus, if formed as intermediates, they must be rapidly oxidized to the nitroarenes. However, reaction of nitrosobenzene by NO_2 in trifluoroacetic acid resulted in a complex tarry mixture. In methylene chloride solvent, nitrobenzene was observed, but with concomitant formation of dinitrobenzenes, chlorobenzene, benzaldehyde, and benzonitrile. These observations are inconsistent with a nitrosoarene as an intermediate in the nitration.

The nitrations of pentamethylbenzene[33,39] and durene[40] with nitronium hexafluorophosphate-containing nitrosonium hexafluorophosphate in nitromethane containing water also show characteristic effects of the N(III) species. In these systems dark-red initial complex formation between the methylbenzenes and nitrosonium hexafluorophosphate was observed. Reaction of the complex with the N_2O_4 formed in the system can give an alternative pathway to direct electrophilic C-nitrosation.

A possible mechanism for the nitration is the oxidation of the initial [ArH·NO$^+$] π-complex to the [ArH·NO$_2^+$] complex, which then rearranges to the Wheland intermediate. This mechanism also accounts for the observed regiochemistry of lower nitrogen oxide-catalyzed nitrations being very similar to that observed in the case of straight N(V) nitrations. Zollinger et al. first proposed this mechanism for the nitration of pentamethylbenzene.[39]

$$[ArH \rightarrow NO]^+ + NO_2 \longrightarrow ArH \overset{\delta+}{\rightarrow} \overset{+}{N} \begin{smallmatrix} O \\ \\ O-N \end{smallmatrix} O^- \longrightarrow ArH^+ \rightarrow \overset{O}{\underset{O}{\overset{\|}{N}}} + NO$$

$$\downarrow$$

nitration products

$$NO\cdot + NO_2\cdot \longrightarrow N_2O_3$$

$$N_2O_3 + HNO_3 \longrightarrow N_2O_4 + HNO_2$$

It appears that this mechanism may be more general than originally considered, and may also be applicable to nitrations in N$_2$O$_4$ medium. Zollinger's original assumption of the nitration proceeding through the initial ArH $\rightarrow$ NO$^+$ π-complex[39] is more probably to be interpreted in Kochi's terms of charge-transfer (CT) complex formation, followed by electron transfer-oxidation leading to the cation radical complex [ArH $\overset{+}{\cdot}$ NO$_2$] collapsing to the arenium ion, as discussed in Section 3.6.8.

3.2 Olah's Modified Mechanism of Two Separate Intermediates

In 1950 when Ingold and his associates published a comprehensive series of papers on their fundamental studies of the mechanism of electrophilic aromatic nitration, for most chemists these studies meant that all significant aspects were resolved and further research on nitration would be of little interest. In fact, however, the Ingold studies acted as a catalyst for further extensive investigations.

The four steps of the Ingold mechanism, as discussed, are (1-2) formation of the nitronium ion, via the nitracidium ion, as the *de facto* nitrating agent; (3) reaction with the aromatic substrate via intermediate complex formation; and (4) proton elimination from the intermediate complex yielding the nitroaromatic product.

The Ingold group firmly established the NO_2^+ ion as the *de facto* nitrating agent in electrophilic nitration. Kinetic studies showing zero order in nitration of reactive aromatics such as benzene, toluene, and alkylbenzenes are best explained with a rate-determining slow formation of the NO_2^+ ion, which then reacts in a very fast reaction with the aromatic.

Ingold's group also isolated nitronium perchlorate, observed its nitrating ability, and spectroscopically characterized it. Due to its instability, however, no detailed studies of nitration were carried out.

3.2.1 Studies with Preformed Nitronium Salts

Olah, Kuhn, and Mlinko, in their studies reported first in 1956[41a] found $NO_2^+BF_4^-$ as a stable and convenient nitrating agent that received subsequently wide use[41b] (see Section 2.24). In Olah's subsequent mechanistic studies[41c] the assumption was pursued that by using preformed $NO_2^+BF_4^-$ and thus eliminating the relatively slow acid- catalyzed transformation of HNO_3 into NO_2^+ ion (i.e., the first two steps of the Ingold scheme), which can give rise to zero-order nitrations, it is possible to study directly the reaction of the NO_2^+ ion with the aromatics. Since the rates of nitration of aromatic hydrocarbons are generally too fast to be measured directly, (techniques for fast-reaction kinetics based on equilibria are not applicable to generally irreversible nitrations) the Wibaut-Ingold method of competitive rate determination was used to gain insight into the mechanism.

The isomer ratios of the product nitroalkylbenzenes obtained were, overall, similar to those observed previously in the acid-catalyzed nitration of alkylbenzenes with HNO_3, showing predominant ortho/para substitution with the meta isomer staying below 5%. The amount of

Table 37. Orientation and Relative Reactivity in the Nitration of Toluene[a]

Conditions	Relative[b] rate	Orientation		
		o-	m-	p-
HNO_3 in CH_3NO_2, 30°C	21	58.5	4.4	37.1
HNO_3 in HOAc, 45°C	24	56.5	3.5	40.0
$NO_2^+BF_4^-$ in $C_4H_8SO_2$, 25°C[c]	1.7	65.4	2.8	31.8
NO_2PF_4 in $C_4H_8SO_2$, 25°C[c]	1.4	67.6	1.4	31.0
NO_2ClO_4 in $C_4H_8SO_2$, 25°C[c]	1.6	66.2	3.4	30.4

[a] Table according to Ridd.[77]
[b] With respect to the rate of nitration of benzene as unity.
[c] Ref. 41c.

ortho substitution was, however, higher and that of para lower than in HNO_3 nitrations. The relative rates of nitration of alkylbenzenes with respect to benzene were considerably reduced, giving $k_{C_6H_5}/k_{C_6H_6}$ values generally only slightly higher than unity. For example, whereas HNO_3 nitration of toluene proceeds about 20 times faster than that of benzene in 68% H_2SO_4–HNO_3, with nitronium salts toluene reacts barely faster than benzene. Results of the nitration of toluene and benzene are shown in Table 37 in comparison with those of conventional HNO_3 nitrations.

3.2.2 The Necessity for Two Intermediates and Their Nature

In order to reconcile the maintained high-positional selectivity despite the loss of substrate selectivity in alkylbenzenes, which is contrary to the Brown's selectivity relationship[43] (based on a Hammett-type treatment[42] of electrophilic aromatic substitution), Olah found it necessary to modify the original Ingold mechanism by invoking two separate intermediates. He argued that the data can be explained only if the substrate and positional selectivities are determined in two distinct steps separated by an intermediate. Since the positional selectivity is clearly determined in the transition states leading to the formation of the isomeric Wheland intermediates,[10] he proposed that substrate selectivity is determined in a transition state leading to the formation of a separate prior intermediate. Olah suggested that the reactive alkylaromatic and the NO_2^+ ion first form a Dewar-type π-complex[42] which then transforms into the Wheland intermediates (σ-complexes) corresponding to the individual regio isomers. This suggestion was made in consideration of the known relative stability of π- and σ- complexes of methylbenzenes.[43,44]

Relying on Dewar's earlier suggestion,[42] Olah assigned π-complex nature to the first intermediate, which was demanded by very low substrate selectivity in nitronium salt nitration of reactive aromatics. π-Complexes of aromatics with halogens, hydrogen halides, and silver salts were firmly established by various physical and spectroscopic (UV, visible) measurements.[43] Hassel's X-ray diffraction studies provided structural evidence for benzene–halogen complexes.[45a] Mulliken gave a thorough theoretical discussion of aromatic π-complexes.[45b]

π-Complex formation of benzene and reactive aromatics with NO_2^+ ion is difficult to observe directly because of their great reactivity, leading very rapidly through the subsequent stages of the nitration reaction. Attempts of their observation therefore were limited to either highly substituted benzenes or deactivated derivatives, and thus are not conclusive. A recent high pressure mass spectrometric study by Holman and Gross,[46] however, provided convincing evidence for the separate existence of arenium ions (σ-complexes) and π-complexes during protonation of arenes in the gas phase, the prototype electrophilic aromatic substitution reaction.

On the basis of the Brown selectivity rule,[47] based on Hammett's linear free-energy relationship,[48] if the reaction were to proceed via a single σ-complex intermediate, it should also have predictably low positional selectivity under these conditions. However, the $o : m : p$ isomer ratio for nitrotoluenes was found to be 65 : 3 : 32%. In other words, the substrate selectivity is low while the positional selectivity remains high. Data in Table 38 show that this trend holds for other alkylbenzenes as well.[41] At the same time, as shown in Table 38 for a typical bromination with Br_2 in 85% AcOH, substrate selectivities are very high.

It should be pointed out that the electronic effects of various substituents arise from the combination of conjugative or inductive effects. Thus, Hammett-type free-energy relationship[48] treatment using single-parameter substituent constants, such as the Brown σ^+-substituent constants,[47] can be only of limited value. Knowles, Norman, and Radda,[49] as well as others have pointed out this limitation and have

Table 38. Competitive Nitration of Alkylbenzenes and Benzene with NO_2BF_4 in Sulfolane Solution at 25°C Comparison with Bromination and π- and σ-complex Stabilities[41]

	$k_{Ar}/k_{benzene}$	Isomer Distribution, %			o-p ratio	Relative π-complex stability (HCl)[45]	Relative σ-complex stability (HF·BF_3)[46]	$k_{Ar}/k_{benzene}$ bromination Br_2, 85% (CH_3CO_2H)
		ortho	meta	para				
Benzene	1.0					1	1	1
Toluene	1.67	65.4	2.8	31.8	2.05	1.5	7.9×10^2	6.05×10^2
o-Xylene	1.75	3-Nitro-o-xylene, 79.7%				1.8	7.9×10^3	5.3×10^3
		4-Nitro-o-xylene, 20.3%						
m-Xylene	1.96	2-Nitro-m-xylene, 17.8%				2.0	10^6	5.1×10^5
		4-Nitro-m-xylene, 82.2%						
p-Xylene	1.96					1.6	3.2×10^3	2.5×10^3
Mesitylene	2.71					2.6	1.9×10^8	1.9×10^8
Ethylbenzene	1.60	53.0	2.9	44.1	1.20	1.5		
n-Propylbenzene	1.46	51.0	2.3	46.7	1.09	–		
i-Propylbenzene	1.32	23.4	6.9	69.7	0.34	1.7		
n-Butylbenzene	1.39	50.0	2.0	48.0	1.04	–		
t-Butylbenzene	1.08	14.3	10.7	75.0	0.19	1.8		

Table 39. Competitive Nitration of Nitrobenzene and Nitrotoluenes with $NO_2^+PF_6^-$ [61]

	Relative Rate		Percent dinitrotoluene				
	CH_3NO_2	96% H_2SO_4	2,3-	2,4-	2,5-	2,6-	3,4-	3,5-
Nitrobenzene	1	1	o : m : p in CH_3NO_2 = 10 : 88.5 : 1.5; in 96% H_2SO_4 = 7.1 : 91.5 : 7.4					
o-Nitrotoluene	384	545	57.4	71.2	1.7	40.9	28.8	
m-Nitrotoluene	91	138	42	28.4	18.6	9.9	35.8	60.1	3.6	1.6
p-Nitrotoluene	147	217		99.8	99.8		0.2	0.2

recommended multi-parameter treatments. Norman et al. took into consideration the electron densities in the ground state of the molecules, the polarizability of the substituent, and the electronic requirements of the reagent. Yukawa and Tsuno developed[50] a particularly useful multi-parameter treatment of electrophilic aromatic substituents. Their discussion is, however, outside the scope of our review.

Kochi succeeded[51] in observing the charge transfer spectrum of $NO_2^+BF_4^-$ with m-toluonitrile in acetonitrile solution, but benzene and toluene were too reactive for a complex to be observed. It is not germane to our discussion at this point whether the electron donor-acceptor complex lies along the reaction coordinate as an intermediate (however, see for further discussion, Section 3.6.8).

Concerning the nature of the second intermediate(s) there is by now general agreement for their arenium-ion nature (also frequently called Wheland intermediates or σ-complexes). Pfeiffer and Wizinger, in 1928, originally suggested their possible carbocationic nature.[52] Wheland later generalized the concept,[10] and the intermediates subsequently are referred to as Wheland intermediates. McCaulay and Lien, studying aromatic-HF-BF$_3$,[53] and Brown, studying aromatic-HCl-AlCl$_3$ systems,[54] used vapor-pressure measurements to substantiate formation of the tertiary 1 : 1 : 1 Wheland-type complexes. Brown named them σ-complexes, indicating that a new σ-bond is formed using 2π electrons from the aromatic.

$$R-\phenyl + HF + BF_3 \rightleftharpoons R-\text{(arenium)} \begin{array}{c} H \\ H \end{array} BF_4^-$$

Olah, Kuhn, and Pavlath, in 1956, first succeeded in preparing a series of stable benzenium ions (Wheland intermediates or σ-complexes) by the reaction of HF-BF$_3$, DF-BF$_3$, alkyl fluoride-BF$_3$, acylfluoride-BF$_3$, as well as NO$_2$F-BF$_3$ with aromatics.[55] The advent of NMR spectroscopy subsequently allowed detailed structural studies of arenium ions.[44,55]

The relative stability of benzenium ions (σ-complexes) derived by protonation of benzene and alkylbenzenes in HF-BF$_3$ was established by the studies of Mackor et al.[48] Olah compared it with the relative reactivities of the nitration of benzene and alkylbenzenes with $NO_2^+BF_4^-$. He observed, as apparent from Table 38, no correlation. At the same time, the known π-stabilities of the same aromatics correlated well, which led Olah to suggest that the initial substrate rate-determining first step of the studied nitrations involve π-type of complexation of the aromatics, i.e., the interaction is with the aromatic π-system as an entity, and not with a single atom, as is the case in arenium ions (σ-complexes).

In the course of criticism raised subsequently, (see Section 3.1.3) that slow mixing rates can render the relative-rate data invalid, Rys[81b] also questioned Olah's conclusion that the small methyl substituent effects on the rate of nitration of benzene derivatives with $NO_2^+BF_4^-$ correlate well only with the relative π- but not with σ-complex stabilities of these systems. When he plotted the relationship between the product ratios for the nitronium salt nitration of methylbenzenes against the equilibrium constants for the formation of π-complexes of the same substrates with hydrogen chloride, the correlation was found "visually" only somewhat poorer with the relative stabilities of the corresponding protonated methylbenzenes (i.e., σ-complexes). He therefore, argued that the transition states of the $NO_2^+BF_4^-$ nitrations do not have π- but σ-complex character and that there is no need to consider two separate intermediates. However, as subsequent studies to be discussed all seem to agree with the necessity of two intermediates in the nitration of reactive aromatics with $NO_2^+BF_4^-$, and the second intermediate, determining positional selectivity, is clearly of arenium-ion (σ-complex) nature, the nature of the first intermediate, determining substrate selectivity (or the lack of it), must involve the aromatics either as some type of π-bonded entity (or as encounter complex with no specific bonding nature as proposed by Schofield).

The π-complex nature of the initial stage of electrophilic nitration was further substantiated by Olah,[39,56] who found that heavy benzene (C_6D_6) reacts faster than its light isotopomer, giving an inverse secondary kinetic isotope effect of $k_H/k_D = 0.86$. In view of the well-recognized effect of deuterium substitution (due to its zero-point energy difference) on the overall nucleophilicity of benzene, the small inverse isotope effect is expected if benzene reacts as a π-donor entity.[56,57] Mixing or diffusion effects hardly seem able to account for heavy benzene reacting faster than its light isotopomer.

At the same time the nitration of heavy benzene or toluene compared to that of their light analogs, as first shown by Melander,[16] gives no primary hydrogen isotope effect. Consequently the proton-elimination step of aromatic nitration is not rate determining and is very fast. Primary kinetic hydrogen isotope effects of the order

k_H/k_D ~3 were observed by Cerfontain[58] only in the nitration of anthracene, indicating that in this case both steric peri-interaction and observed reversibility of the nitration can make proton elimination at least partially rate determining.

Even if the observed low substrate selectivity (or lack of it) in the reaction of reactive aromatics with $NO_2^+BF_4^-$ can, as Schofield et al. pointed out subsequently,[17,27,59] (see Section 3.1.3 and 3.1.4), be due to encounter control, there is also no evidence to exclude that bonding interaction with the aromatic π-system occurs (although as Schofield has pointed out the kinetic results do not necessitate such interaction *per se*). Aromatic hydrocarbons are, however, good π-bases. The NO_2^+ ion is an electrophilic reagent, i.e., an acid. It is thus not unreasonable to expect that a typical acid–base interaction in the initial complex should have bonding nature.

It should be pointed out, as was also done by Schofield, that there is some confusion as to the terminology and the relationship of Dewar's π-complexes[42] to the mechanism of aromatic substitution. Mulliken's "outer" complexes are frequently included in the discussion of π-complexes, together with his more tightly bound molecular "inner" complexes. Keefer and Andrews discussed the topic extensively.[43] They summarized, *inter alia*, heats-of-formation data of molecular π-complexes of aromatics with iodine in carbon tetrachloride and other systems. The observed difference between benzene and alkylbenzenes generally are ~1.5–3 kcal/mol, indicative of weak bonding interaction. Other studies involved solubility of HCl in aromatics and spectroscopic studies. In the discussion of the π-complex nature of aromatic–NO_2^+ "initial" complexes, Olah clearly referred to π-complexes with weak but definite bonding nature.

The energy diagram of nitration of reactive aromatics with NO_2^+ has two energy barriers (Figure 4). The first barrier leads to the formation of a π complex (or encounter complex) determining the overall substrate reactivity, and the second to the corresponding σ-complexes, which determine the positional selectivity during nitration. The overall reactivity is determined by the transition state of highest energy.

For highly reactive (exothermic) systems, the transition state of highest energy lies early and is starting aromatic-like. The positional selectivity is determined in subsequent arenium-ion-like transition states different from the one that dictates the overall reactivity (substrate selectivity); hence, the breakdown of the Brown reactivity-selectivity

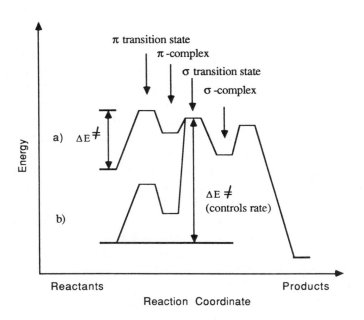

Figure 4. Energy diagram for aromatic nitration of (a) reactive aromatics; (b) less reactive systems.

rule.[60] On the other hand, for deactivated aromatics the transition state of highest energy lies late and is σ-complex-like, thus, positional and substrate selectivities are determined in the same step. These substrates follow the Brown reactivity–selectivity rule even when nitrated with nitronium salts. For example, no loss of substrate selectivity was observed in the nitronium salt nitration of nitrobenzene and nitrotoluenes.[61] High substrate and positional selectivity (Table 39) indicate a single intermediate σ-pattern of the reaction. In these systems it was possible to determine the rates of the relatively slow reactions, also without recourse to the competition method (Table 40).

Masci,[62a] in recent studies, following investigations of Elsenbaumer and Wasserman[62c], as well as Savoie et al.[62b], succeeded in modifying the reactivity of $NO_2^+BF_4^-$ by complexing it with crown ethers such as 18C6 and 21C7. When nitrating toluene and benzene with $NO_2^+BF_4^-$ in CH_2Cl_2 with added excess of 21C7, a high relative $k_T k_B$ rate of 77 was obtained. At the same time the isomer distribution of ortho 19%,

Table 40. Rate of Nitration of Nitrobenzene and Nitrotoluenes with $NO_2^+BF_4^-$ in Nitromethane and Sulfuric Acid at 25°C[61]

ArH	$NO_2^+BF_4^-$ (M)	ArH (M)	k_2 (M^{-1} sec^{-1})	$k_{2(ArB)}/k_{2(NB)}$
		In CH_3NO_2		
Nitrobenzene	0.1-0.2	0.01-0.02	1.19 x 10^{-2}	1
o-Nitrotoluene	0.1-0.2	0.01-0.02	4.25	357 (384)[a]
m-Nitrotoluene	0.1-0.2	0.01-0.02	1.23	103 (91)[a]
p-Nitrotoluene	0.1-0.2	0.01-0.02	1.54	129 (147)[a]
		In 96% H_2SO_4		
Nitrobenzene	0.1-0.2	0.01-0.02	2.30 x 10^{-2}	1
o-Nitrotoluene	0.1-0.2	0.01-0.02	1.41	614 (545)[a]
m-Nitrotoluene	0.1-0.2	0.01-0.02	3.71 x 10^{-1}	161 (138)[a]
p-Nitrotoluene	0.1-0.2	0.01-0.02	5.22 x 10^{-1}	227 (217)[a]

[a] Data in parentheses are from competitive experiments.

effect nitration. Cram and Doxsee have shown[63] that when phenyldiazonium ion, $C_6H_5N_2^+$, is complexed by crown 18C6, it inserts by its cationic end and consequently is dramatically deactivated for coupling reactions with aromatics. The NO_2^+ ion is expected to behave similarly. Consequently, fully complexed (in the internal cavity) NO_2^+ cannot be considered to be the *de facto* nitrating agent. An equilibrium allowing NO_2^+ ion to become free of the crown ether cannot fully explain the data. What we know so far is that a highly selective nitration takes place, concerning both substrate and positional selectivity. Not only is $k_{CH_3C_6H_5}/k_{C_6H_6}$ = 77 the highest reported substrate selectivity in a nitronium salt nitration, but the less than 1% meta isomer is also the lowest value reported for a nitration at room temperature. High para substitution thus may be not only the consequence of steric inhibition of ortho nitration by a complexed NO_2^+ species, but also a reflection of the transition state of a more selective reaction, lying late on the reaction coordinate and thus being of arenium-ion (σ-complex) nature. The para-nitrotoluenium ion is energetically favored over its ortho isomer, thus favoring para substitution. The proposed nature of the reaction is also reflected by the excellent fit of the data with the Brown selectivity relationship,[60] indicating that in this nitration, both substrate and positional selectivity are determined in the same step. The effect of complexing NO_2^+ by the polyether on the "outside" (as contrasted to that in the cavity) should probably be considered in explaining the results.

3.2.3 The Nature of the Nitronium Ion in Nitrations

Ingold's studies[3,8] firmly established the nitronium ion, NO_2^+, as the nitrating agent in HNO_3 nitrations. Olah's studies with stable nitronium salts in organic solvents raised the question of whether it is the "free" NO_2^+ ion or its ion pair or solvated forms that are the *de facto* nitrating agents.[39,41,65] As discussed, similar questions were raised considering NO_2–X type, polarized carriers of NO_2^+, which, as pointed out by Ingold, are nitrating agents closely related to the NO_2^+ ion.

The structure and chemistry of nitronium ion salts are by now well known through extensive studies (see Section 2.24). Whereas HNO_3 is completely ionized in 85% or stronger H_2SO_4, nitronium salts in dilute organic solutions (such as the perchlorate in nitromethane[64] or the tetrafluoroborate in sulfolane),[65] based on electric conductivity and cryoscopic measurements, are ion–paired. Separation of the ion pair in interaction with aromatics thus may involve additional activation energy.

It should be again reminded that the linear nitronium ion, $O=N^+=O$, although a cation, is not a reactive electrophile in its ground state. It has no low-lying LUMO. The cationic nature of the ion is not *per se* rendering it an electrophile. The ammonium ion, NH_4^+, for example, although a nitrogen cation is not a nitrogen electrophile.

The electrophilic nitrating ability of the NO_2^+ ion is due to its polarizability when attracted by π-donor aromatic nucleophiles. The aromatic causes displacement of a π-bonding N–O electron pair to the more electronegative oxygen atom, thus bending the originally linear NO_2^+ ion and allowing bonding interaction with the developing orbital at nitrogen. The bending of the linear nitronium ion necessitates significant energy.

If the NO_2^+ ion is present as an ion pair, as cryoscopic and conductivity measurements in nitromethane or sulfolane indicate,[64,65] solvation initially affects the ion pair itself and does not necessarily alter the linear NO_2^+ ion significantly. On the other hand, more nucleophilic solvents such as acetonitrile, or addition to above solutions nucleophiles such as water, alcohols, dialkyl ethers, and dialkyl sulfides results in nucleophilic solvation through coordinative interaction. Subsequent transfer of the NO_2^+ ion to the aromatic will result in nitration.[66,69] However, the observed regioselectivity of the nitration of toluene by nitronium salts in sulfolane and nitromethane was found to be little affected by such added nucleophiles as dimethyl ether or dimethyl sulfide, showing that the nitrations are still effected by the NO_2^+ ion.

The discussed effect of 18C6 and 21C7 crown ethers on $NO_2^+BF_4^-$ nitration[62] strongly indicates complexation of NO_2^+ by the cyclic ethers, probably however, not in the inner cavity. Para substitution becomes predominant (up to 78%) with $k_{C_6H_5CH_3}/k_{C_6H_6}$ ratios reported as high as 77.

It was frequently observed in preparative nitration with nitronium salts that the presence of small amounts of water is promoting the reactions.[23,66-69] Nucleophilic solvation of NO_2^+ is again indicated, followed by hydrolysis if the amount of water is more excessive.

The hydrolysis of NO_2^+ involves the equilibria.

$$NO_2^+ + H_2O \rightleftharpoons \overset{H}{\underset{H}{\diagdown}}O^+-NO_2 \rightleftharpoons HNO_3 + H^+$$

There never was direct spectroscopic observation of the nitracidium ion, $H_2NO_3^+$, considered by Ingold the first step in the protolytic ionization of HNO_3. It should be pointed out, however, that protonation of HNO_3 can involve the polar nitro group in preference to the OH group, whereas reaction of water with NO_2^+ ion gives only the isomeric nitracidium ion, i.e., the nitroxonium ion. Only the latter is considered as a reactive NO_2^+ ion carrier (hydrated NO_2^+ ion on the other hand may involve only weak association, i.e., solvation).

$$HO-N\overset{O}{\underset{O}{\diagdown\!\!\!/}} \underset{}{\overset{H^+}{\rightleftharpoons}} HO-N^+\overset{OH}{\underset{O}{\diagdown\!\!\!/}} \rightleftharpoons \overset{H}{\underset{H}{\diagdown}}O^+-N\overset{O}{\underset{O}{\diagdown\!\!\!/}} \rightleftharpoons H_2O + NO_2^+$$

The corresponding methylated system was studied by the addition of dimethyl ether to a solution of $NO_2^+BF_4^-$ as well as by the methylation of methyl nitrate by $CH_3F : SbF_5$.[66]

$$CH_3OCH_3 + O{\equiv}N^+{\equiv}O \; SbF_6^- \rightleftharpoons \underset{SbF_6^-}{\overset{H_3C}{\underset{H_3C}{>}}O^+{-}N\overset{\nearrow O}{\underset{\searrow O}{}}} \rightleftharpoons \underset{SbF_6^-}{\overset{CH_3O}{\underset{CH_3O}{>}}N^+{=}O} \rightleftharpoons CH_3ONO_2 + CH_3F \rightarrow SbF_5$$

NMR spectroscopic studies indicate formation of the dimethylnitracidium ion. The isomeric nitrodimethyloxonium ion, however, was not observed, and its equilibrium may lie very much on the solvated NO_2^+ ion side. Regardless, addition of dimethyl ether moderates the nitrating ability of the nitronium ion, but does not change the isomer distribution in the nitration of toluene.

The difference, if any, between ion paired $NO_2^+X^-$ and a highly polarized $^{\delta+}NO_2 \rightarrow {}^{\delta-}X$ complex is difficult to establish. When using performed nitronium salts, a bulk concentration of NO_2^+ is introduced into the system and this can cause, due to its great reactivity with aromatics, mixing difficulties before reaction. *In situ* generated NO_2^+ ion via polarized $NO_2 \rightarrow X$ "carriers" tend to minimize this problem in acid-catalyzed nitrations with HNO_3 and its derivatives.

Kuhn and Olah in 1961 called attention to the fact that the general Friedel-Crafts acylation principle encompasses inorganic acid halides and oxides as well.[64,65] It was consequently suggested that aromatic nitration involving nitryl halides, N_2O_5, and N_2O_4 can be considered as Friedel-Crafts type reactions, as obviously, a very close analogy exists with the corresponding Friedel-Crafts ketone synthesis involving acyl halides and anhydrides. In a generalized sense, acid-catalyzed HNO_3 nitration can also be considered similarly. The scope of Friedel-Crafts nitrations also extended to include acid-catalyzed reactions with mixed anhydrides, and esters of HNO_3. When considering the nature of the nitrating agent, comparison of nitration with acylation is useful and adds to our understanding of the systems.

Both the nitronium and acetylium[70] ions are linear (or close to linear) as established from X-ray studies

$$\overset{+}{O{=}N{=}O} \qquad \overset{+}{CH_3{-}C{\equiv}O}$$

In contrast to the acetylium ion (which can more readily bend through its oxocarbenium ion form developing an empty atomic orbital on carbon), the NO_2^+ ion cannot readily undergo such bending without the polarizing effect of a nucleophile. Consequently, the role of polarized $NO_2 \rightarrow X$ type carriers of the NO_2^+ ion is more important than that of acylation analogs.

Let us compare the proton-catalyzed cleavage of methyl acetate and methyl nitrate. Methyl acetate is protonated on the carbonyl oxygen (observed by NMR in superacidic media),[71] but eventually cleaves to acetyl cation and methyl alcohol (or its protonated form, the methyloxonium ion), probably through the isomeric ether oxygen protonated form (not observed).

$$CH_3-C(=O)OCH_3 \xrightleftharpoons{H^+} CH_3-C(OH)(+)OCH_3 \rightleftharpoons CH_3-C(=O)O^+(H)-CH_3 \rightarrow CH_3\overset{+}{C}O + CH_3OH$$

Protolytic cleavage of methyl nitrate can be visualized similarly

$$CH_3O-N(=O)O \xrightleftharpoons{H^+} CH_3O-N(=O)OH \rightleftharpoons \overset{H_3C}{\underset{H}{O^+}}-N(=O)O \rightleftharpoons CH_3OH + NO_2^+$$

Complexation of acetyl and nitryl halides with Lewis acids, such as $AlCl_3$ is also comparable

$$CH_3COCl + AlCl_3 \rightleftharpoons H_3C-\overset{\delta+}{C}(Cl)(O^{\delta-}-AlCl_3) \rightleftharpoons H_3C-C\equiv O^+ \; AlCl_4^-$$

$$NO_2Cl + AlCl_3 \rightleftharpoons O\overset{\delta+}{=}N(Cl)(O^{\delta-}-AlCl_3) \rightleftharpoons NO_2^+ \; AlCl_4^-$$

In acetylation it is well recognized that O-coordinated donor-acceptor complexes (observed directly by IR and NMR spectroscopy) are capable of acylating aromatics in their own right

[Reaction scheme: benzene + H₃C-C(=O→AlCl₃)Cl ⇌ [arenium intermediate]⁺ → acetophenone + HCl + AlCl₃]

NO_2^+ and CO_2 are isoelectronic and isosteric. Acid-catalyzed carboxylation of aromatics with CO_2 gives even in superacids only very low yields. However, COS and CS_2 were found to be readily methylated (and ethylated) with $CH_3F \rightarrow SbF_5$ and $C_2H_5F \rightarrow SbF_5$, respectively. These S-methyl(ethyl)thio(dithio)carboxonium ions were found to thiolcarboxylate aromatics in high yield.[72a] The reaction is analogous to the $AlCl_3$-catalyzed thiolcarboxylation of aromatics with alkyl(aryl)thiolformates.[72b]

[Reaction scheme: ArR' + [RSC≡X]⁺ SbF₆⁻ → (SO₂, -30°C) arenium intermediate → thiolcarboxylated product]

R' = H, CH₃, OCH₃ or halogen
X = O, S

These reactions serve as further models for Friedel–Crafts nitration of aromatics. It is, thus (in accordance with discussed examples), possible that in certain Friedel–Crafts nitrations of aromatics, the polarized NO_2-X complexes are the nitrating agents

$$\text{C}_6\text{H}_6 + \text{Cl}-\overset{\delta+}{\underset{\text{O}}{\text{N}}}\overset{\delta-}{-}\text{O}\rightarrow\text{AlCl}_3 \longrightarrow \left[\begin{array}{c}\text{H} \quad \text{N}=\text{O} \\ \text{OAlCl}_3 \\ (+)\end{array}\right]\text{Cl}^- \longrightarrow$$

$$\left[\begin{array}{c}\text{H} \quad \text{NO}_2 \\ (+)\end{array}\right]\text{AlCl}_4^- \longrightarrow \text{C}_6\text{H}_5\text{NO}_2 + \text{HCl} + \text{AlCl}_3$$

Differences in isomer distribution of the nitration of excess toluene between a highly polarized donor–acceptor complex and the corresponding NO_2^+ ion-pair salt were observed (see Section 2.22.1). However, it disappears in more polar organic solvents (such as CH_3NO_2) in which the complex ionizes to the NO_2^+ ion

$$NO_2^+AlCl_4^- \rightleftharpoons \overset{\delta+}{NO_2Cl}\cdot\overset{\delta-}{AlCl_3}$$

An extensive compilation[68] of reported isomer distribution data of the nitration of toluene shows remarkable consistency of meta nitration (2–5%), with the ortho and para isomer ratio showing some variation [ortho, 66–50%; para, 30–50% (Table 41)]. Particularly, heterogeneous solid or polymeric acid–catalyzed nitrations affect the ortho substitution, as does complexation with crown ethers and transfer nitration with 9-nitroanthracene.

The data of the isomer distribution in homogeneous nitration is reflected by changes in the ortho/para ratio and the nearly unchanged degree of meta substitution (2–5%). Despite obvious scatter owing to errors in analysis, this is in accord with a common nitrating agent, i.e., the NO_2^+ ion, according to the original Ingold concept. At the same time, the data reflect the effect of the solvation of the NO_2^+ ion, as well as possible involvement of highly polarized NO_2–X type NO_2^+ "carriers". Masci[73] recently reported studies of the effect of organic solvents on the selectivity of aromatic nitration.

Table 41. Isomer Distribution in the Nitration of Toluene[68]

Reagent/solvent[a]	°C	Percent isomer			o/p
		o	m	p	
NO_2Cl, $AgBF_4$/CH_3NO_2	15	69	2	29	2.38
NO_2PF_6/CH_3NO_2	25	68	3	29	2.34
NO_2PF_6/TMS	25	68	3	30	2.27
NO_2AsF_6/CH_3NO_2	25	67	2	29	2.16
NO_2BF_4/CH_3NO_2	25	66	3	31	2.13
NO_2ClO_4/TMS	25	66	3	31	2.13
$CH_3OCH_2ONO_2$/CH_3CN	25	65	4	31	2.10
$C_6H_5ONO_2$/CH_3CN	25	64	5	31	2.06
NO_2BF_4/TMS	25	65	3	32	2.03
NO_2AsF_6/TMS	25	65	3	32	2.03
N_2O_5/TMS	25	65	3	32	2.03
p-$CH_3OC_6H_4ONO_2$/CH_3CN	25	63	6	31	2.03
$Cl_3CNO_2BF_3$/CH_3NO_2	25	64	4	32	2.00
$C_2H_5COONO_2$/CH_3CN	25	64	4	32	2.00
$C(NO_2)_4PF_5$/CH_3NO_2	25	63	5	34	1.97
NO_2Cl,$AlCl_3$/CS_2	0				1.96
2-CH_3-$C_6H_4N^+NO_2BF_4^-$/CH_3CN	25	64	3	33	1.94
2,6-CH_3-$C_5H_3N^+NO_2BF_4^-$/CH_3CN	25	64	3	33	1.94
4-CH_3O-2,6-$(CH_3)_2$-$C_5H_2N^+NO_2BF_4^-$/CH_3CN	25	64	3	33	1.94
$CF_3CO_2NO_2$/CH_3CN	25	63	4	33	1.91
$C_6H_5CO_2NO_2$/CH_3NO_2	25	63	4	33	1.91
$C(NO_2)_4$,BF_3/CH_3NO_2	25	64	2	34	1.88
HNO_3/H_2SO_4	25-50	62	5	33	1.87
2,4,6-$(CH_3)_3C_5H_2N^+NO_2BF_4^-$/$CH_3CN$	25	63	3	34	1.85
HNO_3/Ac_2O	25	63	3	34	1.85
NO_2PF_6,CH_3OH(1:1)/CH_3NO_2	25	63	3	34	1.85
NO_2PF_6,$(CH_3)_2O$(1:2)/CH_3NO_2	25	63	3	34	1.85
NO_2PF_6,$(C_2H_5)_2O$(1:2)/CH_3NO_2	25	62	4	34	1.82
NO_2PF_6,THF (1:2)/CH_3NO_2	25	62	4	34	1.82
NO_2PF_6,$(CH_3)_2O$(1:1)/CH_3NO_2	25	62	4	34	1.82
NO_2PF_6,$(C_2H_5)_2O$(1:1)/CH_3NO_2	25	62	4	34	1.82
CH_3COONO_2/CH_3Cl_2	-25	63	2	35	1.80
NO_2PF_6,$(CH_3)_2S$ (1:2)/$EtNO_2$	-78	62	3	35	1.77
NO_2PF_6,CH_3OH (1:2)/CH_3NO_2	25	62	3	35	1.77
NO_2PF_6,THF (1:1)/CH_3NO_2	25	62	3	35	1.77
Mixed acid (35%)/TMS	25	62	3	35	1.77
HNO_3/TMS	25	62	3	35	1.77
$CH_3ONO_2Ac_2O$	25	62	3	35	1.77
NO_2Cl,PF_5/CH_3NO_2	25	62	3	35	1.77
NO_2PF_6,$(CH_3)_2S$ (1:1)/$EtNO_2$	-78	62	3	35	1.77
$NO_2HS_2O_7$/TMS	25	62	3	35	1.77
HNO_3/CH_3NO_2	25	62	3	35	1.77
HNO_3,H_2SO_4/TMS	25	62	3	35	1.77

Table 41. Isomer Distribution in the Nitration of Toluene (cont.)

Reagent/solvent[a]	°C	o	m	p	o/p
$NO_2^+PF_6^-$,neo-$C_5H_{11}OH$ (1:2)/CH_3NO_2	25	62	3	35	1.77
$NO_2^+PF_6^-$,neo-$C_5H_{11}OH$ (1:1/CH_3NO_2	25	62	3	35	1.77
Fuming HNO_3/Ac_2O	40				1.76
p-$NO_2C_6H_4CO_2NO_2$/CH_3CN	25	61	4	35	1.74
p-$CH_3C_6H_4CO_2NO_2$/CH_3CN	25	61	4	35	1.74
NO_2Cl,$TiCl_4$/CH_3NO_2	25	61	4	35	1.74
NO_2Cl,$AlCl_3$/CH_3NO_2	25	61	4	35	1.74
NO_2Cl,$TiCl_4$/TMS	25	61	4	35	1.74
HNO_3/CH_3NO_2	25	62	2	36	1.72
5-NO_2-$C_9H_6N+NO_2BF_4^-$/CH_3CN	25	62	2	36	1.72
HNO_3/CF_3CO_2H	25	61	3	36	1.69
$CH_3CO_2NO_2$/CH_3NO_2	25	60	4	36	1.67
$CH_3CO_2NO_2$/CH_3CN	25	61	2	37	1.65
HNO_3/Ac_2O	25	61	2	37	1.65
HNO_3/Ac_2O	0	61	2	37	1.65
HNO_3/68.3% H_2SO_4	25	60	3	37	1.62
HNO_3 (94%)/neat	0	60	3	37	1.62
N_2O_5/CH_3CN	0	60	3	37	1.62
$C_6H_5CO_2NO_2$/CH_3CN	0	59	4	37	1.59
HNO_3/CH_3NO_2	30	59	4	37	1.59
HNO_3/H_2SO_4	30	59	4	37	1.59
HNO_3, HNO_2/77% H_2SO_4	30	59	4	37	1.59
HNO_3/H_2SO_4	40	59	4	37	1.59
HNO_3/Ac_2O	30	58	5	37	1.57
$C_6H_5CO_2NO_2$/CCl_4	0	57	6	37	1.54
HNO_3/Ac_2O	0	58	4	38	1.53
HNO_3,H_2SO_4/ArH	25	56	5	39	1.44
HNO_3 (d = 1.47)/ArH	30	57	3	40	1.43
HNO_3/AcOH	25	57	3	40	1.43
$C_5H_{11}ONO_2$/H_2SO_4	50				1.42
HNO_3/90% AcOH	45	56	4	40	1.40
NO_2Cl, PF_5/ArH	25	57	2	41	1.39
NO_2ClBF_3/ArH	25	57	2	41	1.39
HNO_3, H_2SO_4 (75%)/TMS	25	56	3	41	1.37
$CH_3CO_2NO_2$/Ac_2O	25	56	3	41	1.37
HNO_3/Ac_2O	25	56	3	41	1.37
NO_2Cl, $SnCl_4$/ArH	25	57	1	42	1.36
NO_2Cl, $SbCl_5$/ArH	25	56	2	42	1.33
NO_2ClO_4/ArH	25	55	3	42	1.31
NO_2PF_6/ArH	25	55	2	43	1.28
$CH_3CO_2NO_2$/CCl_4	25	55	2	43	1.28
NO_2BF_4/ArH	25	54	3	43	1.26
$CH_3CO_2NO_2$/AcOH	25	54	3	43	1.26
HNO_3/CCl_4	25	53	3	44	1.20

Table 41. Isomer Distribution in the Nitration of Toluene (cont.)

Reagent/solvent[a]	°C	Percent isomer			o/p
		o	m	p	
NO_2BF_4 + 18C6 crown/CH_2Cl_2	25	53	3	44	1.20
NO_2Cl, $TiCl_4$/ArH	25	53	2	45	1.18
HNO_3, PPA/$CHCl_3$	24-40				1.16
NO_2Cl, $AlCl_3$/ArH	25	53	1	46	1.15
CH_3ONO_2, PPA/CH_3NO_2	25	50	3	47	1.06
$NO_2HS_2O_7$/ArH	25	49	4	47	1.04
CH_3ONO_2, PPA/ArH, CH_3NO_2	25	47	3	50	0.94
HNO_3, H_2SO_4/ArH	−15	48	1	51	0.94
HNO_3, PPA/ArH	24-40				0.86
HNO_3, P_2O_5/$CHCl_3$	25	44	4	52	0.85
$EtONO_2$, PPA/ArH	45	42	3	55	0.76
NO_2Cl, $TiCl_4$/ArH	25	41	3	56	0.73
90% HNO_3/Amberlite IR-120	65-190				0.68
n-$BuONO_2$, PPA/ArH	26-35	39	3	58	0.67
$C_5H_{11}ONO_2$, PPA/ArH	50				0.64
CH_3ONO_2, PPA/ArH	25	37	4	59	0.63
sec-$BuONO_2$, PPA/ArH	32-40	36	3	61	0.59
t-$BuONO_2$, PPA/ArH	25-30				0.50
neo-$C_6H_{11}ONO_2$PPA/ArH	30-40				0.49
NO_2BF_4 + 21C7 crown/CH_2Cl_2	25	19	3	78	0.25
9-NO_2-anthracene/TaF_5 or Nafion-H	180-190			>95	0.05

[a]TMS, tetramethylene sulfone; THF, tetrahydrofuran; PPA, polyphosphoric acid.

3.3 Criticism of the Two Intermediate Mechanism

Criticism of Olah's separate π- and σ-complex intermediate mechanism was not long coming.[74a] Tolgyesi first suggested that with more efficient stirring in highly diluted solution, $NO_2^+BF_4^-$ nitration of toluene and benzene gives "regular" substrate selectivity. However, all efforts to change substrate selectivities by more efficient stirring alone were unsuccessful. Tolgyesi's claim[74b] was shown to be a consequence of hydrolysis of the nitronium salt in highly dilute solutions by water impurity. Increased efficiency of mixing had no effect on the regioselectivity of the nitration in any study. Nitration of toluene with nitronium salts under the best mixing conditions, even in stop-flow systems, gave no change in positional selectivity. Similarly, changing the ratio of competing aromatics in competitive nitrations showed no increase of substrate selectivity. These studies, however, did not answer the question as to the possibility of incomplete mixing affecting substrate selectivity in a system where very fast or diffusion-controlled reaction of the NO_2^+ ion with reactive aromatics can take place.

Ridd, studying the nitration of bibenzyl with $NO_2^+BF_4^-$ in sulfolane solution indeed found that the reaction gives a high proportion of dinitration even when a five-molar excess of bibenzyl over nitronium salt was used.[75] He concluded that this product composition is a consequence of nitration during incomplete mixing. He calculated that the NO_2^+ ion was effectively mixed with about 1.2 times the equivalent number of aromatic rings before reaction is complete. When HNO_3 was used in Ac_2O solution to nitrate bibenzyl, the amount of dinitro products was low. Ridd concluded[77] from his studies that nitration of similarly reactive aromatic compounds with nitronium salts should give apparent relative reactivities of near unity irrespective of the true relative reactivity of the substrates to the NO_2^+ ion.

Ridd's studies on the nitration of bibenzyl proved that the fast nitration occurs before sufficient mixing is achieved, thus the reaction is becoming mixing controlled. It is due to the great reactivity of NO_2^+ with the reactive aromatic substrate that such extremely fast reaction takes place.[76,77] As shown by Schofield *et al.*, many nitronium salt nitrations indeed take place at encounter rate (*vide infra*).[76]

The nitration of an excess of monoalkylbenzenes with nitronium salts gave mononitration[68,78] with only very small proportions of dinitro compounds. Mixing control in these cases is not obvious. In contrast, in the nitration of polyalkylbenzenes, such as durene, dinitration predominates, even without preformed nitronium salt.[79] When using $NO_2^+PF_6^-$ in the nitration of durene, 40%–46% dinitrodurene and 5%–8% mononitrodurene is obtained. Mixing control thus clearly is dominant. Similar reaction is indicated with pentamethylbenzene.[80]

Rys *et al.* discussed the question of mixing control in heterogeneous nitrations[81] which can "disguise" the true reactivity. The topic was well reviewed and will not be further elaborated here.

In connection with the question of mixing control, it is instructive to recall Ridd's discussion[77] of the question of microscopic and macroscopic diffusion in nitration. Most nitrations are carried out in such a way that the NO_2^+ ion is formed *in situ* in very small concentrations. For this reason, Ridd pointed out that although with sufficiently reactive aromatics the reaction with NO_2^+ ion may become microscopically diffusion controlled, the experimentally determined rate coefficient (rate = k_2^{obs}[ArH][HNO_3]) may more often than not be measured by conventional methods. If those nitrations that show the phenomenon of a limiting rate coefficient (i.e., loss of substrate selectivity) are correctly regarded as being microscopically diffusion controlled or encounter controlled, then they, and even some nitrations of less reactive aromatics, must become macroscopically diffusion controlled when bulk concentration of nitronium salts are used (as is the case in nitration with nitronium salts).

As discussed, the major criticism raised concerning Olah's competitive-nitration results of benzene and alkylbenzenes in sulfolane solution with $NO_2^+BF_4^-$ was that the fast nitration rates must have exceeded rates of mixing, and therefore the observed low substrate selectivities were caused by differences in the microscopic diffusion of the aromatics. Consequently, the competitive method under these conditions cannot give a true account of the relative reactivity of the aromatics, but nevertheless the data show a practical disappearance of substrate selectivity. The positional selectivities, however, stayed unchanged compared with those resulting from other nitrations.

The possibility of macroscopic diffusion affecting incomplete mixing is thus superimposed on the intrinsic microscopic diffusion of the NO_2^+ ion together with the reactive aromatics. The fact, however, remains that nitration of alkylbenzenes (as well as of halobenzenes) with nitronium salts maintains high positional selectivity, while substrate selectivity is lost. Thus, two distinct intermediates are indicated to account for the results, regardless whether the first step is very fast.

3.4 Schofield's Encounter Pair as the First Intermediate

From their extensive studies on the kinetics of aromatic nitrations, Schofield and co-workers reached the conclusion that the mechanism of nitration of reactive aromatics with NO_2^+ ion must include a separate step prior to the formation of the Wheland intermediate.[78] Whereas they reached their conclusion on independent kinetic grounds, their two-step mechanism – except for the nature of the first intermediate – is similar to that of Olah's.[76] Schofield *et al.* noted, in concert with

Deno's studies, that the observed second-order rate constant for nitration in H_2SO_4 decreased about four orders of magnitude for every 10% drop in the weight % acidity of the H_2SO_4. Since HNO_3 is essentially completely converted into NO_2^+ ion in 90% H_2SO_4, NO_2^+ ion concentration in 68% H_2SO_4 could be estimated to be 10^{-8} of the stoichiometric quantity of HNO_3. Thus, the second-order rate constant for the reaction of the NO_2^+ ion with an aromatic substrate (k_r) in 68% H_2SO_4 will be 10^8 times the observed second order rate constant ($k_2°$). Nitration of benzene in 68% H_2SO_4 at 25°C proceeds with $k_2° = 0.042$ $M^{-1}s^{-1}$. Therefore, k_2[benzene] is 4.2 x 10^6 $M^{-1}s^{-1}$. This value is approaching the rate coefficients for encounter between two species estimated on the basis of the viscosity of the medium ($D = 8RT/3\eta$) 6 x 10^8. Since partial rate coefficients have been calculated using benzene as the reference, it became imperative to determine whether diffusion was at all rate controlling in the nitration of benzene.

Schofield and co-workers studied the kinetics of nitration of a number of reactive aromatics such as toluene, m-xylene, naphthalene, and mesitylene. From the additivity principle, the second-order rate constant was expected to increase steadily along this series. The estimated relative rates (relative to benzene) are: 23, 300, 400, 16,000, respectively. In contrast, the observed relative rates for these substrates in 68% H_2SO_4 were 17, 27, 38, and 38. Evidently, further activation of the benzene nucleus beyond toluene produces a much more attenuated rate enhancement than expected from additivity principle, resulting in a limiting rate constant. When used in conjunction with the estimated concentration of NO_2^+ ion in 68% H_2SO_4, this limiting rate constant leads to a k_r value of 2.2 x 10^8 $M^{-1}s^{-1}$, a value that is very close to the encounter rate. In 61% $HClO_4$, a medium roughly three times less viscous than 68% H_2SO_4, mesitylene reacts about 80 times faster than benzene. Again, the k_r value for mesitylene is almost the same as the encounter rate in the medium.

It was therefore suggested that the observed limiting rates were a consequence of the reaction being limited by the diffusion of the NO_2^+ ion and the reactive aromatic, e.g., mesitylene or m-xylene.

Notwithstanding the loss of substrate selectivity, due to encounter control of the rate of reaction, the isomer distribution for these substrates, e.g., toluene and anisole was far from being statistical, both giving predominantly o- and p-nitro derivatives.

Initially, it was even considered that the reaction at the activated ortho and para sites of toluene occurred at encounter control but not at the less reactive meta position. Nitration of pseudocumene (1,2,4-trimethylbenzene), in which both the activated positions, 5 and 6, should react at encounter control, showed that positional selectivity was nonetheless maintained.

Schofield, considering the question of loss of substrate selectivity but of maintained high-positional selectivity in nitration of reactive aromatics with NO_2^+ ion, concluded that the reaction proceeded via encounter pairs of the aromatic substrate and the NO_2^+ ion (thus loss of positional selectivity) followed by transformation (in a separate step) into the positional selectivity determining Wheland-intermediates.

The encounter pair is considered by Schofield to consist of a NO_2^+ ion and an aromatic molecule held in proximity by a solvent shell. The retention of positional selectivity via encounter-controlled nitration is evidence that a kinetically distinct intermediate proceeds the product-controlling Wheland intermediate. Schofield argued that the magnitude of positional selectivity found in the nitration of monocyclic aromatics does not require the assumption of the presence of stabilizing interactions (i.e., of bonding nature) in the encounter pair formed with NO_2^+ ion, although the data do not necessarily exclude such interactions.[78d] An encounter pair is composed of two reacting molecules in solution that are nearest neighbors and remain as such for some time. During the encounter the reactants involved collide a number of times and may react in one of these collisions. When the number of collisions before reaction is small, in other words, the probability of the reaction approaches unity, the reaction becomes diffusion controlled; the rate of reaction being limited by the rate of encounter.

For nitration to be diffusion controlled, the energy barrier opposing diffusion of the components of the encounter pair must be higher than the energy barrier between the encounter pair and the various possible transition states leading to Wheland intermediates.

The rate constant for diffusion of the components of a noninteracting encounter pair was estimated by Schofield to be of the order of 10^9–10^{10} sec^{-1}. For the reaction to remain diffusion controlled, the rate constant for transformation of the encounter pair into any of the Wheland intermediates must be considerably greater than this. The upper limit, again estimated by Schofield, should have the magnitude of a vibration frequency, i.e., 10^{12}–10^{13} sec^{-1}. If one considers that the encounter pair may be stabilized by a solvent-cage effect, the energy barrier for separation of the encounter complex may be as high as 3 kcal/mol, allowing for some energy range in subsequent formation of arenium ions (σ-complexes).[82] Whereas according to Schofield, observed positional selectivities *per se* do not require the assumption of "stabilizing interactions" in the encounter pair, other considerations strongly suggest such interactions. It must be considered how a strong acid (NO_2^+) and a strong π base (the aromatic, ArH) can come together in an encounter pair without some binding interaction. At the very least, attractive forces of the ion-induced dipole are expected to operate. The fact that the nitration of benzene and toluene, which does not show a primary hydrogen isotope effect (k_H/k_D), but with $NO_2^+BF_4^-$, a small inverse secondary isotope effect ($k_H/k_D = 0.82$–0.85) was observed, strongly indicates π-bonded interaction in the first intermediate. The significant point is that regardless of its nature a separate first intermediate is clearly necessitated by the kinetic data and maintained high positional selectivity.

Evidence for attractive interaction between the NO_2^+ ion and the aromatic is also provided by recent calculations by Politzer and by Cacace's gas-phase nitration studies. *Ab initio* calculations by Politzer on the interaction of NO_2^+ ion with benzene and toluene show two minima on the potential-energy surface.[83] At about 2.9 Å intermolecular distance there is a weakly bound complex in which the NO_2^+ ion is parallel to the plane of the benzene ring. The calculated stabilization energies are 6 kcal/mol for benzene an 7.5 kcal/mol for toluene. At closer distances the strongly bound arenium ion complexes were located.

Cacace *et al.* reported[84] on the gas-phase nitration of benzene and substituted benzenes by irradiating a mixture of the aromatic(s), methyl nitrate, and a large excess of methane with gamma rays (see Section 3.1.6.3). Radiolysis of methane produces CH_5^+ which protonates methyl nitrate and allows subsequent aromatic nitration according to the following sequence of reactions:

$$CH_5^+ + CH_3NO_3 \longrightarrow CH_4 + CH_3O^+(H)NO_2$$

$$C_6H_6 + CH_3O^+(H)NO_2 \longrightarrow C_6H_6NO_2^+ + CH_3OH$$

Nitration is completed by deprotonation of the nitro adduct by any gas-phase base. Addition of ammonia dramatically suppresses nitration by competing with further protonation of methyl nitrate, demonstrating the ionic nature of the reaction.

Unlike other ion–molecule studies of nitration in the gas phase, radiolysis experiments provided isolable product material to allow analysis (including isomer distributions) by gas chromatography. In competitive nitration experiments, the substrate selectivity between benzene and toluene or benzene and anisole were low with $k_{C_6H_5}/k_{C_6H_6}$ ratios of 2.9 and 3.2, respectively. Yet, the amount of meta nitration was also low, ≈4% for toluene and less than the detection limit of 2% for anisole. The observation of high positional selectivity and very low substrate selectivity can only be reconciled by including a "first intermediate" prior to forming the σ complexes, similar to the one discussed in solution chemistry. In gas-phase nitration incomplete mixing cannot be involved and solvent-cage effects cannot exist. Thus, the first intermediate must be held together by attractive forces.

3.5 *Ipso* Attack and Nitration

Nitration of monosubstituted benzenes can give not only ortho-, meta-, and para-substituted isomers, but also can affect the same (*ipso*) position to which a substituent is already attached. Direct effects of the nitration of monosubstituted benzenes, such as toluene, were, however, for long mainly discussed giving ortho, meta, and para nitration.

Ipso nitration (i.e., nitration at a position already carrying a substituent) has been long known in dealkylative nitrations, but not recognized until more recently as a general phenomenon competing with ortho, meta, and para substitution. It has been known for example that nitration of *p*-cymene with HNO_3/H_2SO_4, yields not only nitro-*p*-cymenes, but substantial amounts of 4-nitro toluene.[86]

The nature of such nitrations was recognized and named as *ipso*-nitration by Perrin and Skinner in 1971.[87] An excellent review of the topic was given by Schofield.[85]

Olah *et al.*[88a] as well as Myhre *et al.*,[88b] in their studies of the nitration of alkylbenzenes with $NO_2^+BF_4^-$ in sulfolane solution found generally intact ring nitration, but in case of cymenes, diisopropylbenzenes and *p*-di-*tert*-butylbenzene they also found varying degrees of nitro-dealkylation, indicative of competing *ipso* nitration (Table 42).

In mechanistic terms the reaction goes through an *ipso*-substituted arenium ion intermediate resulting in de-alkylative nitration accompanying intact ring nitration.[87]

Table 42. Intact and Dealkylative Nitration of Alkylbenzenes with Nitronium Tetrafluoroborate in Sulfolane at 25°C[88a,b]

Compound	Intact ring nitration (%)	Dealkylative nitration (%)
Cumene	100	–
p-Cymene	90.5	9.5
m-Cymene	100	–
o-Cymene	59	41
Propylbenzene	100	–
p-Propyltoluene	100	–
m-Propyltoluene	100	–
o-Propyltoluene	100	–
o-Diisopropylbenzene	100	(Traces of *o*-nitrocumene)
p-Diisopropylbenzene	44	46
m-Diisopropylbenzene	100	–
1,3-Dimethyl-5-isopropylbenzene	100	–
1,3,5-Triisopropylbenzene	100	–
t-Butylbenzene	100	–
p-*t*-Butyltoluene	100	–
m-*t*-Butyltoluene	100	–
o-*t*-Butyltoluene	100	–
p-Di-*t*-butylbenzene	80	20
m-Di-*t*-butylbenzene	100	–
o-Di-*t*-butylbenzene	100	–
1,3,5-Tri-*t*-butylbenzene	100	–

Similar nitrodehalogenations, nitrodeacylations, and nitrodecarboxylation[89] as well as nitrodesilylations (see Section 2.26.5) are also known and are assumed to involve *ipso*-attack by the NO_2^+ ion.

When considering *ipso* nitration we must differentiate the attack of the NO_2^+ ion (or its carriers) at *ipso* positions from subsequent possible substitution or nucleophilic–capture reactions (as formation of nitro acetates in reactions with HNO_3 in Ac_2O or other nucleophilic capture reactions). In systems of limited nucleophilicity, such as with $NO_2^+BF_4^-$ in sulfolane solution, no such reactions are possible. The *ipso* nitroarenium, however, can undergo subsequent isomerization leading to ring substitution, or *ipso* nitration can take place if replaceable substituents (such as alkyl or halo substituents) are eliminated; thus the distinction between *ipso* attack and *ipso* substitution.

By comparing the nitration of alkylbenzenes with HNO_3 in Ac_2O and aqueous H_2SO_4 with $NO_2^+BF_4^-$ nitrations in sulfolane, Schofield estimated the relative positional reactivities in a series of alkylbenzenes.[85] Medium dependence of the isomer proportions, however, even when overall nitration yields are quantitative, excludes assignment of truly accurate relative positional selectivities.

The data show that the methyl group in toluene activates its ipso position twice as much as the meta position. Effects on cymenes and polymethylbenzenes are much larger. Pentamethylhalobenzenes with $NO_2^+BF_4^-$ gave at low temperature long-lived nitroarenium ion, formed as a result of *ipso* attack.[91]

Using variable-temperature ^{13}C NMR studies, Olah and co-workers[31] showed that the nitrohexamethylbenzenium ion undergoes a sixfold degenerate isomerization. Thus, the distinct peak due to the sp^3-C coalesced with the other signals at −19°C. Above that temperature only a single peak was observed. From these studies, however, it is not possible to conclude if the rapid isomerization is a result of successive 1,2-shifts or longer-range migration, nor even whether it is an inter- or an intramolecular process. Koptyug et al.[92] used the technique of label saturation transfer to delineate the mechanism of the rearrangement. In this technique, one of the signals is saturated with a strong radio frequency. Molecular rearrangements result in the transfer of the spin to a different location, affecting the signal of that site. In the case of nitrohexamethylbenzenium ion, Koptyug showed that the saturation of the 1-methyl proton signal affected most strongly only the signal due to the 2-methyl protons, indicating that the rearrangement was a result of an intramolecular 1,2-migration and did not involve a long-range rearrangement.

Products of entrapment of the *ipso* Wheland intermediates by nucleophiles are the most frequently used way to study their role in nitrations. For example, Fischer *et al.*[93] isolated a pair of stereoisomeric adducts by low-temperature nitration of *o*-xylene with acetyl nitrate, followed by low-temperature chromatography.

These two nitrocyclohexadienyl acetate adducts represent *cis* and *trans* addition of the NO_2^+ ion followed by nucleophilic capture on the *ipso* intermediate by acetate ion or acetic acid.

Since the isolation of the pair of cyclohexadienyl adducts of *o*-xylene, a large number of similar adducts have been isolated and characterized by Fischer *et al.*[93] Many of these adducts undergo a variety of subsequent reactions. The *ipso* intermediates have also been captured by benzoate, fluoride, and nitrate. Intramolecular capture by an acid has also been reported.

Nitration of *p*-methyl substituted benzenes, bearing Cl, Br, F, −O, and −NH substituents, yield nitrodienones and subsequently nitrophenols via *ipso* substitution.[94]

The yield of the nitrodienones and subsequently nitrophenols is dependent upon the ability of the substituents R^1 and R^2 to stabilize the 1-methyl-1-nitrobenzenium ion.

NMR examination of the reaction mixtures showed that the actual yields of nitrocyclohexadienones were much higher than the yield of isolated nitrophenols.

Myhre first recognized[90] the occurrence of intramolecular 1,2-migrations in the nitration of *o*-xylene with HNO_3 in H_2SO_4 by studying the acidity dependence of the 3- to 4-nitro-*o*-xylene isomers
formed. The *ipso* NO_2^+ ion was suggested to be captured by water at lower acidities, but with increasing acidities 1,2-migrations became increasingly significant, resulting in the change of the isomer ratio.

Ipso nitration adducts, as emphasized by the work of Fischer,[93] as well as Myhre[90] and numerous other investigators, opened up a significant new area of chemistry. Since our discussion, however, is limited to direct methods of nitration, we will not discuss transformation of nitrocyclohexadienyl adducts (except an example of an interesting nucleophilic nitration reaction in Chapter 3, part III).

Ingold's studies in the 1940s and 1950s as well as Olah's nitration studies with nitronium salts in the 1950s and 1960s preceded the realization of the significance of *ipso* nitration and subsequent intramolecular migration processes. The question was consequently raised, that neglect of consideration of possible *ipso* attack in determining reactivities of nitration leaves shortcomings.[27] However, since in most discussions of the mechanistic aspects of aromatic nitration, benzene and toluene are the most frequently considered substrates, and since their nitration is not significantly affected by *ipso* attack, the general mechanistic conclusions reached with them remain substantially valid.

3.6 Electron Transfer in Nitration

3.6.1 Early Suggestions

The possibility of electron transfer (ET) in aromatic nitration was invoked by Kenner[95] and by Weiss[96] as early as the mid-1940s, that is around the same time when the Ingold mechanism was getting established.* Kenner was formulating a scheme for describing organic reactions in a sequence of full electron transfer processes in which the intermediates remain "within the sphere of each other's action." Accordingly, he proposed nitration of benzene as:

$$\left[\begin{array}{c} \bigcirc \\ NO_2^+ \end{array} \right] \longrightarrow \left[\begin{array}{c} \bigcirc^+ \\ \cdot NO_2 \end{array} \stackrel{H}{} \right] \xrightarrow{(SO_4H)^-} \left[\begin{array}{c} \bigcirc \\ \cdot NO_2 \end{array} \right] \longrightarrow \bigcirc\!-\!NO_2$$

Ingold refuted[97] the idea of an intermediate benzene radical–cation $\cdot NO_2$ pair on the grounds that some of the NO_2 should escape and undergo hydrolysis to HNO_2/NO^+ ion. However, these products are not observed in nitrations and hence NO_2 cannot be an intermediate. Of course, if the collapse of the radical cation–radical pair is more rapid than the escape from the solvent cage, Kenner's suggestion would be consistent with the lack of hydrolysis products of NO_2 in the product mixture.

The possibility of electron transfer in aromatic nitrations was subsequently considered by several investigators.[129,130,131] These studies notwithstanding, the possibility of ET in aromatics nitration was not appreciated by most organic chemists until Perrin in 1977 reemphasized its possible significance in his study of the nitration of naphthalene.[98]

---

*It is customary to refer to single-electron or one-electron transfers. Such explicit reference is generally made to draw a distinction between them and the two-electron polar processes. In view of the fact that all electron transfers proceed by one-electron shifts, we have opted to use simply "electron transfer" to refer to single-electron transfers, except when comparing one-electron and two-electron processes.

3.6.2 Perrin's Renewal of the ET Concept

Perrin found that the "electrochemical nitration" of naphthalene involving the quenching of the naphthalene radical cation with NO_2 gave an $\alpha-/\beta-$nitronaphthalene isomer ratio of 9, similar to the ratio of 11 observed in mixed acid nitration. He therefore suggested that the latter reaction also involves single-electron transfer. In his paper,[99] Perrin cites a number of reasons for suggesting the necessity of ET in the nitration of aromatic substrates more reactive than toluene. These include:

(1) The ET mechanism provides a satisfactory answer to the question as to how the NO_2^+ ion, a very reactive electrophile acquires selectivity, which it otherwise lacks. Whereas the mere formation of an initial π complex may alter the nature of the electrophile, ET on the other hand would result in the formation of a totally different species with entirely different characteristics.

(2) Since half-wave potentials for the oxidation of all aromatics more reactive than toluene are less positive than for the oxidation of NO_2, ET from these reactive aromatics to NO_2^+ is considerably exothermic.

(3) Positional selectivity better correlates with the spin densities of the radical cation in the few instances, such as phenanthrene, where a distinction between direct electrophilic substitution and collapse of a radical pair is possible.

(4) The ET mechanism offers a ready explanation of the long-range migration of nitro groups in terms of reversal to the radical-cation–radical pair, which then collapses in a different position rather than sequential 1,2-migrations.

(5) High *ipso* reactivity of polymethylbenzenes is explained in terms of greater spin density at those positions in the radical-cation.

As a test for the ET mechanism, Perrin reported the formation of nitronaphthalenes by controlled potential electrolysis of naphthalene in the presence of NO_2. The electrolysis was carried out on platinum electrodes at +1.3 V versus Ag/0.01 M $AgClO_4$ in acetonitrile. In an independent experiment it was established that this potential is sufficient to oxidize naphthalene, but in order to oxidize NO_2 a potential of +1.8 V was necessary. The experiment was considered to show that the naphthalene radical-cation and NO_2 can react to give observed nitration products. Moreover, the ratio of the 1-nitronaphthalene to 2-nitronaphthalene formed under electrochemical nitration was reported to be the same (≈ 10) as that obtained by nitrating naphthalene with HNO_3 in acetonitrile with added urea to suppress the effect of HNO_2. This evidence was considered to support electron-transfer as the first step in the reaction of NO_2^+ ion and reactive aromatics.

Olah, Schofield, and Perrin all agree insofar as the need of the inclusion of an additional distinct intermediate into the original Ingold mechanism is concerned. They differ in regard to the nature of this intermediate. Moreover, if it is considered that in the formation of this intermediate the aromatic substrate is involved as a whole and not through any specific atom in the molecule, their views appear even more alike. The differences are a matter of degree. At one extreme we have the Schofield model in which the aromatic and the NO_2^+ ion have come together in an encounter complex, but they are not interacting. At the other extreme is Perrin's suggestion in which an electron has been completely transferred to the NO_2^+ ion. Somewhere in between is the π complex suggested by Olah.

3.6.3 Probing the Role of Electron Transfer in Aromatic Nitrations

Perrin's paper attracted considerable interest. Perhaps part of the reason is the increased interest in recent times in electron-transfer reactions in general. Reviews for example by Parker[99] and by Eberson[100] give a good account of the significance of organic radical-cations and electron transfer in organic reactions. Many reactions previously considered to be taking place by two-electron processes have recently been suggested to occur via electron-transfer steps. Some examples are aliphatic and aromatic nucleophilic substitutions, Grignard reactions, the Wittig reaction, anionic polymerization of styrene, and hydride reductions. Whether electron transfer is really as common as suggested by Pross[101] and Shail[102] or in a somewhat related sense by Kochi, relating the significance of charge transfer in electrophilic substitutions and addition,[103] is for the future to decide. It is necessary, however, to caution that experimental proof found in specific cases does not allow generalization to all reactions.

Perrin's publication on ET nitration of naphthalene and related activated aromatics revitalized interest and acted as a catalyst for renewed studies probing the role of electron transfer in aromatic nitration. These studies provided data and answers to the following key questions:

1. Is the NO_2^+ ion capable of oxidizing the aromatic compound under consideration?
2. Does the aromatic radical cation couple with NO_2 to give the corresponding nitroaromatic?
3. Are the rate constants of the individual steps consistent with the observed rates of nitration?
4. Is the product distribution of the reaction of ArH^+ with NO_2 the same as that obtained in electrophilic, acid–catalyzed nitrations?
5. How are the ET and polar mechanisms related?

These questions have been addressed by both experimental and theoretical studies. Of particular relevance are nitrations with preformed aromatic radical cations by electrochemical or chemical oxidation of the aromatics, gas–phase ion–molecule reactions, theoretical examination of the potential energy surface, as well as application of the Marcus theory[104] to estimate the rates of ET.

3.6.3.1 Eberson's Electrochemical Studies

The formation of the same product mixture from two reactions, although not necessarily unequivocal, is a frequently used test for the involvement of a common intermediate. In this regard, Perrin's electrochemical nitration of naphthalene[98] seemed fairly convincing evidence for the ET mechanism of nitration. However, the constant $\alpha-/\beta$-nitronaphthalene isomer ratio of ~10 reported by Perrin was not found by other investigators.[106] Electrochemical nitration by Eberson, Hussey, and Ross, respectively gave much higher α/β ratios (25–65).[107,108] Moreover, acetonitrile used as a solvent is extremely difficult to dry and spectroscopic examination of the NO_2/N_2O_4 (N(IV)) solution in "dry" acetonitrile always showed the presence of HNO_2. N(IV) solutions could be obtained in methylene chloride and even in this medium the ratio of 1- and 2-nitronaphthalenes formed under controlled potential electrolysis was 25.

A detailed examination of the electrochemical nitration by Eberson showed that at least 50% – 60% of the product was formed by acid-catalyzed homogeneous nitration of naphthalene by N_2O_4. In a subsequent study, Eberson and Radner isolated crystalline naphthalene radical-cation hexalfluorophosphate of the composition $(Naph)_2PF_6$. When reacted with N(IV) in methylene chloride at −20°C, 1- and 2-nitronaphthalene were obtained, however, the isomer ratio was 40. At −45°C electrochemical nitration with N_2O_4 gave an α-/β-nitronaphthalene isomer ratio of 65.

In considering changes of isomer ratios, one should keep in mind that they frequently represent only a small change in the less abundant isomer. Further, relatively minor changes in conditions can be reflected in substantial changes in the isomer ratio of acid-catalyzed nitrations. The diversity of the α/β isomer ratio in the nitration of naphthalene is shown in Table 43 to vary from 9 to 65.[106]

Eberson and Radner also investigated the reaction of nitrite ion (NO_2^-) with Naph$^{+\cdot}$.[108c] It was observed that under heterogeneous conditions the naphthalene radical cation and nitrite ion reacted via electron transfer at encounter rate to give naphthalene and NO_2. No nitronaphthalenes were formed in this reaction implying that the NO_2 so generated did not couple with another Naph$^{+\cdot}$, instead it diffused out of the cage and encountered another NO_2 to form N_2O_4. The estimated rate for the coupling of two NO_2 radicals at equilibrium with N_2O_4 under experimental conditions is around 10^8 $M^{-1}s^{-1}$, and since no nitronaphthalenes were formed, the reaction of NO_2 with Naph$^{+\cdot}$ must be even slower. This result is even more striking in view of the heterogeneous nature of the reaction. NO_2 is generated at the interface of the crystalline $(Naph)_2PF_6$ and the solution i.e., it is more likely that the nascent NO_2 will encounter a Naph$^{+\cdot}$ rather than a second NO_2.

According to the ET mechanism of nitration of reactive aromatics, the collapse of the radical radical-cation pair has to be faster than the encounter rate of the NO_2^+ ion and the aromatic, in order for the overall reaction to be encounter limited. The lack of nitro product formation in the reaction of NO_2^- ion with crystalline $(Naph)_2PF_6$ shows that NO_2 and Naph$^{+\cdot}$ react at a rate much slower than the encounter rate. Thus, as pointed out by Eberson, if the reaction of naphthalene and NO_2^+ ion involves ET, there should be a build up of substantial concentrations of naphthalene radical cation and NO_2 pairs, permitting easy spectroscopic detection (which was not the case); thus ET does not take place under these conditions.

Markovnik's studies on single electron transfer in nitration of

Table 43. Nitration of Naphthalene with Various Nitrating Agents [106]

Reagent	Solvent	Temp (°C)	α/β Isomer ratio
$NO_2^+BF_4^-$	Sulfolane	25	10
$NO_2^+BF_4^-$	Nitromethane	25	12
HNO_3	Nitromethane	25	29
HNO_3	Acetic acid	25	21
HNO_3	Acetic acid	50	16
HNO_3	Sulfuric acid	70	22
HNO_3	Acetic anhydride	25	9
CH_3ONO_2/CH_3OSO_2F	Acetonitrile	25	13
$AgNO_3/CH_3COCl$	Acetonitrile	25	12
$AgNO_3/C_6H_5COCl$	Acetonitrile	25	12
N_2O_4	Acetonitrile	25	24
$N_2O_4/Ce(NO_3)_4 \cdot 2NH_4NO_3$	Acetonitrile	65	16
$HNO_3/H_2SO_4/urea$	Acetonitrile	–	11
Electrochemical oxidation + N_2O_4	Methylene chloride	–45	65
$C(NO_2)_4$	Gas phase	300	1
$AgNO_3/BF_3$	Acetonitrile	25	19
4-nitro-1-nitrosopyridinium BF_4^-	Acetonitrile	25	10

reactive aromatics[109] also contributed significantly to our understanding of oxidation by NO_2^+ and NO^+.

3.6.3.2 Use of Chemical Oxidants

An alternative approach to electrochemical oxidation is chemical oxidation. Ceric ammonium nitrate (CAN) is a common oxidant used to generate radical cations from a variety of aromatic hydrocarbons. Suzuki compared the products of direct nitration of 2,4,5-trimethyl-neopentylbenzene and HNO_3 with those obtained using a mixture of CAN and N(IV).[110] With HNO_3, the product mixture consists of ring nitration (30%), two side-chain nitration products (20%), and an alcohol (4%). These products are rationalized as arising from

Wagner–Meerwein shift of the neopentyl side chain that occurs following the formation of the Wheland intermediate. With CAN alone, the substrate gave only a mixture of benzyl nitrates (20 – 50%).

Addition of N(IV) to the reaction mixture gave in addition to the benzyl nitrates some arylnitromethane, but no rearrangement of the neopentyl side chain was observed. The differing reactivities in the two cases, straight nitration with HNO_3 versus CAN/N(IV) nitration, rule out the radical cation as an intermediate in the pathway leading to side-chain substitution. However, the fact that the aromatic radical cation was probably formed in a solvent cage not containing NO_2 obscures any conclusions that could be drawn regarding the relevancy of ET mechanism for nitration of reactive aromatics.

Draper and Ridd[111] sought to test the ET mechanism of nitration by treating mesitylene in acetonitrile with CAN in the presence of NO_2 and to compare the results with the HNO_3 nitration of mesitylene in the same solvent. HNO_3 reacts with mesitylene to give the ring-nitrated product exclusively, whereas CAN alone reacts with mesitylene to give the side-chain nitrate, $ArCH_2ONO_2$, only. This reaction is considered to take place by the loss of a proton from the aromatic radical cation followed by reaction of the benzyl radical with nitrate ion. Addition of N_2O_4 to the CAN reaction with benzene suppressed the overall conversion, but more significantly the product mixture consisted of 74% ring-nitrated derivative and only 26% of the benzyl nitrate.

Subsequently, in a more detailed study, Dinçtürk and Ridd[112] examined the reaction of a range of aromatic hydrocarbons with CAN and with HNO_3 in acetonitrile at 84°C. Several experimental difficulties precluded direct measurement of the rates of nitration with CAN. Therefore, the relative rates were determined using the competitive method. The relative rates of nitration of the hydrocarbons were nearly identical for both methods of nitration (Table 44). Likewise, the isomeric composition of the product from both CAN and HNO_3 nitrations were also the same (Table 45). The identical inter- and intramolecular selectivities can be regarded as a good indication for the involvement of the same reactive intermediate. One simple explanation is that CAN is in equilibrium with a small amount of NO_2^+ ion which then reacts with the substrate. Dinçtürk and Ridd point out the obvious difficulty with this explanation. In order to maintain a small equilibrium concentration, the NO_2^+ ion must react with the small amount of the residual part of the Ce complex in the presence of a large concentration of aromatics. This requirement is generally unreasonable and for substrates such as mesitylene, with which the NO_2^+ ion appears to react at encounter, impossible. The way around the problem is to invoke only a short-lived $NO_2^+CeO(NO_3)_5^{3-}$ pair, and nitration occurs only if the aromatic substrate pre-associates with CAN.

$$ArH + Ce(NO_3)_6^{2-} \rightleftharpoons ArH Ce(NO_3)_6^{2-}$$

$$ArH \cdot Ce(NO_3)_6^{2-} \longrightarrow ArH \cdot NO_2^+ CeO(NO_3)_5^{3-}$$

$$ArHNO_2^+ CeO(NO_3)_5^{3-} \longrightarrow ArNO_2 + Ce(OH)(NO_3)_5^{2-}$$

Table 44. Reactivities Relative to Benzene for Nitration by Cerium(IV) Ammonium Nitrate (CAN)[112] and by Nitric Acid[76b]

Substrate	Relative reactivities		
	CAN (CH_3CN, 84°C)	HNO_3 (CH_3CN, 84°C)	HNO_3 ($C_4H_8SO_2$, 25°C)
Toluene	22.8 ± 0.2	19.2 ± 0.9	20
Ethylbenzene	22.4 ± 1	14.6 ± 0.4	
Isopropylbenzene	15.6	13.3 ± 0.7	
t-Butylbenzene	10.0 ± 0.5	9.6 ± 0.3	
m-Xylene	151 ± 25	122 ± 14	100
Mesitylene	686 ± 94	618 ± 12	350
Anisole	4899	296	175
Naphthalene	7340		33

Table 45. Comparison of the Product Compositions[a] from the Nitration of Aromatic Compounds with Cerium(IV) Ammonium Nitrate in Acetonitrile at 84°C with Those from Nitration by Nitric Acid under the Same Conditions (in parentheses)[112]

Substrate	Isomer proportions (percent)		
	2-	3-	4-
Toluene	55.5 ± 1.5	5.0	39.5 ± 1.5
	(55.7 ± 1.2)	(5.5 ± 6.1)	(39.6 ± 0.8)
Ethylbenzene	46.1 ± 0.3	5.8 ± 0.4	48.1 ± 0.5
	(47.5 ± 1.2)	(6.7 ± 0.4)	(45.8 ± 1.4)
Isopropylbenzene	26.1 ± 0.9	8.8 ± 1.3	65.1 ± 2.0
	(27.8 ± 1.6)	(9.2 ± 0.9)	(63.0 ± 2.0)
t-Butylbenzene	15.3 ± 0.8	14.0 ± 0.3	70.8 ± 0.6
	(15.8 ± 0.5)	(15.5 ± 0.6)	(68.7 ± 1.1)
m-Xylene	19.3		80.7
	(18.8)		(81.2)
Anisole	36.5 ± 0.5		63.7 ± 0.5
	(65.9 ± 0.6)		(34.1 ± 0.6)
Naphthalene	α 93.8	β 6.2	
	(α 91.4)	(β 8.6)	

[a]Excluding products from side-chain substitution.

This scheme does not rule out the formation of the aromatic radical cation by ET to the NO_2^+ ion. However, since the similarity of products extended to substrates such as toluene and ethylbenzene, whose oxidation potentials are higher than that of NO_2, ET could not be a general mechanism.

3.6.3.3 Gas–Phase Ion–Molecule Nitration

The advent of ion cyclotron resonance (ICR) spectrometry and other developments in mass-spectrometric methods greatly facilitated the study of ion-molecule reactions in the gas phase. It is often easier to study the reactions in the gas phase where there are no problems associated with mixing of the reagents, and interpretation of the data is generally more straightforward. Various reactions occurring simultaneously can be easily monitored and quantitative kinetic information can be obtained. These techniques have been applied to many organic reactions and have led to a deeper understanding of the mechanism. While it is recognized that solvent effects can markedly alter the course of reactions, gas-phase studies most conveniently yield the intrinsic reactivity of the species involved. The chemistry observed in the condensed phase can then be resolved into intrinsic chemistry and effects due to solvent or a neighboring atom. For example, the observed order of acidity of alcohols in the condensed phase (primary > secondary > tertiary) had traditionally been explained in terms of +I effect of the alkyl groups. It was not till after the acidities were determined in the gas phase that the decreased acidity of the increasingly bulkier alcohols was correctly ascribed by Brauman to poorer solvation of the corresponding anions.[113]

Bursey and co-workers studied the reaction of perdeuterobenzene (C_6D_6) and a variety of nitrating agents in an ICR instrument operating at a total pressure of $\approx 4 \times 10^{-5}$ torr.[114,115] Ionization was effected using 30 eV electrons. When NO_2 was used as the nitrating agent, product peaks were observed at m/e 130 ($C_6D_6NO_2^+$) and 100 ($C_6D_6O^{+\cdot}$). Using double-resonance techniques, they determined the precursor for the nitrated product to be $C_6D_6^{+\cdot}$; the direct reaction of NO_2^+ with benzene led to the oxygen-transfer product:

$$C_6D_6^{+\cdot} + NO_2 \longrightarrow C_6D_6NO_2^+$$

$$C_6D_6 + NO_2^+ \longrightarrow C_6D_6O^{+\cdot} + NO$$

The observation of the oxygen-transfer product is a demonstration of the ambident nature of the electrophile NO_2^+.

When other nitrating agents such as methyl or ethyl nitrate were used, nitrated product was formed, but in these cases the precursor ion was not the benzene radical cation, but "solvated" NO_2^+ ions formed in these systems: $H_2NO_3^+$ (m/z 64), $CH_2ONO_2^+$ (m/z 76), or $CH_3CHONO_2^+$ (m/z 90). These ions were formed by the fragmentation of the alkyl nitrate ions:

$$C_2H_5ONO_2^{+\cdot} \longrightarrow CH_2=ONO_2^+$$

$$C_2H_5ONO_2H^+ \longrightarrow H_2NO_3^+$$

The differing reactivity of the NO_2^+ ion and "solvated" NO_2^+ ion can be understood in terms of chemical activation. Since the reaction of NO_2^+ with aromatics (ArH) is very exothermic, the $(ArH \cdot NO_2)^+$ complex that is formed will come apart unless the excess energy is dissipated by collision or by loss of a fragment (like NO resulting in the observed O^+-transfer). On the other hand, when reacted with some complexed form of NO_2^+, such as $CH_2ONO_2^+$, aromatic hydrocarbons gave products of the composition $ArHNO_2^+$. In these cases, the excess energy could be dissipated through the loss of the complexing moiety, CH_2O or H_2O.

Dunbar, Shen, and Olah[116] investigated the nitration of benzene and substituted benzenes in the gas phase in an ICR instrument. They used ethyl nitrate as the nitrating agent, which under their conditions of ionization (14 eV electron impact) gave $CH_2ONO_2^+$ and NO_2^+. In accord with the findings of Bursey et al.,[114] they observed the reaction of aromatics and NO_2^+ to lead to oxygen transfer, but from the aromatic and $CH_2ONO_2^+$, formation of nitrated products was observed. Furthermore, by performing experiments with mixtures of aromatics, they were able to obtain relative-rate constants for the nitration reactions. The observed substituent effects were large compared to theoretical predictions and ran contrary to the order of reactivity observed in solution-phase chemistry (Table 46). Thus, nitrobenzene was found to be more reactive than benzene, which, in turn, was more reactive than toluene or ethylbenzene.

Subsequently, Ausloos and Lias also studied the reaction of a "solvated" NO_2^+ ion ($C_2H_5ONO_2NO_2^+$) with a number of substituted benzenes and showed that the charge-transfer reaction dominated the reaction with activated aromatics and hence the absence of the $ArHNO_2^+$ product in those cases.[117]

Table 46. Relative Reaction Rates for Substituted Benzenes and Heteroaromatic Compounds in Nitration with $CH_2ONO_2^{+116}$

Neutral reactant	k_{ArR}/k_{ArH} rate
Benzene	1.0
Benzene-d_6	1.0
Aniline	0
Anisole	0
Nitrobenzene	10
Pyrrole	0
Ethylbenzene	0
Toluene	0.3
Cyclopropylbenzene	0
Chlorobenzene	0.25
α,α,α-Trifluorotoluene	0.3
Pyridine	0.3
Fluorobenzene	0.55
Xylene (o, m, or p)	0
o-Difluorobenzene	0.25
m-Difluorobenzene	0.3
p-Difluorobenzene	0
Pentafluorobenzene	0
p-Bromoanisole	0
1,3,4-Trimethoxybenzene	0

One of the steps, crucial to the ET mechanism of nitration, is the coupling of the aromatic radical cation with NO_2. As mentioned before, Bursey showed the formation of $C_6D_6NO_2^+$ from $C_6D_6^{+\cdot}$ and NO_2. The chemistry of this odd-electron system was further probed by Schmitt, Ross, and Buttrill in the gas phase, using a high-pressure mass spectrometer operating at source pressures of about 10 torr, which was some six orders of magnitude greater than the pressure in the ICR studies.[118] A mixture of the aromatic substrate, NO_2, and argon was led into a pulsed argon- corona discharge ionizer. The photons from the discharge (11.8 eV) were sufficiently energetic to ionize both the aromatic and the NO_2, whose ionization potentials are on the order of 9 and 10 eV, respectively. The discharge causes the ionization of only a small fraction of the molecules, and from the time-resolved mass spectrum the various ion-molecule reactions can be inferred.

In accord with previous investigators, Schmitt *et al.* also reported that the even-electron system reacts to give either electron-transfer or O$^+$-transfer but no nitro adduct. Thus, even at this relatively high pressure the third-body collision frequency was insufficient to dissipate the excess energy of the system. On the other hand, the odd-electron system, ArH $^{+\cdot}$ and NO$_2$, which is less energetic than the even-electron system by about 13 kcal/mol for the case of benzene, did give a nitro complex. The structure of this complex was probed by proton-transfer studies. Addition of bases such as tetrahydrofuran or pyridine resulted in the formation of their conjugate acids. Such proton transfer is expected from a σ-complex, but not from a π complex. Hence the conclusion was made that the ArHNO$_2^+$ complex must be a σ-complex. This study clearly demonstrates that ArH $^{+\cdot}$ and NO$_2$ can react to give nitro products satisfying one of the conditions for the involvement of ET in the nitration of aromatics.

Bimolecular rate constants for the reaction of the radical cation with NO$_2$, determined by Schmitt *et al.*, are given in Table 47. Fluorobenzene reacts slightly faster than benzene. Most of the alkylbenzenes react slower than benzene, and surprisingly, naphthalene reacts very slowly. The reported rate constants are roughly two to three orders of magnitude lower than the collision rates. That is, the reaction of the aromatic radical cation and NO$_2$ has a barrier. The source of this barrier is not yet understood.

Unfortunately, no RRKM-type calculations have as yet been performed on this system. It seems surprising that in going from the high-pressure mass spectrometry study to the ICR study (six orders of magnitude drop in pressure) the odd-electron system still gives the adduct ion, (ArHNO$_2$)$^+$, despite the fact that it is only 13 kcal/mole more stable than the efev-electron system.

In the ET mechanism of nitration, an assumption is made that the odd-electron system will collapse to form a σ-complex. While this turned out to be true for the case of NO$_2$ and ArH $^{+\cdot}$ as demonstrated in the high-pressure mass spectrometric studies by Ross and co-workers, additional studies by the same group also showed that the assumption is not necessarily true for all cases.[119] By substituting NO for NO$_2$, they also studied the chemistry of the systems ArH and NO$^+$, and ArH $^{+\cdot}$ and NO. In both cases, an adduct ArHNO$^+$ was formed. However, in neither case did this complex transfer a proton to tetrahydrofuran, thereby suggesting that the adduct was a π-complex. Reents and Freiser had previously established[120] on the basis of

Table 47. Reactions of Aromatic Radical Cations with NO_2[a][118]

ArH	Nitration via aromatic radical cation	k,[b] cm^3 $molecule^{-1}s^{-1}$
C_6H_6	yes	2.4 x 10^{-11}
$C_6H_5CH_3$	yes	1.2 x 10^{-11}
p-$C_6H_4(CH_3)_2$	yes	1.7 x 10^{-11}
Mesitylene	yes	
1,2,4-Trimethylbenzene	yes	
C_6H_5OH	yes	
C_6H_5F	yes	3.7 x 10^{-11}
C_6H_5Cl	yes	
o-$C_6H_4F_2$	yes	
m-$C_6H_4F_2$	yes	1.2 x 10^{-12}
p-$C_6H_4F_2$	yes	2.9 x 10^{-12}
1,2,4-$C_6H_3F_3$	yes	
1,2,3,4-$C_6H_2F_4$	no	
1,2,4,5-$C_6H_2F_4$	no	
Furan	no	
Pyridine	no	
m-$FC_6H_4CF_3$	very slow	

[a]Reaction by NO_2^+ results in $ArHO^+$ or electron transfer nitration product.
[b]Rate constant for ArH $\stackrel{+}{\cdot}$ includes $ArHNO^+$ (from NO impurity).

proton-transfer and photo-electron spectroscopic studies that the $ArHNO^+$ adduct formed by the reaction of NO^+ and various aromatics was a π-complex. Thus, even when starting with a radical cation (ArH $\stackrel{+}{\cdot}$) and a radical (·NO), a system that on the surface appears to be all set for collapse to a σ-complex, a π-complex was nonetheless formed.

3.6.3.4 Radiolytic Gas-Phase Nitration

Cacace and his associates have developed an integrated approach to structural and mechanistic problems of gas-phase ion chemistry using mass spectrometric and radiolytic techniques.[121,122] The work was applied to the study of gas-phase electrophilic aromatic nitration.[123,124]

As discussed, mass spectrometric approaches to the study of gas-phase ionic nitration have limitations since NO_2^+, the nitrating agent in solution, was found to undergo only charge exchange and oxygen–atom transfer to aromatics under ICR conditions,[114,115,116] while the other cations investigated, $CH_2ONO_2^+$ and $EtO(NO_2)_2^+$, displayed a paradoxical selectivity, reacting at higher rates with deactivated substrates.[117] Cacace et al., have identified protonated methyl nitrate as a suitable reagent, easily obtained in chemical–ionization mass spectrometry via the exothermic process.

$$CH_3ONO_2 + CH_5^+ \longrightarrow (CH_3ONO_2)H^+ + CH_4$$

The cation, essentially a NO_2^+ ion "solvated" by methanol with a bonding energy of $\approx$ 24 kcal/mole, undergoes nucleophilic displacement of methanol by the aromatic substrates.

$$(CH_3ONO_2)H^+ + ArH \longrightarrow ArHNO_2^+ + CH_3OH$$

Significant evidence on this point has been obtained from the nitration of benzylmesitylene, whose rings are characterized by a considerably different activation, roughly corresponding to that of toluene (T) and isodurene (I). In competitive experiments the k_I/k_T ratio is as low as 1.5, but the activated ring of benzylmesitylene reacts at least 20 times faster than the other one, despite the unfavorable (2/5) statistical ratio. This suggests formation of an "early" electrostatic complex, followed by a product–determining step

$$(CH_3ONO_2)H^+ + C_6H_5X \underset{k_{-1}}{\overset{k_1}{\rightleftarrows}} \text{early complex} \overset{k_2}{\longrightarrow} \sigma\text{-complex}$$

Activating substituents increase the interaction in the electrostatic complex thereby making $k_{-1} < k_{-2}$. Under these conditions, the rate–determining step is the formation of the early complex, a step that proceeds at the limiting (collision) rate. The substrate selectivity is thus lost, but not positional selectivity. This represents the gas–phase counterpart of the nitration of reactive aromatics in solution. The early complex is clearly necessitated by the kinetics, irrespective of the factors which stabilize it. As discussed (see Section 3.1.2), π–aromatic–NO_2^+ interaction was suggested in solution chemistry by Olah,[41c] based on Dewar's original π–complex concept,[46] as contrasted by encounter–pair formation, as suggested by Schofield,[59] where the reactants are confined in a solvent cage by a viscosity barrier. In the gas phase electrostatic interactions stabilize the ion molecule complexes.

The influence of the leaving group on the reactivity of different $(RNO_3)H^+$ cations is also of interest. Consistent with the mechanism suggested, inductively stabilizing groups depress the nitrating ability, e.g., the ion is unreactive when $R=i-C_5H_{11}$, and alkylation, rather than nitration, occurs when $R=i-C_3H_7$. On the other hand, electron-withdrawing groups make the reagent more energetic and indiscriminate, eg., the $k_{C_6H_5CH_3}/k_{C_6H_6}$ ratio decreases from 7.6 to 2.1 and to 1.2 in passing from $R=CH_3$ to $R=CF_3CH_2$ and to $R=(CF_3)_2CH$.[125]

Radiolysis of systems containing CH_4, CH_3ONO_2, and ArX in the typical molar ratios 10 : 20 : 1 gives high yields of the corresponding nitrated aromatics of assuredly ionic origin, according to the above sequences, followed by deprotonation of intermediates $ArHNO_2^+$ by a gaseous base. The substrate and positional selectivity, illustrated in Table 48, conform to solution–chemistry trends, characterizing $CH_3(ONO_2)H^+$ as the first well-behaved nitrating cation in the gas phase.[125-127]

The partial-rate factors were reported to fit a Hammett-type plot whose σ value, -3.87, is appreciably less negative than that of conventional nitrations, e.g., $\sigma = -6.53$ in Ac_2O or CH_3NO_2 and -9.7 in H_2SO_4, not unexpected for a free cationic reactant, lacking a solvation shell and a counterion. Another intriguing analogy with liquid–phase nitration emerges from the large negative deviations of the most activated substrates, e.g., PhOMe and Ph–c–C_3H_5 from the Hammett correlation. Further studies have confirmed and generalized such findings, showing that gas–phase nitration of highly reactive aromatic tends toward a limiting rate, unaffected by further activation of the substrate.

The evidence for the existence of an "early complex" of unknown structure preceding formation of a σ–complex for gas phase elctrophilic aromatic substitution occurring under radiolytic conditions was reviewed by Cacace.[123] He pointed out that the kinetic role of the "early complex" is identical with that of the "encounter complex" proposed for solution aromatic substitution, i.e., nitration by Schofield. Holman and Gross[46] recently presented evidence for the coexistence of σ- and π–complexes in the protonation of arenes in the gas phase using high pressure chemical ionization mass spectrometric techniques, including collisionally activated dissociation and tandem MS/MS. The results of their study also revealed that the gas phase "early complex" is a π–complex which exists as a distinct intermediate on the potential energy surface of the arenium ion (σ–complex). Although no direct

Table 48. Selectivity of Nitration of Substituted Benzenes by $(CH_3ONO_2)H^+$ in CH_4 at 37°C, 720 torr [125]

C_6H_5X (X)	$k_{C_6H_5X}/k_{C_6H_6}$	Orientation (%)		
		ortho	meta	para
H	1	53	63	72
CH_3	5.1	59	7	34
C_2H_5	5.6	47	4	49
C_3H_7	7.0	50	4	46
$CH(CH_3)_2$	6.0	31	5	64
C_3H_5	10.6	72	6	22
$C(CH_3)_3$	8.4	17	8	75
C_6H_5	1.5	40	4	56
OCH_3	7.6	41	—	59
F	0.15	14	13	73
Cl	0.19	36	10	54
CF_3	0.0037	—	100	—
Mesitylene	8.1	—	—	—

relevance to conduced phase electrophilic aromatic substitution or nitration exists, the condensed state nitration results are in accord with the gas phase studies.

3.6.3.5 Theoretical Calculations

As mentioned, the concept of ET in electrophilic aromatic substitution reactions has been proposed on several occasions. Electrophilic aromatic substitutions are often accompanied by formation of colored species. Based on Mulliken's formalism[128] for charge-transfer complexes, Brown proposed[129] a mechanism involving charge-transfer complexes as stable intermediates. The geometry and energetics of the species were computed using configuration-interaction molecular orbital method. A reactivity parameter, Z, was calculated for the different positions of various aromatic hydrocarbons such as biphenyl, naphthalene, phenanthrene, and perylene. The results of these calculations were also compared with experimental data for nitration of aromatic hydrocarbons in Ac_2O at 0°C. In general, good qualitative agreement was found between the calculated Z value and the experimentally observed partial-rate constant for nitration at that position. About the same time, Nagakura[130] also developed a similar mechanism for electrophilic aromatic substitution reacts involving charge transfer, but differed in the description of the MO parameters that correlate with the orientation of the substitutions.

In 1973, Pederson et al.[131] correlated the esr hyperfine splittings in the radical cations of the various aromatics with the partial-rate factors. The hyperfine splittings are a measure of the spin density at any given position and the correlation was used to support the intermediacy of the radical cation. However, at best, the data support the ET mechanism only qualitatively. A quantitative treatment using frontier orbital theory[14a] was developed by Elliott, Sackwild, and Richards.[132] According to them, positional selectivity in EAS reaction is related to the frontier charge densities, and they use nitration data to support their contention. Close examination of their data in quantitative terms reveals many inconsistencies. Thus, often centers with very similar frontier charge densities have drastically different extents of nitration, just as centers with sufficiently different frontier charge densities end up having very similar amounts of substitution. For example, the ortho and meta positions of nitrobenzene have very similar calculated frontier charge densities of 0.0061 and 0.0062, yet the relative nitration rates are 6.4 : 93. Likewise, the calculated frontier charge densities for the ortho and para positions of toluene are 0.2507 and 0.2918 but the reported partial rate constants are 29 : 37. Fleming has given an excellent account of the frontier orbital approach to rationalizing electrophilic aromatic substitutions.[132b]

The nature of the stable intermediates in the reaction of benzene and toluene with the NO_2^+ ion was probed by Politzer and co-workers[83,133] with the help of a restricted Hartree Fock *ab initio* approach using GAUSSIAN 70 and GAUSSIAN 80 wave functions. They reported two potential-energy minima. The first is a weak long-range ($\approx$ 3 Å) interaction in which the NO_2^+ ion is parallel to the plane of the aromatic. The stabilization energy is about 6 kcal/mol for benzene and 7.5 kcal/mol for toluene. This structure, which looks like a π-complex, was located by constraining the geometries of the two species to that in their respective ground states. The second minimum corresponds to a σ-complex with a strong interaction between the NO_2^+ ion and specific carbons in the aromatic ring. The structures are shown in Figure 5. The stabilization energies computed for the σ-complexes range from 76 kcal/mol for benzene to 87 kcal/mol for toluene (para). These values are somewhat larger than expected on the basis of simple

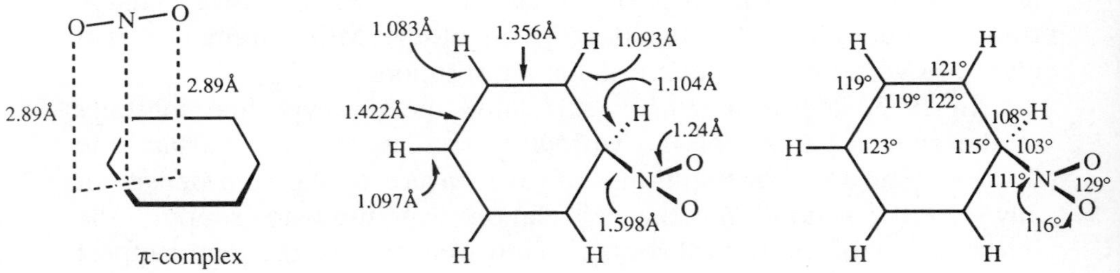

Figure 5. Computed structures for the π- and σ-complexes of nitronium ion and benzene.

bond additivity. The C–NO₂ bond is expected to be around 78 kcal/mol, however, the formation of this bond is accompanied by a loss of a C–C π–bond, worth roughly 50 kcal/mol, and the resonance energy of the benzene ring, worth another 20 kcal/mol. The gain in the resonance energy of the nitro group is only a few kcal and cannot account for the large stabilization shown by the calculations.

Feng, Zheng, and Zerner[134] more recently reported their theoretical investigation of the question of electrophilic nitration versus radical recombination following electron transfer. Semi-empirical MNDO calculations were performed to compare the energetics of reaction via a π–complex versus via ET. The geometries and stabilization energies for the π–complexes are generally in agreement with those reported by Politzer except that the distance between the aromatic and the nitronium ion is about 1 Å longer. Figure 6 shows the computed energetics for π–complex formation and ET for benzene, nitrobenzene, toluene, and xylene. These calculations show that except for nitrobenzene (and possible benzene), ET is the favored pathway for all other aromatics, at least in the gas phase. In further support of this result, the authors correlate the isomeric distribution of the nitration product with the calculated spin densities. In concert with the general finding that the partial–rate constant for nitration of toluene at the para position is greater than that at the ortho position, they report calculated spin densities to be 0.578 and 0.292, respectively. The authors cite the special case of nitration with Nafion–H, which leads to greater than 95% para–nitrotoluene as evidence for ET. However, the calculated spin density at the meta position is also very similar to that at the ortho

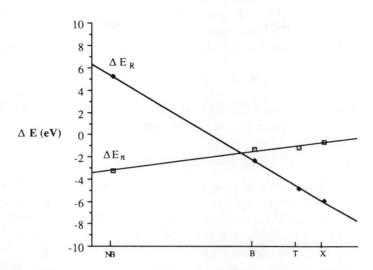

Figure 6. Energy of radical ion-pair formation, ΔE_R, and π-complex formation, ΔE_π, represented as a function of the aromatic. Values plotted are those obtained from the MINDO calculations. (NB = nitrobenzene, B = benzene, T = toluene, and X = o-xylene).[134]

position. Thus, if ET were the operative mechanism one would expect similar amounts of o– and m–nitrotoluene.

The similarity in the predicted spin densities for ortho and meta positions can be understood by simple consideration of the molecular orbitals of benzene. Benzene has a generate pair of highest occupied molecular orbitals. In the case of monosubstituted benzenes, the degeneracy is lifted and results in a pair of orbitals of similar energy, one symmetric and the other antisymmetric with respect to the plane of symmetry. Depending upon the nature of the substituent, either the symmetric or the antisymmetric will be of higher energy and preferentially contain the free electron in the radical cation. However, because the energy difference between the two orbitals is not very large, the contributions of the HOMO and the NHOMO must be taken together when calculating the spin densities. The symmetry of the system is such and the perturbances due to most alkyl and halo substituents are so minor, that the coefficients at ortho and meta positions are nearly the same in both these orbitals as are those of the *ipso* and para positions. Thus, on the basis of spin densities, substitution at ortho and meta positions should be similar and in general different from para. If one includes a further assumption that the initial attack at the *ipso* position results in ortho substitution following a 1,2-migration, then a simple relationship for the positional selectivity follows:

$$o = m + p$$

Of course, it is well recognized that no such simple relationship exists between the extent of substitution at the various positions.

3.6.3.6 Kochi's Studies of Charge Transfer

Kochi and co-workers have observed the appearance of charge-transfer bands during several different electrophilic aromatic substitution reactions including nitration of a deactivated substrate, *meta*-nitrobenzonitrile by $NO_2^+BF_4^-$.[51] They were able to characterize these bands by careful spectroscopic measurements, as resulting from a 1 : 1 complex of the arene and the electrophile. They also showed that the log relative rates of substitution for different arenes correlate linearly with the energy of the CT-transition. Excitation of the charge-transfer is understood in terms of electron-transfer from the donor to the acceptor. Thus, a correlation between the energy of charge-transfer excitation and the relative rates strongly suggests that the chemical activation of the CT complex leading to the σ-complex is tantamount to electron-transfer. Using the CT formulation, Fukuzumi and Kochi[51,135] thus provided a physical interpretation of the chemical activation leading to electrophilic aromatic substitution. They were also able to calculate the slope, ρ, of linear free-energy relationship between substituent constants, σ^+ and relative reactivities of different aromatic substrates.

Kochi was also able to use the CT formulation to predict correctly the orientation of substitution without resorting to ad hoc explanations previously used to explain the abnormal behavior of halo-substituted benzenes and anisoles. Electron-releasing substituents reduce the ionization potential of the arene relative to benzene and give rise to predominantly ortho and para substitution, while electron-withdrawing substituents increase the ionization potential and are predominantly meta orienting. Halobenzenes, which have a lower ionization potential than benzene are therefore ortho and para directing.[136] What the CT formulation does not predict is why halobenzenes are deactivated relative to benzene. The ad hoc explanations that were formerly used were necessary to reconcile their reactivity and orientation behaviors.

In a subsequent study, Kochi and co-workers have reported kinetics of the collapse of aromatic radical cations and NO_2 to give nitro products in solution.[137] The radical-cation/radical pair was formed by photochemical excitation of charge-transfer complexes of arenes and tetranitromethane. They first show that irradiation of the charge-transfer complexes of various substituted dimethoxybenzenes and tetranitromethane leads to the formation of corresponding nitrodimethoxybenzenes.[138] The regioselectivity of the products was found to be identical to that observed upon electrophilic nitration with HNO_3 in Ac_2O.

The details of the charge-transfer nitration were probed with time-resolved spectroscopy. Following the excitation of the charge-transfer complex ($\lambda = 532$ nm), the aromatic radical cation and the trinitromethide ion were observed spectrophotometrically. NO_2 could not be directly observed, but was inferred from the stoichiometry. Photoexcitation of the charge-transfer band leads to electron transfer. The tetranitromethane anion undergoes dissociation to NO_2 and the trinitromethide ion:

$$C(NO_2)_4 \longrightarrow NO_2 + C(NO_2)_3^{\overline{\cdot}}$$

The decay of the absorbance band due to dimethoxybenzene radical cation was shown to follow a simple second-order process. The second-order rate constant in dichloromethane was found to be 7×10^3 A^{-1} s^{-1}. In less polar solvents such as benzene and hexane the rate constant was several orders of magnitude higher. Assuming a molar extinction coefficient of 10^4 for the dimethoxybenzene radical cation, the second-order rate constant in dichloromethane can be estimated to be 7×10^7 M^{-1} s^{-1}, in other words, roughly two orders of magnitude slower than the diffusion rate coefficient. Thus, once again we find that the reaction of the aromatic radical cation with NO_2 is not diffusion limited, implying that, if they were to be formed by outer-sphere electron transfer from an aromatic to NO_2^+, they would also diffuse out of the solvent cage, and we should therefore be able to see the products of subsequent reactions of the aromatic radical cation (and of NO_2) with the solvent.

Recently, Kochi[139] extended his studies to the oxidative nitration of aromatics via charge-transfer complexes of arenes and nitrosonium salts. The brightly colored solutions that are immediately formed from various arenes (including benzene, toluene, methylbenzenes, and naphthalene) with NO^+ ion arise from the charge-transfer absorption bands of 1:1 complexes [ArH, NO^+] that are reversibly formed as persistent intermediates.[40,21,24] The yellow-red charge transfer colors disappear upon introduction of oxygen into the system, and the corresponding nitroarenes can be isolated in excellent yield from the acetonitrile solutions. The oxidative nitration takes place via the initial autoxidation of the charge-transfer complex to a radical ion pair [ArH $\overset{+}{\cdot}$, NO$_2$]. The collapse of the radical ion pair then gives the arenium-ion intermediate and by subsequent proton loss the nitroarene product

$$ArH + NO^+ \rightleftharpoons [ArH, NO^+] \xrightarrow{CT} [ArH \overset{+}{\cdot}, NO] \xrightarrow{\frac{1}{2}O_2}$$

$$[ArH \overset{+}{\cdot} NO_2] \longrightarrow [ArHNO_2]^+ \xrightarrow{-H^+} ArNO_2$$

NO$^+$ ion itself is completely unreactive toward oxygen under the reaction conditions. On the other hand, activated aromatics, such as anthracene, can be directly oxidized by NO$^+$ to their radical ions producing at the same time NO, which is highly susceptible to autoxidation. Subsequent coupling of the anthracene cation radical with NO$_2$ gives 9-nitroanthracene

$$AnH + NO^+ \longrightarrow An \overset{+}{\cdot} + NO$$

$$NO + \tfrac{1}{2} O_2 \longrightarrow NO_2$$

$$AnH \overset{+}{\cdot} + NO_2 \longrightarrow AnHNO_2^+ \xrightarrow{-H^+} 9-AnNO_2$$

No similar path is possible for nonactivated aromatics such as benzene and toluene. Such arenes in contact with NO$^+$PF$_6^-$ show no evidence of undergoing electron transfer. The oxidation potentials of benzene and toluene are too prohibitive for electron transfer to occur with NO$^+$ (even more so, as discussed with NO$_2^+$). However, autoxidation of the arene-NO$^+$ charge-transfer complexes allows a direct pathway for the oxidative nitration to occur. It was also observed that light accelerates oxidative nitration with NO$^+$ (photo-induced nitration).

It is again significant to point out, as emphasized by Kochi,[140] that substrate and positional selectivities in studied oxidative nitrations with NO$^+$ are determined in distinct separate steps. The former is much affected by the oxidation potential of the aromatic, the latter by the spin-density distribution of the radical ion (which generally is similar to the electron distribution pattern of the related arenium ions).

3.6.3.7 Application of Marcus Theory

Eberson and Radner have applied the Marcus Theory of nonbonding electron-transfer to assess the feasibility of oxidation of various aromatic compounds by NO_2^+ and NO^+ ions.[141] Application of the Marcus theory gives the rate constant for the ET step. Since nitration of aromatics, more reactive than toluene, proceeds at encounter rate, the rate constant derived from the Marcus theory must be greater than the diffusion rate constant of $\approx 10^{10}$ M^{-1} s^{-1} in order that ET be kinetically competent to be involved in the reaction scheme. On the other hand, if the calculated value turns out to be significantly smaller than the diffusion rate constant then, barring any serious flaws in the application of Marcus theory, ET may be ruled out.

According to the Marcus theory of non-bonded (outer-sphere) electron transfer, the free-energy change of going from the initial collision complex to the transition state, G, is related to the standard free-energy change of the reaction

$$\Delta G^\ddagger = (\lambda/4)(1 + \Delta G°/\lambda)^2$$

where, λ is the reorganization energy. The rate constant for electron transfer, k_{ET}, is then given by:

$$k_{ET} = \frac{k_d}{1 + (k_{-d}/Z)\exp\{(\lambda/4)(1 + \Delta G°/\lambda)^2/(RT)\}}$$

Here, k_d and k_{-d} are the rate constants of the formation or dissociation of the collision complex and Z is the collision frequency.

In applying Marcus theory to the oxidation of aromatics by NO^+ and NO_2^+, it is necessary to obtain accurate values of the thermodynamic parameters. The standard free-energy change for the electron transfer in the given medium can be obtained from electrochemical data. The NO/NO^+ and NO_2/NO_2^+ systems behave reversibly and their $E°$ values are readily available.[142a] $E°$, for the NO^+/NO and NO_2^+/NO_2 couples are around 1.5 V. However, Boughriet et al. recently found[142b] the values dependent on solvent. In sulfolane $E°'$ (NO^+/NO) and $E°'$ (NO_2^+/NO_2) are 1.3 and 1.9 V (NHE), respectively. In nitromethane the values are 0.88 and 1.62, respectively.[142c]

The electrochemical and thermodynamic behavior of NO_2^+ with aromatics is thus solvent dependent. Eberson and Radner's studies in acetonitrile (*vide infra*) must be consequently considered in view of possible solvent effects, nitromethane being the least solvating one.[142c] However, ArH/ArH$^+$ systems, particularly for benzene and toluene, do not exhibit reversible voltammetric behavior, and estimation of their $E°$ values is difficult. Fortunately, Parker had obtained reliable data on the reversible oxidation potentials of a large number of aromatics[143] and these were used by Eberson and Radner in their treatment.

The reorganization energy may be considered to be a sum of the free-energy changes associated with the changes in the bond lengths and bond angles of reacting species, λ_i, and the free-energy change associated with solvent reorganization, λ_o. λ_i values were estimated by considering the two self-exchange reactions:

$$ArH + ArH^+ \longrightarrow ArH^+ + ArH$$

$$NO_x + NO_x^+ \longrightarrow NO_x^+ + NO_x$$

λ_i values for these self-exchange reactions were determined from a consideration of the IR spectral frequencies of the different species. The mean of the two values was used as an estimate of the reorganization energy for the ET process. λ_o values were obtained by consideration of the Kharkats model.[144]

Using this regimen, Eberson and Radner calculated the rate constant for ET for a large number of aromatic compounds. They recognized that the choice of the various thermodynamic and hydrodynamic parameters strongly dictates the outcome of the application of the Marcus theory. Therefore, they generally chose parameters that would favor ET. In this manner, they would exclude the least number of possible aromatics as possible candidates for ET.

λ-values calculated for NO^+ and NO_2^+ oxidations turned out to be considerably different; 40 and 75 kcal/mol, respectively. The substantially greater reorganization energy required for NO_2^+ is largely a consequence of the fact that NO_2^+ is linear while NO_2 is bent. Thus, oxidations with NO_2^+ will entail some bending of the bonds and hence are relatively inefficient. The rate constants calculated for ET between various aromatics and NO^+ and NO_2^+ for a range of values are summarized in Table 49. For encounter-controlled ET, the log k_{ET} values should be on the order of 10. As can be seen, even though Eberson used parameters most favorable for ET, substrates such as naphthalene and perylene are calculated not to undergo ET with NO_2^+. The NO^+ ion, on the other hand, is a substantially more potent oxidant. Its possible impact on nitrations under ET is considered subsequently.

However, it must be pointed out that recent studies by Ridd *et al.*[148d] point to the fact that nitration of naphthalene (but not mesitylene) could involve limited involvement of electron transfer.

Ridd *et al.* in their studies[148d] of the reaction of naphthalene with $H^{15}NO_3$ initially found no CIDNIP effect indicative of an ET process. However, a recent reinvestigation, using faster reaction rates (permitting a more sensitive test for CIDNIP effects), found slight but significant nuclear polarization (no such effect was, however, found in the case of mesitylene). The weak effect is consistent with the involvement of a small part of the reaction in electron transfer. Even if only a few percent of the reaction occurs by electron transfer, this makes the outer-sphere electron-transfer calculations based on the Marcus Theory less relevant.[105] If the interaction of the NO_2^+ ion with the aromatic occurs through an initial π-complex, this may be sufficient to allow the electron transfer to be classified as an inner-sphere process. If so, then the general application of the Marcus Theory to such nitration processes becomes open to question and such electron-transfer processes may be more important than Eberson and Radner assumed based on their calculations.

3.6.3.8 The Relationship between Ionic and ET Mechanisms

As discussed by Eberson and Radner,[105] the ET and polar pathways of the ArH/NO_2^+ reaction differ in one aspect only, namely, the degree of electronic interaction in the transition state. The ET pathways can be formulated on the basis of the Marcus theory for outer-sphere ET as

Table 49. Calculated log k (M^{-1} s^{-1}) for ET Reactions between ArH and NO$^+$ (NO$_2^+$) in Acetonitrile at 25°C

Compound ($E°$/V)	$G°$/kcal mol^{-1} [a]	log k (M^{-1} s^{-1}) for reaction with NO$^+$ λ(kcal mol^{-1})			log k (M^{-1} s^{-1}) for reaction with NO$_2^+$ λ(kcal mol^{-1})		
		15	25	40	15	25	75
Benzene (3.03)	35.0, 33.8	−18.0	−13.9	−13.3	−16.6	−12.8	−16.4
Toluene (2.61)	25.4, 24.1	−7.4	−6.1	−7.1	−6.3	−5.2	−11.5
Mesitylene (2.43)	21.2, 20.9	−3.5	−3.1	−4.7	−2.5	−2.3	−9.5
Naphthalene (2.08)	13.1, 11.9	2.9	1.9	−0.4	3.7	2.5	−5.9
Hexamethylbenzene (1.85)	7.8, 6.6	6.1	4.6	2.0	6.8	5.2	−3.8
Dibenzo–1,4–dioxine (1.63)	2.8, 1.6	8.6	6.8	4.1	9.1	7.3	−1.5
Anthracene (1.61)	2.3, 1.1	8.8	7.0	4.3	9.3	7.5	−1.7
Perylene (1.30)	−4.8, −6.0	10.3	9.4	6.8	10.3	9.7	0.9
Zn(II) Tetraphenylporphyrin (0.95)	−12.9, −14.1	10.3	10.3	9.1	10.3	10.3	3.4
Phenothioazine (0.71)	−18.4, −19.6	10.3	10.3	10.0	10.3	10.3	5.0
Ferrocene (0.60)	−21.0, −22.2	10.3	10.3	10.2	10.3	10.3	5.7

[a] The first number refers to the NO$^+$ reaction, the second one to the NO$_2^+$ reaction.

$$\text{ArH} + \underset{\text{O}}{\overset{\text{O}}{\underset{\|}{\overset{\|}{\text{N}^+}}}} \rightleftharpoons \left[\text{ArH}\overset{\nearrow\text{O}}{\underset{\searrow\text{O}}{\text{N}}} \longleftrightarrow \text{ArH}\cdot {}^{+}\overset{\nearrow\text{O}\cdot}{\underset{\searrow\text{O}}{\text{N}}} \right]^{\ddagger} \longrightarrow \text{ArH}{}^{+}_{\cdot} + \overset{\nearrow\text{O}\cdot}{\underset{\searrow\text{O}}{\text{N}}}$$

The important feature of this process is the attainment of the transition state by nuclear movement (bond reorganization) and solvent reorganization only; no electronic interaction between the reactants is allowed. This view of the ET transition state is a consequence of the Franck–Condon principle, which says that transfer of an electron is much faster than nuclear movements. The outer–sphere transition state always includes two forms, one with the electronic description of the reactants, and one with that of the products.

The usual polar transition state, where an appreciable electronic interaction is implicit in the dotted bonds of transition state, can be represented as:

$$\text{ArH} + \underset{\text{O}}{\overset{\text{O}}{\underset{\|}{\overset{\|}{\text{N}^+}}}} \rightleftharpoons \left[\text{ArH}^+ \text{---} \overset{\nearrow\text{O}}{\underset{\searrow\text{O}}{\text{N}}} \right]^{\ddagger} \longrightarrow {}^{+}\text{Ar(H)}\overset{\nearrow\text{O:}}{\underset{\searrow\text{O}}{\text{N}}}$$

The structure of the transition states for ET and polar pathways explain the large difference between these two processes. In the ET transition state, the energy expended in bending the initially linear NO_2^+ moiety to a bent species, somewhere halfway to becoming an NO_2 molecule with an O–N–N bond angle of 134.3°, is large.[141] This contribution to the bond–reorganization energy, estimated to be 52 kcal/mol, is the factor responsible for making the ET step energetically difficult. Minor changes take place within the ArH part of the transition state too, but they are negligible in this context (<5kcal/mol).[143] Similar bond reorganization must occur during the attainment of the polar transition state, but here it is compensated by the electronic interaction (partial formation of a C–N bond) in the transition state.

The oxidation of aromatic hydrocarbons by NO^+ can be viewed in the same way.

$$\text{ArH} + \overset{O}{\underset{N^+}{\parallel}} \rightleftharpoons \left[\text{ArHN}^{\overset{O}{\parallel}+} \leftrightarrow \text{ArH}^{+}_{\cdot}\overset{O\cdot}{\underset{N}{|}} \right]^{\ddagger} \longrightarrow \text{ArH}^{+}_{\cdot} + \overset{O\cdot}{\underset{N}{|}}$$

The bond reorganization necessary to reach the transition state is not nearly as extensive as in the case of NO_2^+. The dominant change is the lengthening of the N–O bond, the full change in going from NO^+ to NO being ≈ 0.09 Å. This contribution to the reorganization energy is only on the order of 10 kcal/mol.

In both the nitronium- and nitrosonium-ion effected ET reactions, the total reorganization energy also has a contribution from solvent reorganization, calculated by Kharkats' ellipsoidal model to be 18 and 25 kcal/mol, respectively.[144] This neglects the very small contribution from solvent reorganization around ArH.

Apart from the reorganization energy, the other parameter of interest for the properties of an ET oxidant is its standard potential ($E°$). Reversible potentials for reduction of NO_2^+ and NO^+ in acetonitrile are known to be 1.45 and 1.51 V[145] (vs. normal hydrogen electrode), and can be approximately equated with their $E°$ values. Thus, if one were to look at this parameter only, one would conclude that their powers as ET oxidants are roughly identical. However, the large difference in reorganization energy, qualitatively explained above, makes NO_2^+ a much weaker oxidant. In view of the redox potentials of a series of aromatic compounds[146] listed in Table 50 it is improbable that aromatics with $E°$ higher than that of perylene would be capable of undergoing electron transfer with NO_2^+. Considering, however, the high reorganization energy of NO_2^+ needed for outer-electron transfer, even ferrocene, a very good ET reductant, would be unaffected. Despite of these considerations there are numerous examples of radical cations and their coupling products in treatment of aromatic substrates with NO_2^+. How they are formed, if we exclude the possible presence of NO^+ as an impurity or its formation somehow in catalytic amounts during the reaction, is discussed subsequently. (Formation of NO^+ is certainly not unlikely, unless precautions are taken to add HNO_2 traps in all experiments).

Inspection of energy diagrams for two extreme types of ArH/NO_2^+ reactions gives a reasonable explanation for several phenomena

Table 50. Redox Potentials of Representative Aromatics[146]

Compound	$E°/V°$
Benzene	3.03
Toluene	2.61
Mesitylene	2.43
Naphthalene	2.08
Dibenzo[1,4]dioxin	1.63
Anthracene	1.61
Perylene	1.30
Phenothiazine	0.71
Ferrocene	0.60

mentioned previously.[109] The NaphH/NO_2^+ reaction (Figure 7) is endergonic by 0.52 V in the ET mode and has a very-high-lying transition state (log k = −5.9). The polar mechanism has a very-low lying transition state (reaction proceeds at diffusion-controlled rate), and thus the coupling process between ArH^+ and NO_2 must be exergonic, and conversely, the homolytic cleavage of the Wheland intermediate, endergonic. Thus, it is easily understood why pre-synthesized NaphH$^{+\cdot}$ reacts clearly with NO_2 to give nitro products and why neither the ET nor the polar pathway will give any detectable concentration of the radical cation.

The situation is quite different for an ArH that in principle can undergo exergonic ET to NO_2^+. Figure 8 shows a schematic energy diagram for the perylene/NO_2^+ reaction. Here we still have a fairly high-lying transition state for ET (log k = 0.9) and so for naphthalene, a very low-lying one for the polar pathway. However, the ET process is now overall exergonic, which means that the homolytic cleavage of the Wheland intermediate must necessarily be exergonic too. For the exergonic case we can conclude that (1) the radical cation can be a product from the initially polar pathway, and (2) the nitro product may be formed in a slightly endergonic reaction from PeH$^{+\cdot}$ NO_2, thus creating competition from possible side reactions of PeH$^{+\cdot}$, such as coupling or reaction with solvent.

Based on these conclusions, Eberson and Radner have proposed[142] that the pathway for radical-cation formation from ArH and NO_2^+ follows a Taube-type inner-sphere ET mechanism.[147]

$$ArH + NO_2^+ \rightleftharpoons Ar^+(H)NO_2 \rightleftharpoons ArH^+ + \cdot NO_2$$

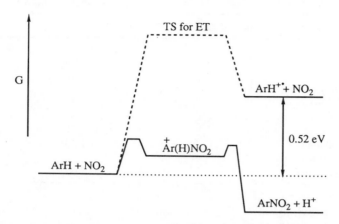

Figure 7. Schematic Free Energy Diagram for a Reaction Between ArH and NO_2^+ where ET is Endergonic. In this Example ArH Corresponds to Naphthalene (Eberson and Radner[105])

The association step to give the Wheland intermediate is equivalent to formation of the bridged intermediate (the precursor complex) and the dissociation step to give a radical cation and NO_2, equivalent to the transfer of the electron from one site to the other.

The simple provision of considering a bond equivalent to the bridging ligand in the inorganic variety of the inner-sphere ET mechanism, immediately makes this useful concept available for use in organic mechanisms. In fact, this is exactly what is implied in Pross' proposal[101] that all organic reaction mechanisms actually consist of one-electron shifts.

The effect of NO^+ on nitration with NO_2^+ (see Section 3.1.3) was demonstrated by Ridd *et al.* who have studied possible ET mechanisms in the ArH/NO_2^+ system using ^{15}N-labeled NO_2^+ ion.[148] For mesitylene in trifluoroacetic acid (TFA, a very good solvent for radical cations), no nuclear polarization is seen in the product from the NO_2^+ ion pathway. This is in agreement with the results described above; mesitylene is a difficult compound to oxidize ($E° = 2.43$ V), so outer-sphere ET is prohibited for kinetic reasons and inner-sphere ET for thermodynamic reasons. On the other hand, the HNO_2-catalyzed pathway gives nitromesitylene with very strong ^{15}N nuclear polarization (emission with a degree of enhancement of more than 200). This CIDNP effect was attributed to the formation of the radical pair, ArH$^{+\cdot}$ NO_2, by diffusion

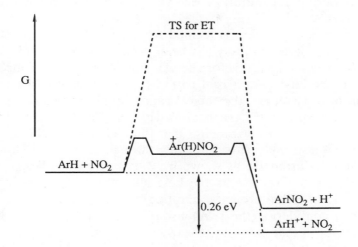

Figure 8. Schematic free energy diagram for a reaction between ArH and NO_2^+ where ET is endergonic. In this example ArH corresponds to perylene (Eberson and Radner[105]).

together of the two components. The effect of HNO_2 catalysis was assumed to be generation of NO^+, which would act as an ET oxidant toward mesitylene. Side products isolated from these experiments, like nitrated diphenylmethane derivatives and a side–chain nitro product, could be rationalized as being derived from the radical cation.

The difficulty with this interpretation is that the mesitylene/NO^+ reaction is endergonic according to estimates based on the standard potentials available. Since these oxidation potentials refer to different solvents (mesitylene in TFA and NO^+ ion in acetonitrile), and since it is a fairly general observation that oxidation potentials tend to increase in acidic media, it might well be that the NO^+ ion has a higher standard potential in TFA than in acetonitrile. Durene was found to behave in a manner similar to mesitylene.

Kochi has discussed recent studies[139] on oxidative nitration of arenes with NO^+ salts showing initial CT–complex formation followed by oxidation (with added O_2, promoted by light) to the radical ion pair [ArH $\overset{+}{\cdot}$ NO_2]. Its subsequent collapse to the arenium ion, giving the nitroarene by proton loss, fit well into the emerging picture of the path of electron transfer nitration.

Considering ET nitration Eberson and Radner drew two major conclusions:[142]

(i) If a radical cation is formed from an ArH/NO_2^+ reaction and the HNO_2 catalyzed pathway has been properly suppressed by addition of an HNO_2 scavenger, it is formed via an inner-sphere ET mechanism.

(ii) If the HNO_2-catalyzed pathway has not been excluded by the addition of a scavenger, the radical cation, in all probability, is formed via an outer-sphere ET mechanism with NO^+ as the oxidant. There exists, however, the possibility that limited inner-sphere electron transfer will also take place, probably through an initial π-complex of NO_2^+ and the aromatic, as shown by Ridd's recent studies.[148d]

Presently this represents the best interpretation of available data on the role of ET in aromatic nitration.

When considering the relationship of ET and polar mechanisms, one must recall that outer-sphere electron transfer is generally defined as occurring via a transition state not exceeding 1 kcal/mol electronic interaction between the reagents. Consequently in the strongly exergonic ArH/NO_2^+ reaction the ET and polar processes merge at this point. Outer-sphere ET thus includes the merging of the ET and the polar mechanism! The same general question arises in the Pross-Shaik treatment of aromatic substitutions in ArH/E^+ reactions. Formerly, one electron is shifted and in the polar process this is accompanied by simultaneous structural changes (group coupling) so that the ArH $\overset{+}{\cdot}$ /E· radical-cation/radical pairs never appear along the reaction coordinate. The two mechanisms again merge in the vicinity of this point.

According to Eberson, the difference between the Marcus (i.e., thermodynamic) approach and the Pross-Shaik single-electron shift idea is that the latter is more difficult to express in quantitative terms, but is probably more attractive for the qualitative reasoning preferred by organic chemists while discussing mechanisms.

It is clear from the foregoing discussion that outer-sphere ET and the polar pathways are two extremes in a spectrum of mechanisms that differ in the degree of concertedness between electron shift and bond rearrangements. The classical Ingold two-electron transfer electrophilic nitration mechanism is in itself just an expression of the same concept with two successive electron transfers following each other so fast as to be expressed in a concerted fashion. The fact that the electrons indeed move not in pairs, but one at a time, but in a concerted although not synchronous fashion, does not affect the continued use of electrophilic nitration to depict the polar mechanism. The reactivity of aromatics towards electrophiles is only formerly related to outer-sphere ET, since outer-sphere ET nitration of aromatics is thermodynamically prohibited.

3.7 Our Present Understanding of the Mechanism of Electrophilic Aromatic Nitration

We now return to the questions we raised earlier, and based on reviewed studies, summarize the conclusions we can draw from them concerning the mechanism of electrophilic aromatic nitration.

3.7.1 The Nitronium Ion Mechanism and Its Modification for Reactive Aromatics

Ingold's and Hughes' conclusions of forty years ago still provide the foundation for our understanding of electrophilic aromatic nitration. The NO_2^+ ion, or its polarized carriers (NO_2–X) formed in acid-catalyzed nitration with HNO_3 and its derivatives, act as a two-electron acceptor in the nitration of the most frequently used benzenoid aromatics. Whereas the NO_2^+ ion is the common reactive nitrating agent in acid-catalyzed aromatic nitrations, its differing solvation (complexation) in various media affects its reactivity, accounting for the observed differences in selectivity. Concerning the "first intermediate" overall discussed considerations favor its π-complex nature, although no direct experimental evidence is yet available. (In gas phase studies of electrophilic aromatic substitution, however, the coexistence of separate π- and σ-complexes was experimentally proven). Fast encounter controlled formation of the "first intermediate" resulting in the loss of substrate selectivity does not contradict such π-bonded nature. Subsequent nitroarenium ion (σ-complex) formation leads to determination of positional selectivity, which always stays high. Single-electron transfer becomes significant with polycyclic aromatic systems with redox potentials substantially lower than that of NO_2^+ to compensate for the bending of the linear NO_2^+ ion. The nitration of reactive benzenoids, however, involves two separate intermediates, modifying the original Ingold mechanism. The nature of the first intermediate is still disputed. Regardless whether it is π-complex, encounter-complex, or radical ion-pair nature, the significant point is its distinct, separate existence from the subsequent arenium ion(σ-complex) intermediates. As stated by Ridd,[77] the formation of Schofield's encounter pair or Olah's π-complex first intermediate both involve the aromatic molecules as a whole and not interaction at a particular carbon atom. The same is the case for Perrin's radical ion pair intermediates. All depict the same concept of a separate initial step involving a "first-intermediate" determining substrate selectivity, followed by arenium ion (σ- complex) formation, determining regioselectivity. As pointed out by Olah et al.,[106] despite differences in interpreting the nature of the "first intermediate", there is general agreement on the necessity of two separate steps determining independently substrate and positional (regio) selectivity. Thus, the original Ingold-Hughes mechanism must be amended for reactive aromatics in the following way:

$$H^+ + HNO_3 \rightleftharpoons H_2ONO_2^+$$

$$H_2ONO_2^+ \longrightarrow NO_2^+ + H_2O$$

$$NO_2^+ + ArH \rightleftharpoons \text{"First intermediate"}$$

$$\text{"First intermediate"} \rightleftharpoons HArNO_2^+$$
(Nitroarenium ion intermediate)

$$HArNO_2^+ \longrightarrow ArNO_2 + H^+$$

With less reactive or deactivated aromatics any interaction before the Wheland intermediate is reversible and thus does not represent a separate kinetic step. These reactions follow the original Ingold reaction scheme and also give a satisfactory Hammett-Brown type σ^+ correlation. When, however, a separate first intermediate determines substrate selectivity (or its absence), followed by σ-complexes that determine positional selectivity, such treatment obviously is not adaptable.

3.7.2 Is the Nitronium Ion Capable of Oxidizing Aromatic Compounds?

The question of outer-sphere electron transfer between aromatics in the condensed phase has been addressed by Radner and Eberson applying Marcus theory. They concluded that because of the substantial free-energy requirement associated with the bending of the NO_2^+ ion to neutral NO_2, the NO_2^+ ion is a rather inefficient oxidizing agent. Electron transfer is indicated for only extremely easily oxidized aromatics such as ferrocene and phenothioazine, but not for perylene, naphthalene, or toluene. Thus, it does not appear likely that nitration of alkylbenzenes and naphthalenes would proceed via ET. Ridd's recent studies[148d] indicate, however, that there exists the possibility of limited inner-sphere electron transfer taking place, probably through an initial π-complex of the aromatic with NO_2^+. Perrin suggested such electron transfer to take place through the first formed Wheland intermediate, but this is less likely as fast deprotonation would lead directly to the nitrated product. Homolysis of the C–N bond in the Wheland intermediate should give an enhanced CIDNIP effect with naphthalene–d_8, which was not observed.[148d] As cautioned by Eberson, it is, however, essential to exclude NO^+ from the systems (impurities can even be formed by side reactions) because it can cause much more efficient oxidation.

3.7.3 Can the Aromatic Radical Cation Couple with NO_2 to Give the Nitroaromatic Product?

The answer to this question is an unequivocal yes. However, as shown in the studies by Ridd and Suzuki, other reactions also accompany the coupling, most notably, reactions at the side chain and formation of biaryls.

3.7.4 Are the Rates of the Individual Steps Consistent with Observed Rates of Nitration?

Nitration of activated aromatics appear to proceed at encounter rates. The ET mechanism for the nitration of such substrates would require that the collapse of the geminate pair of the radical cation and NO_2 be faster than their diffusion from the solvent cage. We have already noted that the ET step is not as rapid for most aromatics even in the gas phase. Eberson's study of the reaction of naphthalene radical–cation hexafluorophosphate with nitrite ion showed that the coupling reaction of NO_2 with naphthalene radical cation is also slower than the diffusion rate. Spectroscopic studies by Kochi lead to the same conclusion. Were the ET mechanism operative, some of the aromatic radical cation and NO_2 would escape from the solvent cage, and we should see products of their respective reactions with the solvent. In questioning the role of electron transfer as proposed by Kenner, Ingold used the same line of reasoning. He cited the lack of HNO_3 as evidence against ET.

3.7.5 Is the Product Distribution of the Reaction of ArH$^{+\cdot}$ with NO_2 the Same as that of Mixed Acid or Nitronium Salt Nitration?

By and large the isomer distributions of the product nitroaromatics are similar in both electrophilic and radical cation nitrations, as charge–distribution patterns in arenium ions and spin–density patterns in arene radical cations are similar. This is not surprising, since structural features in activated aromatics that stabilize a cationic center also stabilize the charge. There are, however, distinct differences in the exact isomer ratios. As these variations require only small differences in activation energies they could result from minor changes in reaction conditions.

3.7.6 How Do the Polar and ET Mechanisms Relate?

Perrin, in a probing paper based on his investigation of the nitration of naphthalene with nitronium salts, reopened the question of

ET in electrophilic aromatic nitration. Detailed investigations, particularly by Eberson and Radner, clarified much of the questions raised, and put ET nitration into a proper perspective. Considering the redox potential of NO_2^+ and a series of aromatics, as well as the reorganization energy needed to bend the linear NO_2^+ ion in the transition state of the reaction, Eberson and Radner concluded that ET of NO_2^+ with aromatics (in the absence of NO^+) can take place only via Taube-type inner-sphere ET from a $ArHNO_2^+$ intermediate. Outer-sphere electron transfer is not feasible, but is operative if NO^+ is present as oxidant. The redox potential of the aromatic must be substantially lower than that of NO_2^+ (1.8 eV) in order to compensate for the reorganization energies. ET nitration, thus, seems limited to polycyclic aromatics with redox potentials lower than that of perylene. Initial π-complex formation of NO_2^+ with aromatics, as pointed out by Ridd, however may lower their redox potentials to allow ET to more readily occur.

II. Free-Radical Nitration

Titov first discussed the mechanism of radical aromatic nitration,[149] which he elaborated subsequently.[150] According to his concept, $NO_2\cdot$ adds to the aromatic π-sextet to form a nitrocyclohexadienyl radical followed by dehydrogenation.

$$ArH + \cdot NO_2 \longrightarrow Ar\underset{NO_2}{\overset{H}{\diagup}} \xrightarrow{NO_2} ArNO_2 + HNO_2$$

Nitration of benzene proceeded at 135–150°C even at low concentration of NO_2 or with dilute HNO_3 containing nitrogen oxides. In accordance with the radical mechanism, the reaction yielded, in addition to nitrobenzene, meta- and para-nitrobenzene, 1,3,5-trinitrobenzene, 2,4-dinitrophenol and 2,4,6-trinitrophenol, and oxalic acid. Formation of dinitrobenzene and trinitrobenzene was considered to be due to successive addition of NO_2 to the initial nitrocyclohexadienyl radical followed by dehydrogenation (by NO_2) of the 1,4-dinitro-2,5-cyclohexadiene.

[Reaction scheme: C_6H_6 + ·NO_2 → nitrocyclohexadienyl radical → dinitro adduct (with O_2N, H at one carbon and O_2N, H at para) + 2 NO_2 → 1,4-dinitrobenzene + 2 HNO_2]

Indeed when anthracene reacts with N_2O_4 in chloroform solution, 9,10-dinitrodihydroanthracene is obtained as a stable crystalline compound. It can be subsequently dehydrogenated to 9,10-nitroanthracene.[151]

[Reaction scheme: anthracene + N_2O_4 → 9,10-dinitro-9,10-dihydroanthracene → 9,10-dinitroanthracene]

The reactions are characterized not only by anomalous isomer distribution (compared to electrophilic reactions) but also by the relative ease of dinitro and trinitro product formation and oxidative side reactions.

The energetics of the radical nitration were also discussed by Titov. The energy of the cleavage of an R–H bond was considered to be 94–102 kcal/mol and the opening of the π-bond (in olefins) as only 58 kcal/mol. The additional energy required to add NO_2 to the benzene π-sextet to form the nitrocyclohexadienyl radical was estimated as 25–30 kcal (i.e., the loss of aromatic stabilizing energy).

In discussing the mechanism of radical (homolytic) aromatic substitution, such as alkylation, Ingold,[152] based on reviews of Hey[153] and Williams,[154] expressed the view that the initial step is the addition of the aryl(alkyl) radicals to the aromatic. The initial adduct radical then achieves some stability by conjugative distribution of its radical center to the isomeric ortho, meta, and para positions. Polar effects of the radicals may be contributing, albeit to a relatively minor degree. The majority of studies on homolytic aromatic substitutions were on arylation, alkylation, hydroxylation and halogenation. Nitration remained relatively unexplored.

The radical nitration of toluene (and other methylbenzenes) was primarily studied with regard to side-chain reactions. At 100°C, NO_2 or dilute HNO_3 (containing nitric oxides) gives a product composition containing about 50% phenylnitromethane, 3% phenyldinitromethane, and 26% benzoic acid (besides unidentified resinous products).[155] The reaction was considered to be affected by hydrogen abstraction by NO_2 and subsequent reaction of the benzyl radical

$$C_6H_5CH_3 \xrightarrow[-HNO_2]{\cdot NO_2} C_6H_5CH_2\cdot \xrightarrow{\cdot NO_2} C_6H_5CH_2NO_2$$

Olah and Overchuck carried out a study comparing electrophilic and free-radical ring nitration of toluene and benzene with regard to regioselectivity (isomer distribution).[156] Nitrations were carried out with N_2O_4 under UV irradiation or with irradiation using a van de Graaff generator. Studies were also extended to thermal nitration with tetranitromethane (>300°C) as a source for NO_2. The high oxidation potential of toluene and fluorobenzene prevents ready initial oxidation to their radical cations and subsequent combination with NO_2. Free-radical nitration of toluene and fluorobenzene gave close to statistical product distributions, i.e., nitration of toluene gave about 40% ortho, 40% meta, and 20% para isomer. The amount of meta-nitrotoluene thus increases from 3-4% in electrophilic nitration to close to 40% in free-radical nitration (Table 51). Similarly fluorobenzene also gave about 35% meta isomer.

Tetranitromethane does not react with toluene up to 60°C in ether, ethanol, nitromethane, or pyridine-ethanol solution. However, when an ethereal solution of toluene is injected through a hot-metal injection port (>300°C) into a gas chromatograph, nitrotoluenes are formed with the relative amount of meta-nitrotoluene being 39%,[156] very close to the theoretical value of 40%. No dinitrotoluenes are formed under the reaction conditions, presumably due to the short contact time of reagents and high dilution. On the other hand, Lewis-acid catalyzed nitra-

Table 51. Free Radical Nitration of Toluene and Fluorobenzene[156]

Compound	Reagent	Percent nitrotoluenes		
		ortho	meta	para
Toluene	N_2O_4 UV Irradiation			
	N_2O_4	37.2	38.1	24.7
	NO_2BF_4	65.4	2.8	31.8
Fluorobenzene	N_2O_4 100 megarads (Van de Graff)			
	N_2O_4	22.2	35.2	42.6
	NO_2BF_4	8.5	<0.2	91.5
Toluene	Tetranitromethane Thermal Reaction (>300°C)			
	$C(NO_2)_4$	42	39	19
	$C(NO_2)_4BF_3$	64	2	34

tion with tetranitromethane exhibits high ortho/para positional selectivity, characteristic of electrophilic nitration and meta substitution is only 2%.

Kurz et al.,[157] while studying the nitration of toluene with benzoyl nitrate, found that at lower temperatures (25–60°C) or with acid catalysts, typical electrophilic nitration takes place with a nitrotoluene isomer distribution of 62% ortho, 4% meta, and 34% para. At higher temperatures (80°C) or in the presence of added peroxides, products characteristic of radical nitration become more evident. The nitrotoluene isomer distribution changed to 34% ortho, 52% meta, and 15% para. At the same time the yield of ring nitration significantly decreased due to predominant side-chain oxidation (giving benzaldehyde and other products). It was considered that at higher temperatures NO_2 and oxygen are formed from decomposition of intermediate N_2O_5 (no evidence for homolytic cleavage of benzoyl nitrate itself was observed), resulting in radical reactions. The trend for statistical ortho, meta, and para–nitrotoluene isomer distribution is again indicated.

In the nitration of naphthalene, electrophilic nitrating agents yield predominantly α-nitronaphthalene with a ratio of α-nitro/β-nitronaphthalene of about 10 (see Section 3.6.4). In contrast, free radical nitration of naphthalene with tetranitromethane at 300°C gives an α : β nitronaphthalene ratio of ~ 1.

III. Nucleophilic Nitration

Mechanistic aspects of nucleophilic aromatic nitration were not specifically studied. The nitro-dehalogenation of activated haloarenes (see Section 2.32), nitro-diazonation of aryldiazonium ions (see Section 3.1) and nitrolysis of diarylhalonium ions (see Section 2.34) (discussed in Chapter 2), however, would seem to fit the established mechanisms of nucleophilic aromatic substitutions.[158]

Nitro-dehalogenation is expected to follow an S_NAr pathway through a Meisenheimer complex.

Nitro-dediazonation of arenediazonium ions could follow the S_N1 pathway

but more probably proceeds through a radical ion mechanism, since copper ion promotes the reaction in neutral or basic media.

An interesting case of nucleophilic aromatic nitration was reported by Feldman and Myhre[159] in the reaction of 4-nitro-3,4,5-trimethyl-4-chlorocyclohexadiene (obtained form the *ipso* 1,4-addition product of 1,2,3-trimethylbenzene and acetyl nitrate[160] and subsequent reaction with dry HCl) with sodium nitrite, which gave a 60% yield of 5-nitro-1,2,3-trimethylbenzene

When 4-methyl-4-nitrocyclohexadienyl acetate was reacted under similar conditions, *p*-nitrotoluene was obtained in 60% yield, with no detectable *o*-nitrotoluene. At the same time, when treating 4-methyl-4-nitrocyclohexadienyl acetate with concentrated H_2SO_4, *o*-nitrotoluene is obtained. The loss of nitrite ion, which rearomatizes the system, is indicated to take place only after the nucleophilic exchange reaction of the halide for the nitro group. Tetraalkylammonium bromide greatly facilitates the reaction by initial fast *in situ* conversion of the nitrodienyl chloride to the corresponding bromide, which is much more reactive in the nucleophilic substitution reaction by nitrite ion.

These results indicate the possibility of regioselective nitration. One can direct the course of reaction of an *ipso*-nitration intermediate in a way to obtain either a pure ortho- or para-nitroalkylbenzene.

IV. Conclusions

The scope of aromatic nitrations ranges from electrophilic to radical cation to free radical to nucleophilic reactions.

After 40 years Ingold's original mechanism of electrophilic (i.e., two-electron transfer) nitration of aromatics by the NO_2^+ ion (or its carriers) still properly depicts, with some modifications, such at the necessity for two separate intermediates in nitration of reactive aromatics, the dominant acid-catalyzed nitrations. Single-electron transfer proceeding through cation-radical intermediates was shown to be significant in nitrations of polycyclic aromatics with oxidation potentials lower than that of NO_2^+, as well as in some oxidative nitrations and nitrations of reactive aromatics effected by lower nitrogen oxides. These nitrations all display high positional selectivity, whereas substrate selectivity can be lost in fast, encounter-controlled reactions.

In contrast, free-radical nitrations with $NO_2 \cdot$ generated via irradiation of N_2O_4 or its thermal dissociation, proceed by statistical product distributions resulting in loss of both regio and substrate selectivity.

Nucleophilic nitrations with the ambident nitrite ion are less significant and tend also to give isomeric nitrites and, through them, phenolic products.

References

1. Euler, E. *Liebig Ann. Chem.* **1903**, *330*, 280; *Z. Angew. Chem.* **1922**, *35*, 580.

2. (a) Walden, P. *Z. Angew. Chem.* **1924**, *37*, 390; (b) Ri, T., Eyring, H. *J. Chem. Phys.* **1940**, *8*, 433; (c) Price, C. C. *Chem. Revs.* **1941**, *29*, 51.

3. (a) Hughes, E. D., Ingold, C. K., Reed, R. I. *Nature* **1946**, *158*, 448; (b) Gillespie, R. J., Graham, J., Hughes, E. D., Ingold, C. K., Peeling, E. R. A. *Nature* **1946**, *158*, p. 480; (c) Ingold, C. K., Millen, D. J., Poole, H. C. *Nature* **1946**, *158*; Goddard, D. R., Hughes, E. D., Ingold, C. K. *Nature* **1946**, *158*; (d) Gillespie, R. J., Millen, D. J. *Quart. Rev.* **1948**, *2*, 1277.

4. (a) Holleman, A. F. "*Die direkte Einführung von Subtituenten in den Benzolkern*" Veit: Leipzig, 1910; *Chem. Rev.* **1925**, *1*, 187; (b) Wibaut, J. P. *Rec. Trav. Chim.* **1915**, *34*, 241.

5. (a) Hantzsch, A. *Ber.* **1925**, *58*, 958; (b) Hantzsch, A., Berger, K. *Ber.*, **1928**, *61*, 1328.

6. (a) Bunton, C. A., Halevi, E. A., Llewellyn, D. R. *J. Chem. Soc.* **1952**, p. 4913; (b) Bunton, C. A., Halevi, E. A. *J. Chem. Soc.* **1952**, p. 4917; (c) Bunton, C. A., Stedman, G. *J. Chem. Soc.* **1958**, p. 2420.

7. Westheimer, F. F., Kharash, M. S. *J. Am. Chem. Soc.* **1946**, *68*, 1871.

8. For a discussion, Ingold, C. K. "*Structure and Mechanism in Organic Chemistry*", Bell: London, 1953; (b) de la Mare, P., Ridd, J. H. "*Aromatic Substitution: Nitration and Halogenation*", Butterworth: London, 1959.

9. Ingold, C. K., Hughes, E. D. *et al. J. Chem. Soc.* **1950**, p. 2400 and subsequent papers.

10. Wheland, G. W. *J. Am. Chem. Soc.* **1942**, *64*, 900; Wheland, G. W. "*The Theory of Resonance*", Wiley: New York, 1944.

11. Brown, H. C., Brady, J. D. *J. Am. Chem. Soc.* **1952**, *74*, 3570.

12. Olah, G. A. *et al. Nature*, **1956**, *178*, 694, 1344; *J. Am. Chem. Soc.* **1958**, *80*, 6535, 6540, 6541.

13. Olah, G. A. *Accts. Chem. Res.* **1971**, *4*, 240.

14. (a) Klopman, G. *J. Am. Chem. Soc.* **1968**, *90*, 223; Fleming, I. "*Frontier Orbitals and Organic Chemical Reactions*", Wiley-Interscience: New York, 1976; (b) Elliott, R. J., Sackwild, V., Richards, W. G. *Theochem.* **1982**, *3*, 301.

15. Benford, G., Ingold, C. K. *J. Chem. Soc.* **1938**, p. 929.

16. Deno, N. C., Stein, R. *J. Am. Chem. Soc.* **1956**, *78*, 587; Deno, N. C., Jaruzelski, J. J., Schrieshiem, A. *J. Am. Chem. Soc.* **1955**, *77*, 3044; Deno, N. C., Berkheimer, H. E., Evans, W. L., Peterson, H. J. *J. Am. Chem. Soc.* **1959**, *81*, 2344.

17. Coombes, R. G., Moodie, R. B., Schofield, K. *J. Chem. Soc. B*, **1968**, p. 800.

18. (a) Melander, L. *Arkiv. Kemi.* **1950**, *2*, 211; (b) Melander, L. *"Isotope Effects on Reaction Rates"*, Ronald Press: New York, 1960.

19. Vinnik, M. I., Grabovskaya, Zh. E., Arzamaskova, L. N. *Russ. J. Phys. Chem.* **1967**, *41*, 580; Grabovskaya, Zh. E., Vinnik, M. I. *Russ. J. Phys. Chem.* **1966**, *40*, 1221.

20. (a) Chedin, J., Pradier, J. C. *Cr. hebd. Seanc. Acad. Sci., Paris,* **1936**, *203*, 722; Chedin, J. *Cr. hebd. Seanc. Acad. Sci., Paris,* **1935**, *201*, 552; (b) Zamen, M. B., Ph.D. Thesis, University of Bradford, England, 1972.

21. (a) Marziano, N. C. *J. Chem. Soc. Chem. Comm.* **1976**, p. 873; (b) Marziano, N. C., Traverso, P. G., Cimino, G. G. *J. Chem. Soc. Perkin II,* **1984**, p. 574; (c) Marziano, N. C., Sampoli, M., Pinna, F., Passerini, A. *J. Chem. Soc. Perkin II,* **1984**, p. 1163.

22. Ross, D. S., Kuhlmann, K. F., Malhotra, R. *J. Am. Chem. Soc.* **1983**, *105*, 4299.

23. Olah, G. A., Prakash, G. K. S., unpublished results.

24. Moodie, R. B., Schofield, K., Taylor, R. *J. Chem. Soc. Perkin II,* **1979**, p. 133.

25. Sheats, G. F., Strachan, A. N. *Can. J. Chem.* **1978**, *56*, 1280.

26. Katritzky, A. R., Terem, B., Scriven, E. V. *J. Chem. Soc. Perkin II,* **1975**, p. 1600.

27. Schofield, K. *"Aromatic Nitration"*, Cambridge University Press: London, 1980.

28. (a) Bayliss, N. S., Dingle, R., Watts, D. W., Wilkie, R. J. *Austr. J. Chem.* **1963**, *16*, 933; (b) Singer, K., Vamplew, P. A. *J. Chem. Soc.* **1956**, p. 3971.

29. Angus, W. R., Leckie, A. H. *Proc. Royal Soc. A,* **1935**, *150*, 615.

30. Goulden, J. D., Millen, D. J. *J. Chem. Soc.* **1950**, p. 2620; Millen, D. J., Watson, D. *J. Chem. Soc.* **1952**, p. 1369.

31. Pinck, L. A. *J. Am. Chem. Soc.* **1927**, *49*, 2536.

32. Bunton, C. A., Hughes, E. D., Ingold, C. K., Jacobs, D. I. H., Jones, M. H., Minkoff, G. J., Reed, R. I. *J. Chem. Soc.* **1950**, p. 2628.

33. Veibel, S. *Ber.,* **1930**, *63*, 1577.

34. (a) Giffney, J. C., Ridd, J. H., *J. Chem. Soc. Perkin II,* **1979**, p. 618; (b) Al-Omran, F., Fujiwara, K., Giffney, J. C., Ridd, J. H. *J. Chem. Soc. Perkin II,* **1980**, p. 518.

35. Ross, D. S., Hum, G. P., Blucher, W. G. *J. Chem. Soc. Chem. Comm.* **1982**, p. 532.

36. Challis, B. C., Lawson, A. J. *J. Chem. Soc. (B),* **1971**, p. 770.

37. (a) Main, L., Moodie, R. B., Schofield, K. *J. Chem. Soc. Chem. Comm.* **1982**, p. 48; (b) Ross, D. S., Moran, K. D., Malhotra, R. *J. Org. Chem.* **1983**, *48*, 2118; (c) Ross, D. S., unpublished results.

38. Milligan, B. *J. Org. Chem.* **1983**, *48*, 1495.

39. Hunziker, E., Myhre, P. C., Penton, J. R., Zollinger, H. *Helv. Chim. Acta,* **1975**, *58*, 230.

40. Hanna, S. B., Hunziker, E., Saito, T., Zollinger, H. *Helv. Chim. Acta,* **1969**, *52*, 1537; (b) Hunziker, E., Penton, J. R., Zollinger, H. *Helv. Chim. Acta,* **1971**, *54*, 2043.

41. (a) Olah, G. A., Kuhn, S., Mlinko, A. *J. Chem. Soc.* **1956**, p. 4257; (b) Olah, G. A., Kuhn, S. *J. Am. Chem. Soc.* **1961**, *83*, 4564; (c) Olah, G. A., Kuhn, S., Flood, S. H. *J. Am. Chem. Soc.* **1961**, *83*, 4571, 4581.

42. Dewar, M. J. S. *Nature,* **1946**, *156*, 784; *J. Chem. Soc.* **1946**, *406*, 777; *The Electronic Theory of Organic Chemistry*", Oxford University Press: London, 1946.

43. Andrews, L. J., Keefer, R. M., *Molecular Complexes in Organic Chemistry,"* Holden-Day: San Francisco, 1964.

44. (a) Mackor, E., L., Hofstra, A., van der Waals, J. H. *Trans. Faraday Soc.* **1958**, *54*, 66, 187; (b) Dallinga, G., Mackor, E. L., Verrijn Stuart, A. A. *Mol. Phys.* **1958**, *1*, 123, 247; for a review see: Brouwer, D. M., Mackor, E. L., MacLean, C. In: "*Carbonium Ions*", Vol. II, Olah, G. A., Schleyer, P. v. R., eds. Wiley: New York, 1970, p. 837.

45. (a) Hassel, O., Stromme, K. O. *Acta Chem. Scand.* **1958**, *12*, 1146; **1959**, *13*, 1781; Hassel, O. *Mol. Phys.* **1958**, *1*, 241. (b) Mulliken, R. S. *J. Am. Chem. Soc.* **1950**, *72*, 600.

46. Holman, R. W., Gross, M. L. *J. Am. Chem. Soc.* **1989**, *111*, 3560.

47. Brown, H. C., Okamoto, Y. *J. Am. Chem. Soc.* **1957**, *79*, 1931; **1958**, *80*, 4979.

48. Hammett, L. "Physical Organic Chemistry", McGraw Hill: New York, 1940.

49. Knowles, J. R., Norman, R. O. C., Radda, G. K. *J. Chem. Soc.* **1960**, *4*, 885; Norman, R. O. C., Taylor, R. *"Electrophilic Substitution in Benzenoid Compounds"*, Elsevier: New York, 1965.

50. Yukawa, T., Tsuno, Y. *Bull. Chem. Soc. Japan* **1959**, *32*, 971.

51. Fukuzumi, S., Kochi, J. K. *J. Am. Chem. Soc.* **1981**, *103*, 7240.

52. Pfeiffer, P., Wizinger, R. *Ann. Chem.* **1928**, *461*, 132.

53. McCaulay, D. A., Lien, A. P. *J. Am. Chem. Soc.* **1952**, *74*, 6246.

54. Olah, G. A., Kuhn, S., Pavlath, A. *Nature,* **1956**, *178*, 693; *J. Am. Chem. Soc.* **1958**, *80*, 6535.

55. Olah, G. A. et al. *J. Am. Chem. Soc.* **1972**, *94*, 2034; **1978**, *100*, 6299.

56. Olah, G. A. *J. Tenn. Acad. Sci.* **1965**, *40*, 77.

57. Halevi, E. H. *Tetrahedron* **1957**, *1*, 174.

58. Cerfontain, H., Telder, A. *Rec. Trav. Chim.* **1967**, *86*, 371.

59. Hoggett, J. G., Moodie, R. B., Penton, J. R., Schofield, K. *"Nitration and Aromatic Reactivity",* Cambridge University Press: London, 1971.

60. Stock, L. M., Brown, H. C., *Adv. Phys. Org. Chem.* Vol. I, Gold, V. ed. p. 35 and references therein.

61. Olah, G. A., Lin, H. C. *J. Am. Chem. Soc.* **1974**, *96*, 549.

62. (a) Masci, B. *J. Chem. Soc. Chem. Comm.* **1982**, p. 1262; *J. Org. Chem.* **1985**, *50*, 4081; (b) Savoie, R., Pigeon-Gosselini, M., Rodrique, A., Chénevert, R. *Can. J. Chem.* **1983**, *61*, 1248; (c) Elsenbaumer, R. L., Wasserman, E. 2nd *Chem. Congr. North Am. Cont. Abst. Papers*, **1980**, p. 77.

63. Cram, D. J., Doxsee, K. M. *J. Org. Chem.* **1986**, *51*, 5068.

64. Foster, P. W. ACS 136th National Meeting, Atlantic City, *1959*, Abstr. Paper 12p and personal communication quoted by Olah, G. A., Kuhn, S. In *"Friedel-Crafts and Related Reactions"*, Olah, G. A., ed. Vol. 3, pp. 1432-33, Wiley: New York, 1964.

65. Kuhn, S. J., Olah, G. A. *J. Am. Chem. Soc.* **1961**, *83*, 4567.

66. Hanna, S. B., Hunzinker, E., Saito, T., Zollinger, H. *Helv. Chim. Acta,* **1969**, *52*, 1537.

67. (a) Olah, G. A., Lin, H. C. *Synthesis*, **1974**, p. 49; (b) Olah, G. A. In: *"Industrial and Laboratory Nitrations"*, ACS Symposium Series No. 22; Albright, L. F., Hanson, C., eds., 1975.

68. Olah, G. A., Lin, H. C., Olah, J. A., Narang, S. C. *Proc. Natl. Acad. Sci. USA* **1978**, *75*, 1045.

69. Olah, G. A., Narang, S. C., Olah, J. A., Lammertsma, K. *Proc. Natl. Acad. Sci. USA* **1982**, *79*, 4487.

70. Boer, P. *J. Am. Chem. Soc.* **1966**, *88*, 1572; **1968**, *90*, 6706.

71. Olah, G. A., O'Brien, D. H., White, A. M. *J. Am. Chem. Soc.* **1967**, *89*, 5694.

72. (a) Olah, G. A., Bruce, M. R., Clouet, F. L. *J. Org. Chem.* **1981**, *46*, 438.
    (b) Olah, G. A., Shilling, P. *J. Liebigs Ann. Chem.* **1972**, *67*, 77.

73. Masci, B. *Tetrahedron Lett.* **1989**, *45*, 2719.

74. (a) Ritchie, C. D., Win, H. *J. Org. Chem.* **1964**, *29*, 3093; (b) Tolgyesi, W. S. *Can. J. Chem.* **1965**, *43*, 343 (for a rebuttal, see Olah, G. A. and Overchuck, N. *Can. J. Chem.* **1965**, *43*, 3279); (c) Ridd, J. D. *"Studies on Chemical Structure and Reactivity"*, Methuen: London, p. 152; (d) Brown, H. C., Wirkkala, R. *J. Am. Chem. Soc.* **1966**, *88*, 145; (e) Cerfontain, H., Telder, A. *Recl. Trav. Chim. Pays-Bas*, **1967**, *86*, 370; (f) Caille, S. W., Corriu, J. P. *Chem. Comm.* **1967**, p. 1251; *Tetrahedron* **1969**, *25*, 2005; (g) Ingold, C. K. "Structure and Mechanism in Organic Chemistry", 2nd ed., Cornell University Press: Ithaca, New York, 1969, p. 290.

75. Christy, P. F., Ridd, J. H., Stears, N. S. *J. Chem. Soc. (B)*, 1970, p. 797.

76. (a) Coombes, R. D., Moodie, R. B., Schofield, K. *J. Chem. Soc. (B)* **1968**, p. 800; (b) Hoggett, J. G., Moodie, R. B., Schofield, K. *J. Chem. Soc. (B)* **1969**, p. 1; (c) Moodie, R. B., Schofield, K., Tobin, G. D. *J. Chem. Soc. Perkin II* **1977**, p. 1688; (d) Schofield, K. *"Aromatic Nitration"*, Cambridge University Press: London, 1980, pp. 44-53; (e) Moodie, R. B., Schofield, K., Weston, J. B. *J. Chem. Soc. Chem. Comm.* 1974, p. 382.

77. Ridd, J. H. *Accts. Chem. Res.* **1971**, *4*, 248; *Adv. Phys. Org. Chem.* **1968**, *16*, 1.

78. Olah, G. A., Overchuck, N. A. *Can. J. Chem.* **1965**, *43*, 3279.

79. Hanna, S. B., Hunziker, E., Saito, T., Zollinger, H. *Helv. Chim. Acta* **1969**, *52*, 1537.

80. Hunziker, E., Penton, J. R., Zollinger, H. *Helv. Chim. Acta* **1971**, *54*, 2043.

81. (a) Ott, R. J., Rys, P. *Helv. Chim. Acta* **1975**, *58*, 2074; Rys, P. *et al. Helv. Chim. Acta* **1977**, *60*, 2926, 2937; Rys. P. *Angew. Chem. Int. Ed.* **1977**, *16*, 807.

82. Stock, L. M. *"Industrial and Laboratory Nitrations"* Albright, L. F., Hanson, C., eds., American Chemical Society: Washington, D. C. 1976, p. 45.

83. Politzer, P., Jayasuriya, K., Sjoberg, P., Laurence, P. R. *J. Am. Chem. Soc.* **1985**, *107*, 1174.

84. Attina, M., Cacace, F. *J. Am. Chem. Soc.* **1986**, *108*, 318.

85. Schofield, K. *"Aromatic Nitration"*, Cambridge University Press: London, 1980, pp. 171-235.

86. Wheeler, A. S., Smithy, I. W. *J. Am. Chem. Soc.* **1921**, *43*, 2611; Wheeler, A. S., Harris, C. R. *J. Am. Chem. Soc.* **1927**, *49*, 494; Doumani, T. F., Kobe, K. A., *J. Org. Chem.* **1942**, *7*, 1.

87. Perrin, C. L., Skinner, G. A. *J. Am. Chem. Soc.* **1971**, *93*, 3389.

88. (a) Olah, G. A., Kuhn, S. J. *J. Am. Chem. Soc.* **1964**, *86*, 1067; (b) Myhre, P. C., Beug, M. *J. Am. Chem. Soc.* **1966**, *88*, 1568, 1569.

89. Nightingale, D. V. *Chem. Rev.* **1947**, *40*, 117; de Lange, M. P. *Rec. Trav. Chim.* **1926**, *45*, 19; *Methoden der Organischen Chemie* (Houben-Weyl, ed.), 4th ed., 1971, Vol. **101**, G. Thieme Verlag: Stuttgart.

90. Myhre, P. C. *J. Am. Chem. Soc.* **1972**, *94*, 7921.

91. Olah, G. A., Lin, H. C., Mo, Y. K. *J. Am. Chem. Soc.* **1972**, *94*, 3667; Olah, G. A., Lin, H. C., Forsyth, D. A. *J. Am. Chem. Soc.* **1974**, *96*, 6908.

92. Mamatyuk, V. I., Derendyaev, B. G., Detsina, A. N., Koptyug, V. A., *J. Org. Chem. (USSR)* **1974**, *10*, 2506; Detsina, A. N., Mamatyuk, V. I., Koptyug, V. A. *J. Org. Chem. (USSR)* **1977**, *13*, 122; Koptyug, V. A. *"Arenium Ions"* (in Russian), Acad. Nauka: Siberian Dept. Novosibirsk, 1983.

93. Blackstock, D. J., Fisher, A., Richards, K. E., Vaughan, J., Wright, G. J. *J. Chem. Soc. Chem. Comm.* **1970**, 641; Blackstock, D. J., Cretney, J. R., Fischer, A., Hartshoth, M. R., Richards, K. E., Vaughan, J., Wright, G. J. *Tetrahedron Letters* **1970**, *32*, 2793; Fischer, A., Henderson, G. N., Iyer, L. M. *Can. J. Chem.* **1985**, *63*, 2390.

94. Barnes, C. G., Myhre, P. C. *J. Am. Chem. Soc.* **1978**, *100*, 973, 975.

95. Kenner, J. *Nature (London)*, **1945**, *156*, 369.

96. Weiss, J. *Trans. Faraday Soc.* **1946**, *42*, 116.

97. Benford, G. A., Bunton, C. A., Holbenstadt, E. S., Hughes, E. D., Ingold, C. K., Minkoff, G. J., Reed, R. E. *Nature (London)* **1945**, *156*, 688.

98. Perrin, C. L. *J. Am. Chem. Soc.* **1977**, *99*, 5516.

99. Parker, V. D. *Adv. Phys. Org. Chem.* **1983**, *19*, 13.

100. Eberson, L. *"Electron Transfer Reactions in Organic Chemistry"*, Springer: Berlin, 1987.

101. Pross, A. *Accts. Chem. Res.* **1985**, *18*, 212; *Adv. Phys. Org. Chem.* **1985**, *21*, 99.

102. Shaik, S. S. *Progr. Phys. Org. Chem.* **1985**, *15*, 197.

103. Fukuzumi, S., Kochi, J. K. *J. Am. Chem. Soc.* **1981**, *103*, 7240; **1982**, *104*, 7599.

104. Marcus, R. A. *Ann. Rev. Phys. Chem.* **1964**, *15*, 155.

105. Eberson, L., Radner, F. *Accts. Chem. Res.* **1987**, *20*, 53.

106. Olah, G. A., Narang, S. C., Olah, J. A. *Proc. Natl. Acad. Sci. USA* **1981**, *78*, 3298.

107. (a) Achord, J. M., Hussey, C. L., *J. Electrochem. Soc.* **1981**, *128*, 2556; (b) Ross, D. S., Malhotra, R., Schmitt, R. *"Study of Nitration and Oxidation in Oxynitrogen Systems"* Final Report, U. S. Army Res. Off. Contract No. DAAH29-80-C-0046, 1983.

108. (a) Eberson, L., Jonsson, L., Radner, F. *Acta Chem. Scand.* **1978**, *B32*, 749; (b) Eberson, L., Radner, F. *Acta Chem. Scand.* **1980**, *B34*, 739. (c) Eberson, L., Radner, F. *Acta Chem. Scand.* **1985**, *B39*, 357.

109. Markovnik, A. S., *et al. Zh. Obshch. Khim.* **1985**, *55*, 692; *Zh. Org. Khim.* **1982**, *18*, 378.

110. Suzuki, J. *J. Chem. Soc. Chem. Comm.* **1980**, 1245.

111. Draper, M. R., Ridd, J. H. *J. Chem. Soc. Chem. Comm.* **1978**, 445.

112. Dinçtürk, S., Ridd, J. H. *J. Chem. Soc. Perkin II*, **1982**, 965.

113. Lowry, T. H., Richardson, K. S. *"Mechanism and Theory in Organic Chemistry"*, 3d ed., Harper and Row: New York, 1987.

114. Benezra, S. A., Hoffmann, M. K., Bursey, M. M. *J. Am. Chem. Soc.* **1970**, *92*, 7501.

115. Hoffmann, M. K., Bursey, M. M. *Tetrahedron Lett.* **1971**, 2539.

116. Dunbar, R. C., Shen, J., Olah, G. A. *J. Am. Chem. Soc.* **1972**, *94*, 6862.

117. Ausloos, P., Lias, S. G. *Int. J. Chem. Kinet.* **1978**, *10*, 657.

118. Schmitt, R. J., Buttrill, S. E., Jr., Ross, D. S. *J. Am. Chem. Soc.* **1981**, *103*, 5265.

119. Schmitt, R. J., Buttrill, S. E., Jr., Ross, D. S. *J. Am. Chem. Soc.* **1984**, *106*, 926.

120. Reents, W. D., Jr., Freiser, B. S. *J. Am. Chem. Soc.* **1980**, *102*, 271.

121. Cacace, F. *Radiat. Phys. Chem.* **1982**, *10*, 99.

122. Cacace, F. In: *"Structure' Reactivity in Thermochemistry of Ions"*, (Ausloos, P. and Lias, S. G., eds.), D. Reidel Co. Dordrecht: Netherlands, 1987, p. 467.

123. Cacace, F. *Accts. Chem. Res.* **1988**, *21*, 215.

124. Attina, M., Cacace, F., Yanez, M. *J. Am. Chem. Soc.* **1987**, *109*, 5092.

125. Bernardi, F., Cacace, F., Grandinetti, F. *J. Chem. Soc. Perkin Trans. II* **1988** (in press).

126. Harrison, A. G. *"Chemical Ionization Mass Spectrometry"*, CRC Press: Boca Raton, Florida 1983.

127. Attina, M., Cacace, F., de Petris, G. *Angew. Chem.* **1987**, *99*, 1174.

128. Mulliken, R. S., Person, W. B. *"Molecular Complexes: A Lecture and Reprint Volume"*, Wiley: New York, 1969.

129. Brown, R. D. *J. Chem. Soc.* **1959**, 2224, 2232.

130. Nakagura, S., Tanaka, J. *J. Chem. Phys.* **1954**, *22*, 563; Nakagura, S. *Tetrahedron* **1963**, *19* (suppl. 2), 361.

131. Pederson, E. B., Petersen, T. E., Torsell, K., Lawsson, S. O. *Tetrahedron* **1973**, *29*, 579.

132. (a) Elliott, R. J., Sackwild, V., Richards, W. G. *Theochem.* **1982**, *3*, 301; (b) Fleming, I. *"Frontier Orbitals in Organic Reactions"*, Wiley: New York, 1976.

133. Politzer, P., Sjoberg, P. Nitration Conference, SRI International, 1983, Abstr., p. 26.

134. Feng, J., Zheng, Z., Zerner, M. C. *J. Org. Chem.* **1986**, *51*, 4531.

135. Fukuzumi, S., Kochi, J. K. *J. Org. Chem.* **1981**, *46*, 4116.

136. Sankararaman, S., Haney, W. A., Kochi, J. K. *J. Am. Chem. Soc.* **1987**, *109*, 7824.

137. Masnovi, J. M., Kochi, J. K., Hilinski, E. F., Rentzepis, P. M. *J. Am. Chem. Soc.* **1986**, *108*, 1126.

138. Sankararaman, S., Haney, W. A., Kochi, J. K. *J. Am. Chem. Soc.* **1987**, *109*, 5235.

139. Kim, E. K., Kochi, J. K. *J. Org. Chem.* **1989**, *54*, 1692.

140. Kochi, J. K. *Angew. Chem.* **1988**, *100*, 1331.

141. Eberson, L., Radner, F. *Acta Chem. Scand.* **1984**, *38B*, 861.

142. (a) Ram, M. S., Stanburg, D. M. *J. Am. Chem. Soc.* **1984**, *106*, 8136; (b) Boughriet, A., Wartel, M., Fischer, J. C. *J. Electroanal. Chem.* **1985**, *186*, 201; **1985**, *190*, 103; (c) Boughriet, A., Wartel, M. *Chem. Comm.* **1989**, 809.

143. Parker, V., *Adv. Phys. Chem.* **1982**, *18*, 79.

144. Kharkats, Yu I. *Soc. Electrochem. (Engl. Transl.)* **1976**, *12*, 1176; **1974**, *10*, 588.

145. Bontempelli, G., Mazzocchin, G. A., Magno, F. *J. Electroanal. Chem.* **1974**, *55*, 91.

146. Todres, Z. V. *Russ. Chem. Rev. (Engl. Transl.)* **1978**, *47*, 260; *Tetrahedron*, **1985**, *41*, 2771.

147. Taube, H. *"Electron Transfer Reactions of Complex Ions in Solution"*, Academic Press: New York, 1970.

148. (a) Clemens, A. H., Ridd, J. H., Sandall, J. P. B. *J. Chem. Soc. Perkin Trans. II* **1984**, 1659; (b) Clemens, A. H., Ridd, J. H., Sandall, J. P. B. *J. Chem. Soc. Perkin Trans. II* **1984**, 1667; (c) Clemens, A. H., Helsby, P., Ridd, J. H., Omran, F., Sandall, J. P. B. *J. Chem. Soc. Perkin Trans. II* **1985**, 1217; (d) Clemens, A. H., Ridd, J. H., Sandall, J. P. B. *J. Chem. Soc. Perkin Trans. II* **1985**, 1227; (e) Johnston, J. F., Ridd, J. H., Sandall, J. P. B. *Chem. Comm.* **1989**, 244; and Ridd, J. H., personal communication.

149. Titov, A. I. *Tetrahedron* **1963**, *19*, 558; Titov, A. I. Dissertation, Moscow, 1941; Titov, A. I. *Zh. Obshch. Khim.* **1948**, *18*, 190.

150. Titov, A. I. *Zh. Obshch. Khim.* **1947**, *17*, 382; **1948**, *18*, 190; **1952**, *22*, 1329, 1335.

151. Meisenheimer, J. *Liebigs Ann. Chem.* **1902**, *323*, 205.

152. Ingold, C. K. *"Structure and Mechanism in Organic Chemistry"*, 2nd ed., Cornell University Press: 406-417, 1969.

153. Hey, D. H. *"Vistas in Free Radical Chemistry"*, Pergamon Press: London, 1959, p. 209.

154. Williams, G. H. *"Homolytic Aromatic Substitution"*, Pergamon Press: London, 1960.

155. Titov, A. I. *Zh. Obshch. Khim.* **1940**, *10*, 1878.

156. Olah, G. A., Overchuck, N. A. *Can. J. Chem.* **1965**, *43*, 3279.

157. Kurz, M. E., Yang, L. T. A., Zahora, E. P., Adams. R. C. *J. Org. Chem.* **1973**, *38*, 2271.

158. Lowry, T. H., Richardson, K. S. *"Mechanism and Theory in Organic Chemistry"*, 3d ed., Harper and Row: New York, 1987, pp. 640-649.

159. Feldman, K. S., Myhre, P. C. *J. Am. Chem. Soc.* **1979**, *101*, 4768.

160. Barnes, C. E., Feldman, K. S., Lee, H. W. H., Johnson, M. W., Myhre, P. C. *J. Org. Chem.* **1979**, *44*, 3925.

CHAPTER 4.
Aliphatic Nitration

4.1 General Aspects

In contrast to the nitration of aromatic hydrocarbons, which dates back to Faraday[1] and Mitscherlich,[2] nitration of saturated aliphatic hydrocarbons ("paraffins") developed only later.[3] Beilstein and Kurbatov, in 1880, first showed[4] that certain fractions of Caucasian petroleum (recognized by its aliphatic nature), when allowed to react with HNO_3, gave small amounts of nitro compounds of alicyclic nature along with predominant oxidation products. Subsequently, Konovalov[5] and Markownikov[6] carried out extensive studies of the nitration of aliphatic hydrocarbons. Konovalov studies particularly proved that paraffinic hydrocarbons (i.e., alkanes) can be nitrated comparatively readily at high temperatures (in sealed tubes) with dilute HNO_3.

The nitration of saturated aliphatic hydrocarbons was not followed up from an academic or an industrial point of view till the 1930s when Hass took up such investigations[7] and developed a process for gas-phase nitration of the lower homologous alkanes including propane, butanes, and pentanes, with HNO_3 vapors at 350–450°C. The process has been used in industry since the 1940s. All these nitrations of alkanes can be considered as radical nitrations, in contrast to the predominantly electrophilic nature of aromatic nitrations. It was only in the 1970s that Olah *et al.* were able to show[8] that purely electrophilic nitration of alkanes is also possible using nitronium salts.

Nitroaliphatic compounds are frequently prepared by nucleophilic substitution from alkyl halides with metal nitrites, a reaction first discovered by Victor Meyer in 1872 in the case of alkyl iodides with silver nitrite.[9] The reaction has since been improved and extended to other alkyl halides and metal nitrites, most notably by Kornblum using sodium nitrite in dimethyl formamide solution.[10]

We will discuss aliphatic nitrations according to their three major reaction types, i.e., radical, electrophilic, and nucleophilic nitrations. Compared with aromatic nitrations, the aliphatic counterparts are much less investigated and more limited in scope. Our discussion will combine preparative and mechanistic aspects. We will also discuss nitration not only of alkanes, but of alkenes and alkynes, as well as O, N and S nitration of heteroaliphatic compounds.

Nitroalkanes, particularly nitromethane, are used primarily as solvents,[11] although they also find applications as intermediates for nitro alcohols, hydroxylamines, and polynitro compounds, as stabilizers, explosives, and high-energy fuels. A now-abandoned process for the manufacture of ϵ-caprolactam[11] used the nitration of cyclohexane.

Polynitrates (such as nitroglycerine) and nitramines (RDX,HMX) play a significant role as explosives. Nitroolefins are gaining significance as synthons.[12]

4.2 Nitration of Alkanes, Cycloalkanes, and Arylalkanes

4.2.1 Free-Radical Nitration

Saturated aliphatic hydrocarbons are most frequently nitrated by free-radical reactions. We will begin our discussion of aliphatic nitration with the development and early studies of nitration with dilute HNO_3 in the liquid phase, which subsequently led to high-temperature gas-phase nitration.

4.2.1.1 Liquid-Phase Nitration with Dilute Nitric Acid and Nitrogen Oxides (Konovalov Reaction)

Nitration of alkanes and saturated aliphatic hydrocarbon derivatives is normally carried out with dilute HNO_3. Concentrated HNO_3 tends to extensively oxidize alkanes.

Konovalov first succeeded in nitrating saturated hydrocarbons, such as trimethylcyclohexane, with dilute HNO_3.[13] Prior to his studies, attempts to nitrate saturated hydrocarbons with concentrated HNO_3 (by Bayer, Markownikov, Beilstein, Konovalov, and others[14]) resulted mainly in oxidation with only small yields of nitro products.

Konovalov found that nitration of saturated aliphatic hydrocarbons, such as n-hexane with dilute HNO_3 of specific gravity 1.075 ($\approx$ 13% by wt.) at 140°C for 4-6 hr gave nitrohexane in about 40% yield. n-Heptane and n-octane under similar conditions gave the corresponding nitro compounds in yield of 47% and 49-52%, respectively.

The nitration of isobutane by Konovalov's method with HNO_3 in a sealed tube at 150°C gave nitroisobutane, showing the much higher reactivity of the tertiary methine hydrogen over the primary methyl hydrogens.[15]

Konovalov also showed that his nitration method with highly dilute HNO_3 serves to introduce the nitro group into the side chain of alkylbenzenes when the two are heated in a sealed tube at 100-110°C.[16] Propylbenzene was found to be more reactive than ethylbenzene and when reacted with HNO_3 of 1.075 specific gravity at 105°C for 4-5 hr gave a 74% yield of phenylnitropropanes. Propylbenzene was even nitrated with 1% HNO_3 at 100-105°C for 90 hr.

The main factors affecting Konovalov's nitration were the concentration of HNO_3, temperature, and duration of the reaction. Nitration under pressure gave generally superior results compared with reactions in open vessels. The reactivity of C-H bonds in saturated hydrocarbons or in the side chain of alkylbenzenes generally showed the reactivity in the order 3°>2°>1°. Nitrations were always accompanied by oxidative side reactions. Konovalov considered these as secondary reactions of the nitroalkane (arylalkane) products.

Konovalov initially suggested that extensive oxidation observed during nitration of alkanes with HNO_3 in the liquid phase is entirely due to subsequent secondary oxidation of the nitro products formed initially. Later, however, when the stability of nitroalkanes to oxidation became clear he concluded that the nitro products underwent oxidation in the moment of their formation. Clearly his explanation was unsatisfactory. Nametkin subsequently suggested[17] that initially aci- or iso-nitro compounds are formed in the reaction of alkanes with HNO_3,

$$\diagdown_{CH_2}\diagup + HONO_2 \longrightarrow \diagdown_{C}\diagup = N\diagup^{O}_{\diagdown OH}$$

which then either isomerize to nitro compounds or undergo decomposition (via Nef's reaction) into ketones (aldehydes) and N_2O, with subsequent oxidation. Ingold[20] and Titov[18] later clearly established (*vide infra*) that HNO_3 itself does not react with alkanes in the absence of nitrogen oxides, thus the hypothesis of nitration of alkanes by HNO_3 via iso-nitro compounds cannot be correct.

Nitration of saturated acyclic and cyclic aliphatic hydrocarbons was also studied by Markownikov.[1] In his extensive studies he found that open-chain *n*-alkanes nitrated relatively slowly as compared to iso-alkanes containing C–H bonds.

Markownikov's[19] studies found that dilute HNO_3 nitrated not only compounds which contained alkyl substituted methylene groups but also unsubstituted polymethylenes. In the latter case, however, nitration occurred less readily and therefore more concentrated HNO_3 had to be used. Under these conditions, however, oxidation that occurs concurrently became a strongly competing reaction.

Evidence that the nitration of alkanes is not effected by HNO_3 itself seems to come first from the studies of Ingold.[20]

When an optically active alkane such as (+)-3-methylheptane, is nitrated by 50% HNO_3 at 100°C, the product, 3-methyl-3-nitroheptane, is racemic. Nitrations of *cis*- and *trans*-decalins and *cis*- and *trans*-hydrindanes are also not stereospecific. Therefore, a mechanism involving bimolecular substitution with a transition state involving HNO_3 is not favored, but a mechanism involving a free carbon radical, which has escaped the solvent cage, is probable.

$$(+)\text{-} \underset{H}{\overset{CH_3}{\diagup}}\diagdown\diagup\diagdown \xrightarrow[-HNO_2]{NO_2} \underset{}{\overset{CH_3}{\diagup}}\diagdown\diagup\diagdown \xrightarrow{NO_2}$$

$$(rac)\text{-} \underset{O_2N}{\overset{CH_3}{\diagup}}\diagdown\diagup\diagdown$$

Alkyl nitrites, which are byproducts, are easily decomposed to form carbonyl compounds, alcohols, and carboxylic acids.

Although Konovalov's, Markownikoff's, and Nametkin's pioneering investigations of liquid–phase nitration with HNO_3 were carried out around the turn of the century, the nature of these nitrations was not understood until the extensive studies by Titov,[21] who established that HNO_3 itself is not a nitrating agent for alkanes, and it is the NO_2 present in the acid system that is the *de facto* nitrating agent.

Titov's studies in the 1940s and 1950s established the free–radical mechanism of saturated hydrocarbons[21] (as well as related reactions of unsaturated and aromatic systems). As the original work was published exclusively in the Russian literature, it received wide-spread recognition only after an English review paper was published in 1963.[21]

In nitrations with N_2O_4, particularly in those at elevated temperatures, the reagent is dissociated into NO_2 radicals, which are the active species. As in the case of vapor–phase nitrations, HNO_3 does not react as such with the hydrocarbons but rather is a source of NO_2 radicals.[22,23,24] For example, neither *n*-heptane nor 2,7-dimethyloctane reacts with HNO_3 even during prolonged contact, if NO_2 is absent.[21] Similarly, in the absence of NO_2, cyclohexane can be kept with dilute HNO_3 for prolonged periods of time without any reaction, but NO_2 reacts with cyclohexane slowly at room temperature and rapidly at 100°C.[25]

Nitrations with NO_2 and with dilute HNO_3 give the same mixture of products. When HNO_3 is added to a mixture of cyclohexane and NO_2, the oxidizing and nitrating effect of NO_2 is lowered. An increase in the NO_2 concentration increases the yield of products of oxidation and at the same time lowers the yield of the nitro compounds.[25]

As in the vapor–phase processes, the reaction of NO_2 with hydrocarbons in the liquid phase produces either nitro products or alkyl nitrites. As in the vapor–phase reactions, the first step in the nitration in solution is abstraction of a hydrogen atom of the substrate by NO_2.

$$RH + NO_2 \longrightarrow R\cdot + HNO_2$$

The hydrocarbon radical $R\cdot$ then reacts with NO_2 to give the nitro compound or nitrite.

$$R\cdot \xrightarrow{NO_2} \begin{cases} RNO_2 \\ RONO \end{cases}$$

The nitro product is stable and can be isolated as such. The nitrite, on the other hand, may react with water in an equilibrium reaction.[25]

$$RONO + H_2O \rightleftharpoons ROH + HNO_2$$

In liquid–phase nitrations the order of reactivity of the hydrogen atoms of the substrate is the same as that observed for other free-radical systems namely, tertiary hydrogen atoms are substituted more easily than secondary, which in turn are more readily substituted than primary ones. For example, the nitration of methylcyclopentane with HNO_3 at 115°C yields a mixture of the tertiary and secondary nitro compounds in a 2 : 1 ratio.[26,27]

(and regioisomers)

Alkyl benzenes and cycloalkylbenzenes are preferentially nitrated and oxidized at the benzylic position. For example, ethylbenzene and cyclohexylbenzene give mainly α–nitroethylbenzene and α–nitrocyclohexylbenzene, respectively.[28]

The ease of nitration of saturated carbon atoms is generally in the order: tertiary>secondary>primary. Exceptions, however, are observed in some cases. For example, camphane and isocamphane are nitrated in the liquid phase to yield only secondary nitro compounds.[28]

Ogata *et al.*[29] rationalized the acid dependence of the oxidation of arylalkanes by HNO_3 being due to the rate-determining hydrogen abstraction by the radical cation $HNO_2{}^+$, i.e., protonated NO_2.

Nitration of alkanes using N_2O_5 has received little attention. Titov and Shchitov proposed that N_2O_5 is in equilibrium with NO_2 and NO_3 radicals.[30]

$$N_2O_5 \rightleftharpoons \cdot NO_2 + NO_3 \cdot$$

The more reactive NO_3 radical abstracts a hydrogen atom from the hydrocarbon. The hydrocarbon radical then undergoes further reaction with N_2O_5 and with nitrogen oxides to give nitro compounds and a nitrate, as well as some oxidized products ($R'CO_2H$). The formation of these products is shown in Scheme 1.

The reaction of N_2O_5 with paraffins is retarded by NO_2. This effect is in contrast to that observed with NO_2 in nitration processes using dilute HNO_3. This is explained by the shifting of the equilibrium into the direction of the undissociated N_2O_5.

Cyclohexane and N_2O_5 in carbon tetrachloride react at 0° to give nitrocyclohexane in 30% yield and cyclohexyl nitrate in 41% yield. In boiling carbon tetrachloride, adipic acid is also obtained.[30] The reaction with *n*-heptane under similar conditions produces mainly *n*-heptyl nitrate and secondary nitroheptanes. The reaction with *n*-octane produces *n*-octyl nitrate and secondary nitrooctanes.[30]

4.2.1.2 Gas-Phase Nitration with Nitrogen Dioxide

Shorygin and Topchiev in 1934 found that the reaction of *n*-hexane with N_2O_4 diluted with CO_2 at temperature up to 80°C gave about 15%[31] of mononitrohexane.

Urbański and Slon reported in 1936 that the nitration of straight-chain alkanes, ranging from methane to nonane can be substantially improved when carried out with NO_2 in the gas phase at 200°C, and they obtained mixtures of mono- and dinitroalkanes.[31] The mononitro compounds were said to contain predominantly primary terminal derivatives, whereas the dinitro compounds were assumed to

$$N_2O_5 \rightleftharpoons NO_2\cdot + NO_3\cdot$$

$$RH + NO_3\cdot \longrightarrow R\cdot + HNO_3$$

$$R\cdot + N_2H_5 \longrightarrow RONO_2 + NO_2$$

$$2R\cdot + 2NO_2 \longrightarrow RNO_2 + RONO$$

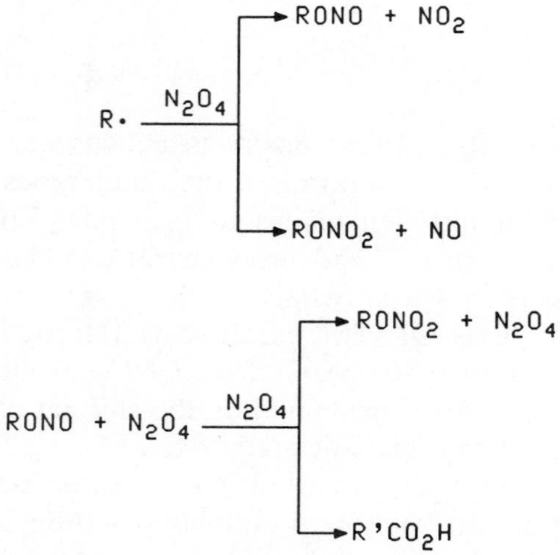

$$R\cdot \xrightarrow{N_2O_4} \begin{array}{l} RONO + NO_2 \\ \\ RONO_2 + NO \end{array}$$

$$RONO + N_2O_4 \xrightarrow{N_2O_4} \begin{array}{l} RONO_2 + N_2O_4 \\ \\ R'CO_2H \end{array}$$

Scheme 1. Free Radical Nitration of Alkanes with N_2O_5

be mainly ω, ω-derivatives.

$$2\,RCH_3 + 4NO_2 \longrightarrow 2\,RCH_2NO_2 + H_2O + N_2O_3\,(NO_2 + NO)$$

In addition to the nitro derivatives, aldehydes and carboxylic acids, formed by oxidative side reactions, were also found in the products.

Reinvestigation of these results by Hass *et al.* in 1941 showed,[32] however, that the mononitro products were mixtures of isomers similar in composition to those obtained independently by Hass in gas-phase nitration with HNO_3. The dinitro compounds could not be obtained as well-defined isomers from the mixtures.

The mechanism of the reaction based on the original suggestion by Titov is now generally accepted to involve a free-radical path with $NO_2\cdot$ (formed from dissociation of N_2O_4) reacting with the alkane via initial hydrogen atom abstraction.

$$RH + NO_2\cdot \longrightarrow R\cdot + HNO_2$$

$$R\cdot + \cdot NO_2 \longrightarrow RNO_2$$

HNO_2 itself is a further source of NO_2 via is gas-phase decomposition with oxygen present in the system. Under these circumstances NO_3 can also be involved in the reaction.

A breakthrough in the development of aliphatic nitration came when Hass *et al.* developed a process for the ready nitration of aliphatic hydrocarbons such as methane, ethane, propane, and butane with nitric acid in the vapor phase around 400°C with a few seconds of reaction time.[34] Hass discovered his reaction in 1935 and subsequently extensive studies appeared in the technical literature.

$$RH + HNO_3 \xrightarrow{400°C} RNO_2 + H_2O$$

Initially, the nitration of propane, butane, and pentane was studied and nitro products included nitromethane and nitroethane, indicative of chain cleavage during the reaction. This is a significant characteristic of the Hass reaction. Subsequently, nitration of methane itself to nitromethane was found to be feasible under more forcing conditions (460°C with only 0.1 sec residence time). Nitroaliphatics are used predominantly as solvents (e.g., for paints) and to some degree as intermediates in synthesis. With the rise of the petroleum industry and abundant availability of gaseous hydrocarbons, vapor-phase nitration gained importance. The reaction may be conducted at atmospheric pressure, although for large-scale manufacturing the reaction is carried out under pressure in the vicinity of 10 atm. The vapor-phase nitration is not practical for the production of pure nitro products from any alkane except methane. With other alkanes the reaction gives mixtures of mono-, di-, and polynitrated alkanes at every combination of positions, and is also accompanied by extensive chain-cleavage reactions.[35] For example, nitration of *n*-butane produces besides

1-nitrobutane and 2-nitrobutane, 1-nitropropane and 2-nitropropane, nitroethane, and nitromethane. The extent of chain cleavage (cracking) increases with temperature. Consequently, if higher molecular weight products are desired, the reaction is conducted at lower temperatures.

The free-radical mechanism for vapor-phase nitration was first established by Hass and Bachman[36] following an earlier suggestion by Titov for liquid-phase nitration.[18] In accordance with their mechanism, the nitration can also be achieved with a mixture of NO_2 and steam. Use of NO_2 allows the reaction to be conducted at lower temperatures. Generally, about half of the added HNO_3 is converted into the nitro product. The remainder ends up as a mixture of nitrogen oxides. In industry[37] this mixture is reconverted to HNO_3 and recycled. The conversion of HNO_3 into nitro products can also be improved by including oxygen and/or halogens. These gases help increase the concentration of the alkyl radicals. Other variations include use of silent electric discharge and gamma rays.[37]

The free-radical chain mechanism of the Hass nitration was extensively studied in subsequent work and excellent reviews have appeared.[38,39] As already mentioned, Titov has demonstrated[21] that HNO_3 has no effect on alkanes in the absence of NO_2, which suggests that the reaction proceeds by $NO_2\cdot$ (a free radical) abstracting hydrogen from the alkane. HNO_3 only serves as a source for NO_2.

$$RH + NO_2\cdot \longrightarrow R\cdot + HNO_2$$

$$R\cdot + \cdot NO_2 \longrightarrow RNO_2$$

Thermal decomposition of HNO_3, producing $\cdot OH$ and $NO_2\cdot$, initiates a chain reaction. Both of these radicals can abstract hydrogen from the hydrocarbon to give alkyl radicals. Coupling of the alkyl radical with NO_2 gives the nitroaliphatic product, while decomposition of the alkyl radical, generally by scission reactions, leads to cracked products. Coupling of the alkyl radicals to produce higher hydrocarbons is also possible. Under the reaction conditions employed, however, such coupling is not favored.

The Hass reaction has been reviewed extensively. The reader is referred to the literature for further details. The relative brevity of our discussion should not detract from the fact that this method is presently the most significant nitration method available for aliphatic saturated hydrocarbons, although as discussed, it generally does not give pure individual nitro compounds.

4.2.2 Electrophilic Nitration

In contrast to free-radical nitration, electrophilic nitration of saturated aliphatics is much less developed, except for the nitration of active methylene compounds.

4.2.2.1 Active Methylene Compounds

The nitration of active methylene compounds generally proceeds through the reaction of carbanionic intermediates with an electrophilic nitrating agent.[40] The nitration of arylacetonitriles and arylacetic esters by alkyl nitrates was discovered by Wislicenus,[41a] who also converted fluorene to 9-nitrofluorene in high yield.[41b] The reaction of benzyl cyanide, sodium ethoxide, and methyl nitrate provided a route to phenylnitromethane.[42] The overall yield was 50–55%.

$$C_6H_5CH_2CN + CH_3ONO_2 \xrightarrow{NaOC_2H_5} \left[\begin{array}{c} C_6H_5C=NO_2 \\ | \\ CN \end{array} \right]^- Na^+ \xrightarrow[H_2O]{NaOH}$$

$$\left[\begin{array}{c} C_6H_5CNO_2 \\ | \\ CO_2 \end{array} \right]^{2-} 2Na^+ \xrightarrow{HCl} C_6H_5CH_2NO_2 + CO_2$$

Wieland applied the reaction to the nitration of ketones.[43] Cyclohexanone reacted with potassium ethoxide and ethyl nitrate in a mixture of ethyl alcohol and diethyl ether to form a mixture of α-nitrocyclohexanone and α,α'-dinitrocyclohexanone.

The Wieland procedure generally leads to nitro ketones in low yields. Feuer and co-workers reported[44] a thorough study to determine the optimum experimental conditions. For example, cyclopentanone was treated with excess sublimed potassium t-butoxide in tetrahydrofuran at a low temperature (−30°C to minimize condensation of the ketone) and amyl nitrate was added. A small amount of alcohol lowers the yield drastically, and a substantial excess of base is essential. Cyclohexanone, cyclopentanone, and cyclooctanone react similarly and the yields range from 35–72%.[45]

$$\text{cyclopentanone} \xrightarrow[\text{RONO}_2]{(CH_3)_3COK} \left[O_2N \overset{O}{=}\!\!\!\underset{}{\bigcirc}\!\!\!= NO_2 \right]^{2-} 2K^+ \xrightarrow{HCl} O_2N\!-\!\!\underset{}{\overset{O}{\bigcirc}}\!\!-NO_2$$

An excellent method was found by Feuer and Anderson[46] for the synthesis of α, ω -dinitroalkanes in the cleavage of dinitrocycloalkanones. The potassium salt of 2,6-dinitrocyclohexanone is converted in acidic medium to 1,5-dinitropentane in 78% yield. In a comparable manner, 1,6-dinitrohexane (75%) and 1,4-dinitrobutane (72%) are prepared. The synthesis of terminal dinitroalkanes is a general reaction for which the requisite cyclic ketones are readily available.

The method developed for the preparation of α,α'-dinitrocyclic ketones is applicable to aliphatic esters, nitriles, α, ω, aliphatic ketones and aralkyl ketones.[47] *N,N*-Dialkylamides undergo alkyl nitration, and the products are isolated as their bromo derivatives.[48] Modifications in the procedure of the alkyl nitrate nitration facilitate the synthesis of mononitro ketones.[49] The α-nitro ketones are susceptible to cleavage reaction. Cyclooctanone reacts with potassium *t*-butoxide and amyl nitrate to form α-nitrocyclooctanone and amyl 8-nitrooctanoate.

In the competitive nitration and cleavage reactions, nitration predominates with the C_5, C_6, and C_7-cyclic ketones, but cleavage dominates with medium ring ketones. Tertiary nitro ketones, which cannot form a stable anion, are cleaved. Dinitration does occur in some instances, and the dinitro ketone and unreacted ketone complicate the isolation of the desired mononitro ketone. Nevertheless, nitration of ketones with alkyl nitrates is the most effective method for the synthesis of α-nitro ketones. Nitration of enol silyl ethers with nitronium salts is an excellent alternative for the synthesis of α-nitro ketones without any subsequent ring cleavage.

In the reaction, a carbanion is first formed, which then reacts with the electrophilic nitrating agent in an S_E1 fashion.

$$-\overset{|}{\underset{|}{C}}^- + CH_3ONO_2 \longrightarrow -\overset{|}{\underset{|}{C}}-NO_2 + CH_3O^-$$

Under the basic reaction conditions, the conjugate base of the nitro compound is the reaction product.

C–H bonds activated by only one electron-withdrawing group such as at the α-positions of ketones, nitriles, sulfones, or N,N-dialkylamides tend to form carbanions readily with bases such as $t\text{-}C_4H_9O^-$ or $NaNH_2$ and thus can be nitrated subsequently with alkyl nitrates.[40,50,51] Aliphatic nitro compounds can be also α-dinitrated by reaction of the conjugate bases with NO_2^- and $K_3Fe(CN)_6$.[52]

$$R_2CHNO_2 \longrightarrow R_2\bar{C}NO_2 \xrightarrow[K_3Fe(CN)_6]{NO_2^-} R_2C(NO_2)_2$$

Diethyl nitromalonate is obtained in 45% yield from diethyl malonate, sodium hydride, and acetone cyanohydrin nitrate in tetrahydrofuran. α-Nitro esters are prepared by the nitration of monosubstituted acetoacetic or malonic esters with acetone cyanohydrin nitrate by the use of excess sodium hydride.[53] The yields range from 40 – 70%.

$$\text{RCH(CO}_2\text{C}_2\text{H}_5)_2 \text{ or } \underset{\underset{\text{COCH}_3}{|}}{\text{RCHCO}_2\text{C}_2\text{H}_5} + \underset{\underset{\text{CN}}{|}}{\text{(CH}_3)_2\text{CONO}_2}$$

$$\downarrow 2\text{NaH}$$

$$\underset{\underset{\text{NO}_2}{|}}{\text{RCHCO}_2\text{C}_2\text{H}_5}$$

HNO_3 reacts with acetone cyanohydrin in acetic anhydride to form acetone cyanohydrin nitrate in 65–68% yield.[54a] Sodium alkoxides destroy the nitrating agent.[54b] Sodium n-amylate reacts with acetone cyanohydrin nitrate to form n-amyl nitrate and n-amyl hydroxyisobutyrate. Therefore, the choice of base is limited, and sodium hydride is effective.

Feuer gave an excellent review of nitration of active methylene compounds with alkyl nitrates in the presence of bases.[55] The reader is referred to this review for further discussion and details.

Cyclic β-diketones are effectively nitrated with HNO_3. For example, indane-1,3-dione is converted to 2-nitroindane-1,3-dione in 78% yield.[56] The nitroindanedione synthesis also furnishes a method for the preparation of primary nitro compounds.[57]

$$\text{indane-1,3-dione} \xrightarrow{HNO_3} \text{2-nitroindane-1,3-dione}$$

The nitroindane-1,3-diones formed are cleaved by dilute aqueous sodium hydroxide to sodium phthalate and the primary nitro compound.

$$\text{2-R-2-nitroindane-1,3-dione} \xrightarrow{OH^-} RCH_2NO_2$$

For example, 1-naphthalynitromethane[58] as well as other analogs of phenylnitromethane may be prepared. The preparations of 2-nitro-5,5-dimethylcyclohexane-1,3-dione[59] and 2-nitro-2-carbethoxyindane-1,3-dione[60] are additional illustrations of the nitration of cyclic β-diketones with HNO_3.

The nitration of 3-bromocamphor with boiling concentrated HNO_3 resulted in a mixture of the corresponding 3-bromo-3-nitro epimers. The mixture was treated with sodium ethoxide, and 3-nitrocamphor[61] was obtained in 17% yield.[62] Obviously, few ketones will survive these experimental conditions, but the method provided an α-nitro ketone[63] for study at a time when these compounds were not readily available.

Nitration of diethyl malonate provides diethyl nitromalonate in 92% yield.[64] Nitrations of monoesters of malonic acid with nitric acid affords the α,α-dinitro esters in low yields.[65]

Diphenylcyanonitromethane[66] is prepared from diphenylcyanomethane and NO_2. Diphenylnitromethane is obtained by the action of NO_2 on diphenylmethane in the presence of copper sulfate and oxygen.[67] Nitrogen oxides convert malonate to diethyl oxomalonate and no nitration is observed.[68]

Activated C–H bonds of alkanes are in some cases also nitrated by fuming HNO_3 in AcOH and by acetyl nitrate and an acid catalyst.[69]

The reaction of alkyllithiums with $NO_2^+BF_4^-$ generally does not yield the corresponding nitro compounds, as complex reactions including electron transfer seem to dominate, probably also involving the formed nitro compound.

$$\text{RLi} + \text{NO}_2^+\text{BF}_4^- \longrightarrow \text{RNO}_2 + \text{LiBF}_4$$

An interesting example of nitration of a carbanionic system with $\text{NO}_2^+\text{BF}_4^-$ is that of the sulphonium dinitroylide to form a trinitromethylsulphonium salt.[70]

$$(\text{CH}_3)_2\text{S}^+\text{-C(NO}_2)_2^- + \text{NO}_2^+\text{BF}_4^- \xrightarrow{\text{CH}_3\text{CN}} [(\text{CH}_3)_2\text{S}^+\text{-C(NO}_2)_3]\text{BF}_4^-$$

Related derivatives of trinitromethylselenonium salts are less stable and all attempts to synthesize such compounds have led to the formation of tetranitromethane.[71]

$$(\text{C}_4\text{H}_5)_2\text{Se}^+\text{-C(NO}_2)_2^- + \text{NO}_2^+\text{BF}_4^- \xrightarrow{\text{CH}_3\text{CN}} \text{C(NO}_2)_4$$

4.2.2.2 Nitration with Nitronium Salts

Mixtures of HNO_3 with H_2SO_4 (mixed acid) used extensively in nitrations of aromatic hydrocarbons, are generally unsuitable for the nitration of alkanes, since primary nitroalkanes are rapidly hydrolyzed by hot H_2SO_4 and secondary and tertiary nitroalkanes form tars (in all probability via rapid alkene–forming elimination and subsequent polycondensation and polymerization). However, it is significant to point out that it is not necessarily the lack of reactivity of "paraffins" with mixed acid that makes the nitration of saturated hydrocarbons unsuitable, but the fast secondary reactions of any nitro products formed (as well as oxidative side reactions). This difficulty can be, at least in part, overcome by using preformed nitronium salts as nitrating agents.

Whereas electrophilic aromatic nitration is one of the most thoroughly studied reaction, electrophilic aliphatic nitration remained for a long time unrecognized. It was only after the development of stable nitronium salts as effective nitrating agents that *bona fide* electrophilic nitration and nitrolysis of alkanes and cycloalkanes was achieved.[72] [The terms are defined as substitution (of hydrogen for the nitro group) and nitrolytic cleavage (of C–C bonds), respectively.]

A solution of a stable nitronium salt (generally the hexafluorophosphate $NO_2^+PF_6^-$ but also the hexafluoroantimonate $NO_2^+SbF_6^-$ or tetrafluoroborate $NO_2^+BF_4^-$) in methylene chloride–sulfolane or nitromethane solution was allowed to react with the alkane (cycloalkane), with usual precautions taken to avoid moisture and other impurities. Reactions were carried out at room temperature (25°C) in order to avoid or minimize the possibility of radical side reactions and/or protolytic cleavage reactions (tertiary nitroalkanes particularly undergo ready protolytic cleavage reactions, even if the system initially is acid free, but nitration forms acid). Data obtained are summarized in Table 52.

At 25°C only 0.1% of nitromethane was obtained in the nitration of methane. Substantially (at least tenfold) increased yields were obtained in HF and HSO_3F (or other superacid) solutions (see subsequent discussion). Higher alkanes and isoalkanes gave yields of 5–10% and adamantane was nitrated in 30% yield. The data indicate that nitration (nitrolysis) of alkanes with nitronium salts proceeds in accordance with the generalized concept of electrophilic reactions of single bonds[73] involving two–electron three–center–bond carbocationic intermediates (transition states) as illustrated in the case of methane.

$$CH_4 + NO_2^+PF_6^- \rightleftharpoons \left[H_3C \cdots \begin{smallmatrix} H \\ NO_2^+ \end{smallmatrix} \right]^+ PF_6^- \xrightarrow{-H^+} CH_3NO_2$$

In cases of higher hydrocarbons, nitrolysis of the C–C bond also takes place in competitive reactions (Scheme 2).

The nitronium ion nitration (and nitrolysis) of alkanes and cycloalkanes follows the same pathway as protolytic reactions and alkylations.[74] The reactions proceed via two–electron three–center– bound five–coordinate carbocationic transition states formed by the NO_2^+ ion attacking the two–electron covalent σ–bonds, forcing them into electron–pair sharing. It should be reminded that the linear NO_2^+ ion $O=N^+=O$ has no vacant orbital on nitrogen (similar to the ammonium

Table 52. Nitration and Nitrolysis of Alkanes and Cycloalkanes with $NO_2^+PF_6^-$ in CH_2Cl_2–Sulfolane Solution at 25°C[72a]

Hydrocarbon	Nitroalkane products and their mol ratio
Methane	CH_3NO_2
Ethane	$CH_3NO_2 > CH_3CH_2NO_2$, 2.9 : 1
Propane	$CH_3NO_2 > CH_3CH_2NO_2 > 2\text{-}NO_2C_3H_7 > 1\text{-}NO_2C_3H_7$, 2.8 : 1 : 0.5 : 0.1
Isobutane	$tert\text{-}NO_2C_4H_9 > CH_3NO_2$, 3:1
n-Butane	$CH_3NO_2 > CH_3CH_2NO_2 > 2\text{-}NO_2C_4H_9 \sim 1\text{-}NO_2C_4H_9$, 5 : 4 : 1.5 : 1
Neopentane	$CH_3NO_2 > tert\text{-}C_4H_9NO_2$, 3.3 : 1
Cyclohexane	Nitrocyclohexane
Adamantane	1-Nitroadamantane > 2-nitroadamantane, 17.5 : 1

ion) and therefore *per se* can act only as a polarizable electrophilic nitrating agent. In contrast to π–donor aromatics, σ–donor alkanes are, however, weak electron donors to bring about ready polarization. The interacting NO_2^+ ion thus must be at least partially bent with a developing empty p–orbital on nitrogen. The bending of the NO_2^+ ion can be facilitated by the ability of the oxygen non–bonded electron pairs to coordinate with strong acid present in the nitration systems. Proto–solvation should result in at least partial bending and starting of the development of an empty bonding orbital on nitrogen. If full protonation would be achieved (for which there is no evidence) the protonated nitronium ion NO_2H^{2+}

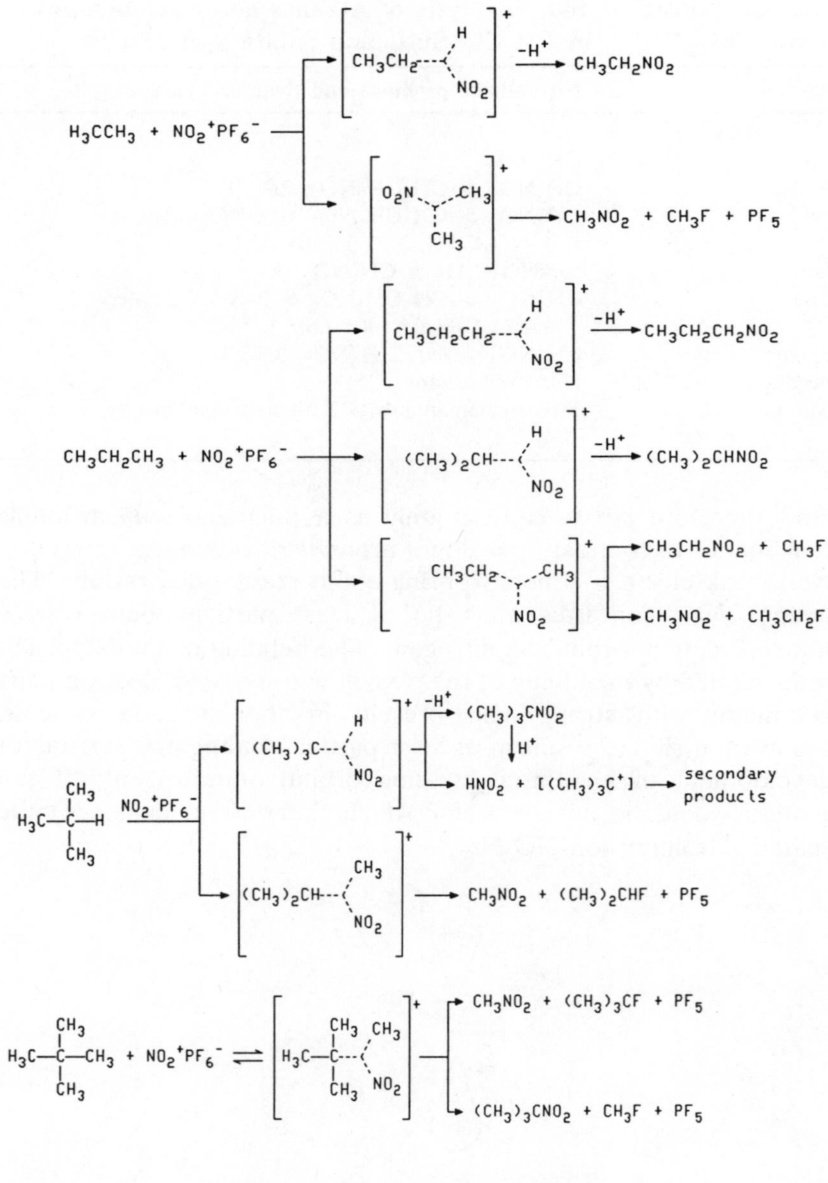

Scheme 2. NO_2^+ Nitration of Alkanes

would be fully bent with an empty atomic orbital on nitrogen, and could be considered as the ultimate electrophilic nitrating agent. The proto-solvated NO_2^+ ion may indeed be the reactive species formed in electrophilic nitrations (hence the higher reactivity of nitronium salts in superacid solutions). Whereas reaction at the C–H bond results in substitution of the nitro group for hydrogen, reaction on C–C bonds causes nitrolysis as shown in the reactions of ethane, propane, isobutane, and neopentane.

C–C bonds are generally more reactive than secondary or primary C–H bonds, leading to preferential nitrolysis of n-alkanes (see Table 52). Side products of the nitrolysis are methyl, ethyl, and isopropyl fluoride (formed by the reaction of PF_6^- with the cleaved alkylcarbenium ions) or secondary alkylation products, which by themselves are capable of undergoing reaction with the nitronium salt.

Tertiary C–H bonds show the highest reactivity. However, protolytic cleavage of tertiary and secondary nitroalkanes is a major side reaction, and can lead to the formation of a variety of by-products. Protolytic denitration was demonstrated by reacting 2-nitro-2-methylpropane with FSO_3H–SbF_3, HF–SbF_3, and HF–PF_5 at −80°C. The protolytic cleavage reaction yields t-butyl cation and HNO_2 (or subsequently, NO^+ ion). No NO_2^+ ion is formed, as shown by the fact that upon quenching of the reaction mixtures with benzene and toluene, no nitroaromatics were obtained. At the same time, products of t-butylation were observed.

$$(CH_3)_3CNO_2 \underset{}{\overset{H^+}{\rightleftharpoons}} [(CH_3)_3CNO_2H]^+ \longrightarrow (CH_3)_3C^+ + HNO_2$$

The steric requirements for the reaction with tertiary C–H bonds in alkanes with NO_2^+ have not yet been sufficiently studied. Adamantane is nitrated with nitronium salts in methylene chloride–sulfolane mixtures, to form 1-nitroadamantane in 10–30% yield (accompanied by 2-nitroadamantane in <1% yield). In purified nitromethane solution recently a 65% yield was obtained with $NO_2^+BF_4^-$[72b]. Because of the cage structure of the adamantane, geometric constraints indicate front-side attack by the NO_2^+ ion. The reaction therefore proceeds via an S_E2 like electrophilic substitution involving the σ-electron pair of the involved C–H bonds.

$$\text{adamantane} + NO_2{}^+PF_6{}^- \underset{-k_1}{\overset{k_1}{\rightleftharpoons}} \left[\text{adamantane-H-NO}_2 \right]^+ \xrightarrow[k_2]{-H^+} \text{1-nitroadamantane}$$

Probing the mechanism of the reaction a kinetic hydrogen isotope effect study of the rate of nitration of 1,3,5,7-tetradeuterioadamantane compared with that of light adamantane showed no primary kinetic hydrogen isotope effect (k_H/k_D)=1.06). This seems to rule out a single step reaction, indicating a carbocationic discrete intermediate. The formation of the intermediate (k_1) must be slow (or $-k_1$ comparable), while the rate of the proton elimination step (k_2) is fast and thus not rate determining.[126]

As discussed previously, substantially increased yields of nitromethane can be obtained by carrying out the nitration in anhydrous HF or HSO_3F. As no protolytic reaction of nitromethane occurs in HF under the reaction condition, no protolytic by-products are formed. However, during the nitration of isobutane in HF solution, 90% of the nitroalkane obtained consisted of 1-nitro-2-methylpropane. Since only traces of this isomer are formed in nitronium salt nitration in methylene chloride-sulfolane solution, this product must be formed from isobutylene, which in turn is formed either by the direct reaction between isobutane and HF/PF_5 or more probably from the protolytic cleavage of 2-nitro-2-methylpropane.

$$H_3C-\underset{CH_3}{\underset{|}{\overset{CH_3}{\overset{|}{C}}}}-H \begin{array}{c} \xrightarrow{NO_2{}^+} (CH_3)_3CNO_2 \xrightarrow{H^+} (CH_3)_3C^+ + HNO_2 \\ \\ \xrightarrow{H^+} (CH_3)_3C^+ + H_2 \xrightarrow{-H^+} (CH_3)_2C=CH_2 \end{array}$$

$$(CH_3)_2C=CH_2 \xrightarrow{NO_2{}^+} O_2NCH_2C^+(CH_3)_2 \xrightarrow{+H^-} O_2NCH_2CH(CH_3)_2$$

As mentioned the high reactivity of the NO_2^+ ion in FSO_3H solution is attributed to its protosolvation.

$$O=\overset{+}{N}=O \xrightarrow{FSO_3H} O=\overset{+}{N}=O\text{---}HOSO_2F \rightarrow \left[O=\overset{+}{N}=\overset{+}{O}\text{---}H\right] \longleftrightarrow \left[O=N\diagdown_{OH}\right]^{2+}$$

Fully protonated NO_2^+ ion, however is unstable, as indicated by Simonetta's quantum mechanical calculations.[75] Evidence for proto-solvation of the NO_2^+ ion by FSO_3H comes from infrared studies of these mixtures in arsenic trifluoride solution. The O–H stretching frequency of FSO_3H in AsF_3 is shifted from 3300 to 3265 cm^{-1} upon addition of 10% w/w of $NO_2^+PF_6^-$. Simultaneously, the O–H band broadens. The N=O stretching vibration at 2380 cm^{-1} does not shift or broaden significantly under these conditions.

As discussed nitronium salts react extremely readily with π-donor aromatic compounds, as well as with alkenes and alkynes (*vide infra*). They also show high reactivity towards n-donors, such as alcohols, ethers, amines, amides, imides, carbodiimides, oximes, hydrazones, sulfides, sulfoxides, halides, and phosphines. However their reactivity towards σ-donors is understandably much lower.

There exists also another reaction mode of the NO_2^+ ion towards σ-donors, where NO_2^+ acts as an oxidant, resulting in the formation of carbocations via hydride abstraction. Rate constants and efficiencies ($k_r/k_{collision}$) of the gas-phase hydride-transfer reactions from alkanes to NO_2^+ have been measured.[76]

$$NO_2^+ + RH \longrightarrow R^+ \quad HONO$$

The efficiency of hydride transfer to NO_2^+ is very low. However, gas-phase studies indicated disproportionately high efficiency for the abstraction of a tertiary hydrogen. The efficiency of hydride abstraction by NO^+ is much higher.

It has been demonstrated in solution-phase studies that carbenium ions might be formed directly from alkanes via formal hydride abstraction, instead of via nitration followed by proto-denitration.

$$RH + NO_2^+ \longrightarrow R^+ + HNO_2$$

$$RH + NO_2^+ \longrightarrow R\text{-}NO_2 + H^+ \longrightarrow R^+ + HNO_2$$

If the reaction is carried out in acetonitrile then the carbocation is converted into the corresponding amide by the Ritter reaction. 2-Methylbutane reacts with $NO_2^+BF_4^-$ in acetonitrile solution to provide the corresponding amide in moderate yield.

Similarly, bicyclo[2.2.2]octane yielded a mixture of acetamides in 75% yield on reaction with $NO_2^+BF_4^-$ in acetonitrile.

When the nitration of adamantane with $NO_2^+BF_4^-$ was carried out in acetonitrile N-(1-adamantyl)acetamide was obtained in 88% yield. Similarly, norbornane yielded N-(exo-2-norbonyl)acetamide in 77% yield. Bicyclo[2.2.2]octane gave, in 73% yield, the corresponding amides.

The reaction clearly proceeds via initial formation of a bicyclooctyl carbocation, which undergoes subsequent rearrangement. It is striking that the ratio of isomers formed is approximately the same as in the similar isomerization of the carbocation obtained in the solvolysis of bicyclo[2.2.2]octyl-2-p-bromobenzenesulphonate, which supports the carbocationic mechanism.[77,78]

From these studies, it is clear that the reaction medium plays a significant role in the nature of products formed in the reactions of alkanes with the NO_2^+ ion. In methylene chloride–sulfolane, nitromethane, or in FSO_3H, the pentacoordinate transition state leads, via deprotonation, to nitration. In the sufficiently nucleophilic acetonitrile, the solvent reacts with the incipient carbocation.

This dichotomy is quite evident in the reaction of cyclooctane with nitronium salts in trifluoroacetic acid. The products obtained were cyclooctyl trifluoroacetate, cyclooctyl nitrate, and nitrocyclooctane. Conversion of adamantane to 1-fluoroadamantane in 95% yield on reaction with $NO_2^+BF_4^-$ in pyridine polyhydrogen fluoride indicates that formation of the adamantyl cation by formal hydride abstraction is a significant alternative to the nitration–protodenitration pathway.

4.2.2.3 Nitroalkanes via Nitro–Desilylation and Nitro–Destannylation

Nitro–desilylation with $NO_2^+BF_4^-$ is a mild and efficient way to prepare aliphatic nitro compounds from readily available alkylsilanes.

Olah *et al.* have studied[79] the scope of the reaction. Tetramethylsilane reacts readily with $NO_2^+BF_4^-$ in sulfolane solution to give nitromethane in 80% yield.

$$(CH_3)_4Si + NO_2^+BF_4^- \longrightarrow CH_3NO_2 + (CH_3)_3SiF + BF_3$$

At higher temperature and excess $NO_2^+BF_4^-$, a further equivalent of nitromethane can be obtained.

$$(CH_3)_4SiF + NO_2^+BF_4^- \longrightarrow CH_3NO_2 + (CH_3)_3SiF_2 + BF_3$$

Nitrodesilylation probably proceeds via a pentacoordinate two-electron three-center bonded siliconium ion.

$$H_3C-\underset{CH_3}{\underset{|}{\overset{CH_3}{\overset{|}{Si}}}}-CH_3 + NO_2^+BF_4^- \longrightarrow \left[H_3C-\underset{CH_3}{\underset{|}{\overset{CH_3}{\overset{|}{Si}}}}\cdots\underset{CH_3}{\overset{NO_2}{\diagup}} \right]^+ BF_4^- \xrightarrow{-BF_3}$$

$$\left[H_3C-\underset{CH_3}{\underset{|}{\overset{F\diagdown CH_3\ NO_2}{\overset{|\ \diagup}{Si}}}}\cdots CH_3 \right] \longrightarrow CH_3NO_2 + (CH_3)_3SiF$$

Ethylsilanes react similarly but higher alkylsilanes gave only low yields of nitroalkanes accompanied by products of elimination and subsequent polymerization.

Allylsilanes react readily with nitronium salts to yield nitroalkenes.

$$CH_2=CH-CH_2-Si(CH_3)_3 \xrightarrow{NO_2^+BF_4^-} \underset{80\%}{CH_2=CH-CH_2-NO_2}$$

$$CH_2=C(CH_3)-CH_2-Si(CH_3)_3 \xrightarrow{NO_2^+BF_4^-} \underset{65\%}{CH_2=C(CH_3)-CH_2-NO_2}$$

$$CH_3CH=CH-CH_2-Si(CH_3)_3 \xrightarrow{NO_2^+BF_4^-} \underset{75\%}{CH_3-CH(NO_2)-CH=CH_2}$$

The reactions follow the addition–elimination course

$$R_3Si\text{-CH}_2\text{-CH=CH}_2(R^1) \xrightarrow{NO_2^+} R_3Si\text{-CH}_2\text{-CH}(R^1)\text{-CH}_2\text{-NO}_2^+ \xrightarrow{F^-} (R^1)\text{CH=CH-CH}_2\text{NO}_2$$

On the other hand, benzylsilanes do not undergo nitro–desilylation, but instead ring nitration takes place.

Although alkylsilanes undergo nitro–desilylation with $NO_2^+BF_4^-$, organostannanes, in preference, undergo one-electron oxidation.

$$RSn(CH_3)_3 + NO_2^+BF_4^- \longrightarrow [RSn(CH_3)_3^+ NO_2 \cdot BF_4^-\ ] \longrightarrow$$

$$R\cdot\ +\ \cdot NO_2 + (CH_3)_3SnBF_4$$

Tetramethylstannane reacts with nitronium salts in dichloromethane to form chloromethane, chloroethane, methane, and ethane. The products are indicative of the intermediacy of free radicals.[80]

Corey et al. have reported[81] facile nitro–destannylation of allyltrimethylstannane with tetranitromethane in DMSO.

$$\text{CH}_2\text{=CH-CH}_2\text{-Sn(CH}_3)_3 + C(NO_2)_4 \xrightarrow[17\,°C]{DMSO} \text{CH}_2\text{=CH-CH}_2\text{-NO}_2$$

56%

4.3 Nitration of Alkenes

Nitration of π–systems can be carried out easily not only in the case of aromatics, but unsaturated C–C bonds in aliphatic and cycloaliphatic systems can be readily nitrated by a variety of methods.

4.3.1 Nitric Acid

Nitration of the lower olefins with concentrated HNO_3[82,83] was considered to take place through an addition–elimination process.

$$\diagup\!\!\!\!\diagdown C=C \diagup\!\!\!\!\diagdown^H + HNO_3 \longrightarrow HO-\underset{|}{\overset{H}{C}}-\underset{|}{\overset{H}{C}}-NO_2 \longrightarrow \diagup\!\!\!\!\diagdown C=C \diagup\!\!\!\!\diagdown_{NO_2} + H_2O$$

It is also possible that initial attack by the NO_2^+ ion is followed by elimination

$$\diagup\!\!\!\!\diagdown C=C \diagup\!\!\!\!\diagdown^H + NO_2^+ \longrightarrow -\underset{|}{\overset{+}{C}}-\underset{|}{\overset{H}{C}}-NO_2 \xrightarrow{-H^+} \diagup\!\!\!\!\diagdown C=C \diagup\!\!\!\!\diagdown_{NO_2}$$

In the reaction of 1,1-diphenylethylene with HNO_3, some of the intermediate addition product (or its hydrolysis product) was indeed observed.[84]

$$\underset{H_5C_6}{\overset{H_5C_6}{\diagdown}}C=CH_2 \xrightarrow{HNO_3} HO-\underset{C_6H_5}{\overset{C_6H_5}{\underset{|}{\overset{|}{C}}}}-CH_2NO_2 + \underset{H_5C_6}{\overset{H_5C_6}{\diagdown}}C=CHNO_2$$

The reaction of cinnamic acid with HNO_3 gives, via decarbonylation, β-nitrostyrene or ring-nitrated products.[85]

$$\text{Ph}-CH=CH-CO_2H \xrightarrow{HNO_3} \text{Ph}-CH=CH-NO_2$$

It should, however, be emphasized that the double bonds in olefins are very readily oxidatively cleaved by HNO_3 in the liquid phase. Propylene for example gives oxalic acid with 70% HNO_3 and oxygen at

25–30°C. 2–Butylene gives 80% propionic acid (as well as some acetic and oxalic acid). Thus, mild reaction conditions as well as absence of oxygen are needed to minimize oxidations.[86]

Under mild conditions the nitration of carbon–carbon double bonds with HNO_3 is also applicable to the synthesis of nitro steroids.[87,88] Nitration of cholesterol acetate is effected by fuming HNO_3, and the yield of 6–nitrochloroesteryl acetate is 79%. In a similar way, Δ^4–cholestene affords the unsaturated nitro steroid in 75% yield.[87,89,90] The nitration products of Δ^7–, $\Delta^{9(11)}$–, $\Delta^{7,9(11)}$–unsaturated steroids are reported by Anagnostopoulos and Fieser.[91]

6-nitrocholesteryl acetate

nitro Δ^4-cholestene[1]

Nitration studies of steroids have been extended to the preparation of 6α– and 6β–nitro derivatives of testosterone, progesterone, and androstene–3,17–dione.[91] Nitration of pregnenolone acetate affords 6–nitropregnenolone acetate in 61% yield.

The nitration of unsaturated compounds with dilute HNO_3 is also known but tends to give substitution nitration, in contrast to thus-far discussed nitrations, which seem to proceed through addition-elimination pathway.

The nitration mechanism of olefins with dilute HNO_3 has been investigated by Petrov and Bulygina.[92] They concluded that these nitrations proceed by a radical mechanism similar to that suggested by Titov for the nitration of saturated compounds. As in the nitration of saturated compounds, the HNO_3 is only a source of the NO_2 radical. They also observed that nitration of olefins may be accompanied by a shift of the double bond (see subsequent discussion of nitration with NO_2).

When 1-octene is nitrated with 20% or 80% HNO_3 at 90–100°C, 1-nitro-2-octene is formed in 83% yield.[92]

$$CH_3(CH_2)_5CH=CH_2 \longrightarrow CH_3(CH_2)_4CH=CHCH_2NO_2$$

Nitration of 2-ethyl-1-hexene (possibly contaminated with some 2-ethyl-2-hexene with HNO_3 at 70–75°C resulted in an 86% yield of a nitro-2-hexene derivative.[92]

Isobutylene, nitrated with HNO_3 (specific gravity 1.52) forms the nitro derivative in 6–10% yield.[93]

$$(CH_3)_2C=CH_2 \longrightarrow (CH_3)_2C=CHNO_2$$

Butadiene reacted with HNO_3 (specific gravity 1.50) at −30°C to form a mononitro product, which could be the 1-nitro derivative.[94]

The nitration of 1,1-dichloro-1-pentene or 1,1,5-trichloro-1-pentene with HNO_3 (specific gravity 1.495) at about 60° resulted in a mixture of saturated compounds.[95] On the basis of the products formed, these reactions fit the general picture of radical reactions.

Cyclohexene does not react at 60–65° with HNO_3, having a specific gravity of 1.075 ($\approx$ 13 wt%).[96] With more concentrated acid (specific gravity 1.2 $\approx$ 33 wt%),) 1-nitrocyclohexene, secondary nitrocyclohexene (which is believed to be the allylic 3-nitrocyclohexene), cyclohexene-pseudonitrosite, and some adipic acid are obtained. The combined yield of nitro products is 16%.

The nitration of camphene with dilute HNO_3 at 105-110° gives a low yield of α- and ω-nitrocamphene.

4.3.2 Nitric-Sulfuric Acid and Nitric Acid-Anhydrous Hydrogen Fluoride

Nitration of highly fluorinated olefins was studied with strong nitrating acid mixtures, such as HNO_3/H_2SO_4 and HNO_3/HF.

Fluoronitroacetyl chloride, which can readily be converted to other derivatives, was prepared by the nitration of 1,1-dichloro-2-fluoroethylene.[86]

$$HCF=CCl_2 \xrightarrow[H_2SO_4]{HNO_3} HCFNO_2C(O)Cl$$

Knunyants[97a] and Titov[97b] achieved nitro-fluorination of some unbranched alkenes with HNO_3 in anhydrous HF solution. It is, however, usually inconvenient to use HF as a solvent, especially in the reaction of reactive, branched alkenes, which readily polymerize and undergo other side reactions in the system.

Hexafluoropropene reacts with HNO_3/HF to give a mixture of products[97c]

$$CF_2=CF-CF_3 \xrightarrow{HNO_3/HF} F_3C-\underset{\underset{NO_2}{|}}{C}F-CF_3 + F_3C-\underset{\underset{NO_2}{|}}{C}F-CF_2NO_2$$

$$\left(+ F_3C-\underset{\underset{NO}{|}}{C}F-CF_3 \right)$$

4.3.3 Acetyl Nitrate

In the reaction of olefins with HNO_3-Ac_2O acetyl nitrate is the active regent (caution: explosive). The reactions can also lead to secondary products, but in the presence of tertiary amines these are absent.

Nitration of many olefins have been studied,[98] and that of cyclohexene is only illustrative.[99,100]

$$\text{cyclohexene} \xrightarrow{ArONO_2} \text{(1-ONO}_2\text{, 2-NO}_2\text{-cyclohexane)} + \text{(1-OAc, 2-NO}_2\text{-cyclohexane)} + \text{(3-NO}_2\text{-cyclohexene)} + \text{(3-NO}_2\text{-cyclohexene isomer)}$$

The separation of products is rather difficult in these reactions. The reaction of acetyl nitrate with cyclopentene gives products that are analogous to those obtained from cyclohexene. Separation of the complex mixture of products, produced in the reaction of acetyl nitrate with olefins, poses a serious limitation to the method.

In the nitration of stilbenes and styrenes, mainly β-nitroacetates are obtained.[100]

$$Ph-CH=CH-Ph \xrightarrow{ArONO_2} Ph-CH(OAc)-CH(NO_2)-Ph$$

The yield of threo-1-acetoxy-1,2-diphenyl-2-nitroethane from the nitration of trans-stilbene is 70%, and α-methylstyrene is converted to 1-nitro-2-acetoxy-2-phenylpropane in 70% yield. High yields of β-nitroacetates are also obtained in the reactions of acetyl nitrate with 1,1-diarylalkenes.[101] The nitration of 1,1-diphenylethene gives 1-acetoxy-1,1-diphenyl-2-nitroethane in 70% yield. β-Nitroacetates, which are accessible by the nitration reactions, can undergo elimination to form allenes.

The reaction of acetyl nitrate with 1-phenylcyclohexene results in a mixture of β-nitroacetates in 65% yield.[102]

Acetyl nitrate is prepared *in situ* by the reaction of Ac_2O with HNO_3 at 15°C according to Bordwell and Garbisch.[98] Careful control of the exothermic reaction is essential, but acetyl nitrate is not formed at lower temperatures. A mole ratio of Ac_2O to 70% HNO_3 of about seven is necessary in order to avoid crystallization of AcOH at lower temperatures during the nitration. The olefin is introduced at −20 to −30°C. The reaction is complete within a brief period of time, and, with some olefins, acid catalysis is helpful.

The preparation of β-nitroacetates, the principal products of the nitration reaction, may be formulated as the result of Markownikov addition of $AcO-NO_2$[98] to carbon–carbon double bonds. The reaction is catalyzed by H_2SO_4, which suggests that the nitrating species is $(AcOHNO_2)^+$ or the NO_2^+ ion. Products that are the result of both cis and trans additions are obtained from a single olefin. In the case of cyclohexene, the structure of the numerous products imply the intermediacy of carbocations. The addition of acetyl nitrate to 1,1-diarylalkenes yields products that are characteristically derived from carbocations.[99] It was also suggested that the addition of acetyl nitrate to simple acyclic olefins involves a cyclic transition state.[98]

Acetyl nitrate also reacts with enol esters to produce nitro compounds. The reaction of 2-buten-2-yl acetate with acetyl nitrate yields 3-nitro-2-butanone (13%).[103] Acetyl nitrate and 1-phenyl-1-propen-1-yl acetate react to form β-nitropropiophenone in 21% yield. Though the initial report indicates that the yields of nitro ketones are low, acetyl nitrate and the enol acetate of cyclohexanone react to give 2-nitrocyclohexanone in 40% yield.[99]

4.3.4 Dinitrogen Tetroxide/Nitrogen Dioxide

N_2O_4 adds to olefins below room temperature.[104-106] The reaction, which is carried out under oxygen to eliminate NO, gives dinitro compounds and nitro–nitrites, which are further hydrolyzed to nitro alcohols in the presence of water or are oxidized to nitro–nitrates.

The reaction of olefins such as ethylene, propylene, isobutylene, and 1–butene with N_2O_4 were described as giving a mixture of nitro products. The reaction of cyclohexene with N_2O_4 at 0°C can serve as a representative example[59,109]

The orientation of addition observed with terminal olefins leading to nitro–nitrites shows that the C–N bond is formed at the olefinic carbon, which bears more hydrogen atoms and the C–O bond is formed at the carbon atom, which bears fewer hydrogen atoms, irrespective of the polarity of the R group.[104,105]

$$R-CH=CH_2 + N_2O_4 \longrightarrow R-CH(ONO)-CH_2NO_2$$

$$R = CH_3,^{104} \quad CF_3,^{105} \quad CO_2CH_3,^{107,108}$$

In some cases, the reaction of olefins with N_2O_4 also gives nitroso-nitrates.[109] Isobutylene, for example, reacts in the following way (further oxidation and hydrolysis can give methacrylic acid):

$$(CH_3)_2C=CH_2 + N_2O_4 \longrightarrow (CH_3)_2C(ONO_2)CH_2NO$$

The above data can be rationalized in accord with an ionic mechanism, i.e., N_2O_4 adding as $NO_2^+NO_2^-$ or $NO^+NO_3^-$ in an electrophilic fashion to the olefin.

Shechter and Conrad on the other hand showed that the addition of N_2O_4 to carbon–carbon double bonds can be a free–radical reaction.[110a,b] N_2O_4 adds to methyl acrylate to give three distinct products after hydrolysis and neutralization: methyl 3–nitroacrylate (13%), methyl 2–hydroxy–3–nitropropionate (27%), and oxalic acid dihydrate (up to 80%), and also nitrogen containing polymers of methyl acrylate. The reaction of N_2O_4 with olefins in the presence of excess iodine provides additional evidence for the free–radical nature of the reaction.[111] Iodine traps the intermediate β–nitroalkyl radical and β–nitroalkyl iodides are obtained in yields that range from 50–70%.

$$CH_3CH=CH_2 + \cdot NO_2 \rightarrow CH_3\dot{C}HCH_2NO_2 \rightarrow CH_3CHICH_2NO_2 + I\cdot$$

It should be pointed out, however, that the formation of β–nitro-iodoalkanes can also be explained on the basis of addition of NO_2^+ followed by reaction of the carbocation with I_3^-. Hassner's work points in the direction of an ionic mechanism.[110c] Similarly N_2O_4 also adds to fluoroethylenes to give the corresponding 1,2–dinitroethanes. Tetrafluoroethylene yields 1,2–dinitro–1,1,2,2–tetrafluoroethane.[110]

The above modification of the N_2O_4 reaction with olefins results in a reaction that is useful in organic synthesis. Nitroolefins are obtained from the dehydro–halogenation of β–nitroalkyl iodides by sodium acetate. 1,3–Butadiene gives 1,4–dinitro–2–butene.[112]

$$CH_2=CH-CH=CH_2 + N_2O_4 \longrightarrow O_2NCH_2CH=CH-CH_2NO_2$$

In the above discussed reaction, depending on the reaction conditions, both free-radical and polar (ionic) addition reacts of NO_2^+ to olefins are possible.[113]

It is clear from the complexity of the reaction products obtained that ionic and free-radical reactions can be superimposed on each other, with free-radical reactions[110,114,118] generally predominating.

More recently, Pryor et al.[119] carried out a detailed study of the reaction of NO_2 with alkenes, particularly cyclohexene, in a carrier gas (nitrogen, oxygen, or air) changing the concentration of NO_2 from 70 ppm to 50%. Based on product studies, they have found that NO_2 adds, at ambient temperature, to the double bonds of alkenes at high NO_2 concentrations (>1%). These results are in accord with previous investigations.[115,116] However, at low concentrations, NO_2 appears to abstract allylic hydrogen atoms in competition with the addition reaction. NO_2 is in equilibrium with its dimer with an equilibrium constant, K_{eq}, equal to about 10^{-4} M. Because dissociation and recombination are both rapid, compared with the reaction of NO_2 with alkenes, the equilibrium is maintained during the reaction; it is thought that the reaction involves the NO_2 radical.[115] Scheme 3 shows the reaction involved int he addition of NO_2 to olefins. The initial reaction is the reversible[120,121] addition to the double bond to form the carbon-centered radical 1. Radical 2 then reacts with more NO_2, N_2O_4,

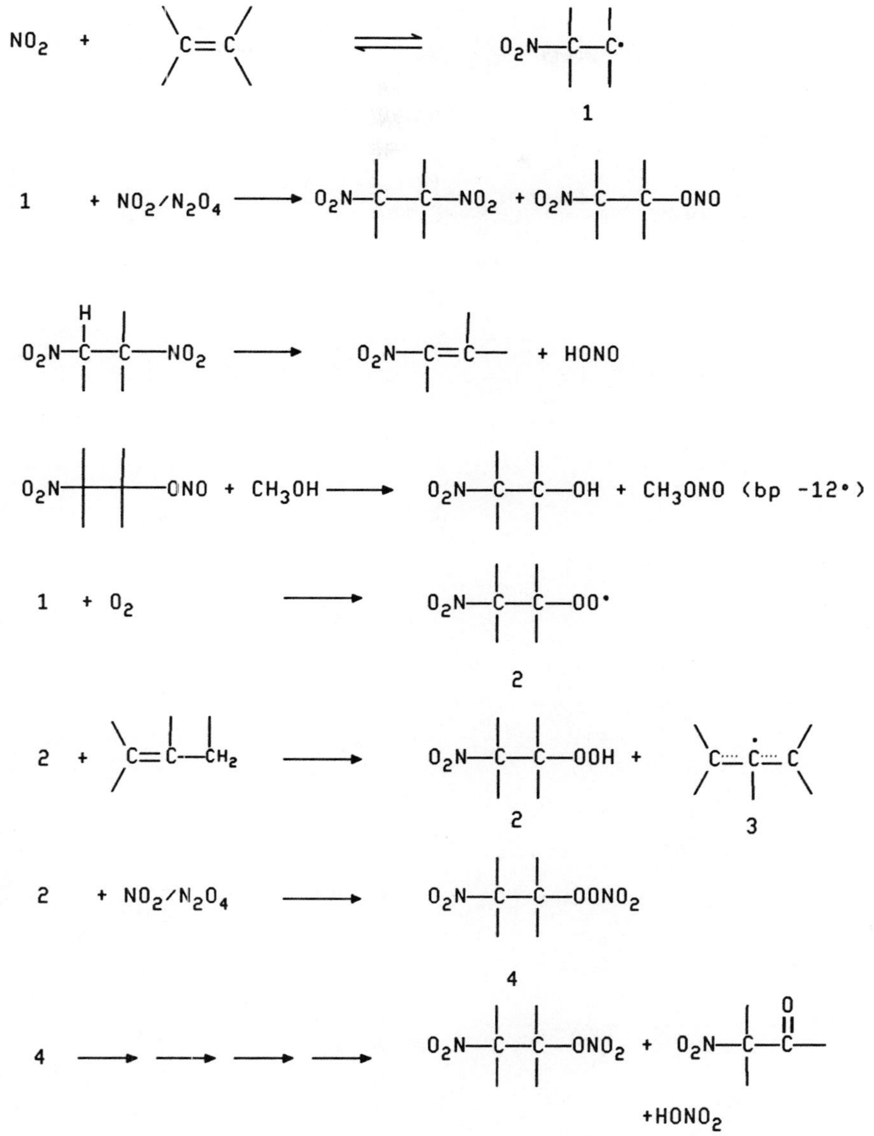

NO_2 + C=C ⇌ $O_2N-C-C\cdot$ **1**

1 + NO_2/N_2O_4 ⟶ $O_2N-C-C-NO_2$ + $O_2N-C-C-ONO$

$O_2N-\underset{|}{\overset{H}{C}}-C-NO_2$ ⟶ $O_2N-C=C-$ + $HONO$

$O_2N-C-C-ONO$ + CH_3OH ⟶ $O_2N-C-C-OH$ + CH_3ONO (bp $-12°$)

1 + O_2 ⟶ $O_2N-C-C-OO\cdot$ **2**

2 + $\text{C=C}-CH_2$ ⟶ $O_2N-C-C-OOH$ + $\text{C}\cdots\overset{\cdot}{\text{C}}\cdots\text{C}$
                                                                      2   **3**

2 + NO_2/N_2O_4 ⟶ $O_2N-C-C-OONO_2$ **4**

4 ⟶ ⟶ ⟶ ⟶ $O_2N-C-C-ONO_2$ + $O_2N-C-\overset{O}{\underset{||}{C}}-$ + $HONO_2$

Scheme 3. Addition mechanism of reaction of N_2O_4/NO_2 with alkenes.[119]

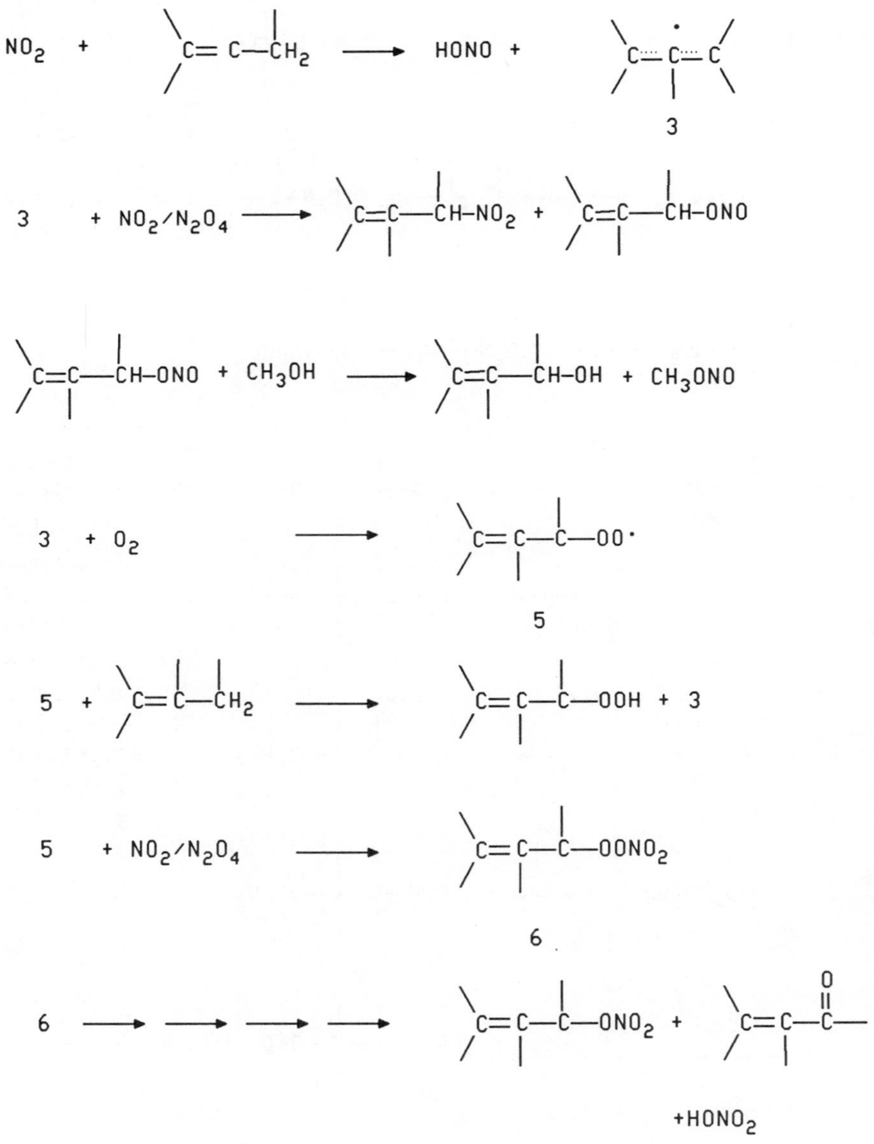

$$NO_2 + \;\;\diagdown\!\!\!\!\!\!\diagup\!\!\text{C}\!=\!\text{C}\!-\!\text{CH}_2 \longrightarrow HONO + \;\;\diagdown\!\!\!\!\!\!\diagup\!\!\text{C}\!\cdots\!\text{C}\!\cdots\!\text{C}\diagdown$$
                                                                                                                    3

$$3 + NO_2/N_2O_4 \longrightarrow \;\;\diagdown\!\!\!\!\!\!\diagup\!\!\text{C}\!=\!\text{C}\!-\!\text{CH}\!-\!NO_2 + \;\;\diagdown\!\!\!\!\!\!\diagup\!\!\text{C}\!=\!\text{C}\!-\!\text{CH}\!-\!ONO$$

$$\diagdown\!\!\!\!\!\!\diagup\!\!\text{C}\!=\!\text{C}\!-\!\text{CH}\!-\!ONO + CH_3OH \longrightarrow \;\;\diagdown\!\!\!\!\!\!\diagup\!\!\text{C}\!=\!\text{C}\!-\!\text{CH}\!-\!OH + CH_3ONO$$

$$3 + O_2 \longrightarrow \;\;\diagdown\!\!\!\!\!\!\diagup\!\!\text{C}\!=\!\text{C}\!-\!\text{C}\!-\!OO\cdot$$
                                                                                                                     5

$$5 + \;\;\diagdown\!\!\!\!\!\!\diagup\!\!\text{C}\!=\!\text{C}\!-\!\text{CH}_2 \longrightarrow \;\;\diagdown\!\!\!\!\!\!\diagup\!\!\text{C}\!=\!\text{C}\!-\!\text{C}\!-\!OOH + 3$$

$$5 + NO_2/N_2O_4 \longrightarrow \;\;\diagdown\!\!\!\!\!\!\diagup\!\!\text{C}\!=\!\text{C}\!-\!\text{C}\!-\!OONO_2$$
                                                                                                                     6

$$6 \longrightarrow \longrightarrow \longrightarrow \;\;\diagdown\!\!\!\!\!\!\diagup\!\!\text{C}\!=\!\text{C}\!-\!\text{C}\!-\!ONO_2 + \;\;\diagdown\!\!\!\!\!\!\diagup\!\!\text{C}\!=\!\text{C}\!-\!\overset{O}{\overset{\|}{\text{C}}}\!-$$

$$+HONO_2$$

Scheme 4. Hydrogen-abstraction mechanism of reaction of N_2O_4/NO_2 with alkenes.[119]

or O_2 (if air is present). In the absence of O_2, radical **1** reacts with NO_2 to give a nitro-nitrite, 1,2-dinitroalkane, or 1-nitroalkene, (subsequent workup of the reaction mixture with methanol converts the nitrite ester to the nitro alcohol[115]). If oxygen is present, radical **1** can give the peroxy radical **2**. Radical **2** either abstracts an allylic hydrogen atom to give a 2-nitro hydroperoxide and the allyl radical **3**, or is trapped by NO_2 to form an alkyl peroxynitrate **4**. Alkyl peroxynitrates are short-lived intermediates at ambient temperature and ultimately from a nitro-nitrate or nitro ketone, depending on structure and reaction conditions.[122]

In contrast, below 1% NO_2 concentration, hydrogen abstraction from the alkene by NO_2 dominates as reflected by product analyses. The initial products of hydrogen abstraction by NO_2 form an alkene are HNO_2 and allyl radical **3**. (The HNO_2 decomposes to form water and nitrogen oxides.[123]) The reactions of **3** (and workup with methanol) produce allylic substitution products, as shown in Scheme 4.

Competition is possible between the thermodynamically favored but reversible addition reaction and the irreversible hydrogen abstraction (the HNO_2 formed either decomposes or is removed by the carrier gas from the reaction mixture). At higher concentration of NO_2 it acts as a radical trapping agent to give the addition product. At low concentration the initial carbon radical is trapped more slowly and a larger fraction of the radical can revert to olefin. Therefore, less addition occurs and hydrogen abstraction can become favored. This can be illustrated with oxygen as a radical trapping agent.

$$NO_2 + \mathrm{-C=C-} \underset{k_2}{\overset{k_1}{\rightleftharpoons}} \mathrm{-C-C\cdot} \xrightarrow[O_2]{k_3} O_2N\mathrm{-C-C-OO\cdot}$$

The situation is similar to the well-known reaction of bromine atoms with olefins.[124-128]

4.3.5 Dinitrogen Trioxide

Nitro products are also obtained by the addition of N_2O_3 to carbon-carbon double bonds. Cinnamaldehyde reacts with N_2O_3 to form mainly 4-nitro-3-phenylisoxazole.[129]

$$C_6H_5CH=CHCHO \xrightarrow{N_2O_3} \left[\begin{array}{c} C_6H_5CH-CHCHO \\ | \quad | \\ ON \quad NO_2 \end{array} \to \begin{array}{c} C_6H_5CH-CNO_2 \\ \| \quad \| \\ HON \quad CHOH \end{array} \right]$$

$$\longrightarrow \underset{\substack{\text{isoxazole ring with } H_5C_6 \text{ and } NO_2}}{}$$

Benzalacetone reacts to form a nitro–nitroso dimer.

$$C_6H_5CH=CH-\overset{O}{\overset{\|}{C}}CH_3 \xrightarrow{N_2O_3} \left[\begin{array}{c} C_6H_5CH-CH-\overset{O}{\overset{\|}{C}}CH_3 \\ | \quad | \\ ON \quad NO_2 \end{array} \right]_2$$

p-Anisalacetophenone reacts with N_2O_3 to give a nitro–nitroso dimer in 15% yield.[130]

$$H_3CO-C_6H_4-CH=CH-\overset{O}{\overset{\|}{C}}-C_6H_5 \xrightarrow{N_2O_3}$$

$$\left[H_3CO-C_6H_4-\underset{\substack{| \quad | \\ NO \quad NO_2}}{CH-CH}-\overset{O}{\overset{\|}{C}}-C_6H_5 \right]_2$$

The addition of N_2O_3 to the anethole analog also forms a nitro–nitroso dimer.

$$\underset{\substack{| \\ OCH_3}}{C_6H_4}-CH=CHCH_3 \xrightarrow{N_2O_3} \left[CH_3O-C_6H_4-\underset{\substack{| \quad | \\ NO \quad NO_2}}{CH-CH}-CH_3 \right]_2$$

Styrene reacts with N_2O_3 to form a nitroso dimer,[131] which is readily converted to ω-nitroacetophenone oxime and ω-nitroacetophenone.[132] Camphene reacts with oxides of nitrogen to form ω-nitrocamphene in 40% yield.[133]

4.3.6 Dinitrogen Pentoxide

N_2O_5 can add to olefins according to its ionic nature, i.e., $NO_2^+NO_3^-$ giving nitrates as well as nitro compounds.[134]

In contrast the thermolysis of N_2O_5 can form NO_2 and NO_3 and thus involve radical nitrations.

4.3.7 Nitryl Halides

Nitro–fluorination of alkenes is obtained on reaction with nitryl fluoride.

$$R\text{-}CH=CH_2 + NO_2F \longrightarrow R\text{-}CHF\text{-}CH_2NO_2$$

Addition of nitryl chloride (NO_2Cl) to olefins gives, besides chloronitro products, dichloroalkanes and chloroalkyl nitrites.[135]

$$H_2C=CH_2 \xrightarrow{NO_2Cl} ClCH_2\text{-}CH_2\text{-}NO_2 + ClCH_2\text{-}CH_2\text{-}NO + ClCH_2\text{-}CH_2\text{-}Cl + NO_2$$

Products indicate both heterolytic and homolytic (i.e., free–radical) reactions.

Nitroketones are obtained by the reaction of NO_2Cl with enol esters. The yields vary considerably, and competing chlorination may occur[136] (Table 53). NO_2Cl appears to be less useful than acetyl nitrate in the conversion of enol esters to nitro compounds.

Table 53. Nitration of Enol Acetates with Nitryl Chloride[136]

Enol acetate	Product	Yield (%)
$(CH_3)_2C=CHOCOCH_3$	$(CH_3)_2CClCHO$	21
	CH_3CNO_2CHO	12
$(CH_3)CH=C(CH_3)OCOCH_3$	$CH_3CH(NO_2)CH_3$	36
$CH_2=C(C_6H_5)OCOCH_3$	$C_6H_5COCH_2NO_2$	36
$CH_3CH=C(C_6H_5)OCOCH_3$	$C_6H_5COCH(NO_2)CH_3$	28

NO$_2$Cl adds to unsymmetrical terminal olefins to yield 1-nitro-2-chloroalkanes.[137] Vinyl bromide and NO$_2$Cl react to give 1-bromo-1-chloro-2-nitroethane in 85% yield.[138] NO$_2$Cl converts cyclohexene[139] to 1,2-dichlorocyclohexane, cyclohexene pseudonitrosite, and 1-chloro-2-nitrocyclohexane (33%).

A mixture that consists of methyl 2,3-dichloropropionate (7%), methyl 2-chloro-2-nitropropionate (75%), and dimethyl 2-chloro-4-nitromethylpentanedioate (5–10%) is obtained from the reaction of NO$_2$Cl and methyl acrylate at 0°C.[135] The suggestion was made that the addition of NO$_2$Cl to a carbon–carbon double bond is a homolytic process in which NO$_2$, formed from NO$_2$Cl, attacks the terminal position of the double bond, forming a C–N bond exclusively. The intermediate radical reacts with excess NO$_2$Cl to form a chloronitro compound, or with methyl acrylate and another radical to complete the reaction. At the same time in the presence of Lewis acid catalyst, electrophilic addition can take place (see nitration with nitronium salts).

Cholesteryl acetate was reported to react with nitrosyl chloride[140,141a] to give the 5-chloro-6-nitro compound in 92% yield. This surprising result can be probably explained by the observation[141b] that the reaction with pure NO$_2$Cl is very slow.

Thus reaction with NOCl and subsequent oxidation is indicated. In the presence of NO_2 the nitrochloride is formed rapidly, probably by a free-radical oxidative mechanism. Two additional examples of the reaction with unsaturated steroids were reported[140] with yields over 80%.

1-Chloro-2-nitro-1,1,2,2-tetrafluoroethane is obtained from the reaction of tetrafluoroethylene with NO_2Cl or with NOCl.

$$\diagdown\!\!\!\!C\!\!=\!\!C\!\!\diagup + \cdot NO_2 \longrightarrow \diagdown\!\!\!\!\overset{\cdot}{C}\!\!-\!\!\underset{NO_2}{C}\!\!- \xrightarrow{NOCl} -\!\!\underset{Cl}{C}\!\!-\!\!\underset{NO_2}{C}\!\!-$$

Recently two significantly improved methods of nitration of alkenes via halonitration with *in situ* formed nitryl halides were reported.

Jew et al.[142] found an improved way for the preparation of nitroalkenes by reacting alkenes with sodium nitrite and iodine in ethyl acetate (or ether) and water in the presence of ethylene glycol (or propylene glycol). The yield of the conjugated nitroalkenes obtained, such as 1-nitrocyclohexene, $C_6H_5CH=CHNO_2$, $C_6H_5OCOCH=CHNO_2$ were 72, 81, and 82%, respectively. Intermediate iodonitroalkane formation was detected in the reaction. In the absence of diols, iodohydrins are obtained as the main product. The diol prohibits formation of hypoiodous acid, but accelerates that of nitryl iodide, which then adds to the alkene.

Vankar and Kumaravel[143] have found that ammonium nitrate in trifluoroacetic anhydride (Crivello's reagent), in the presence of ammonium bromide followed by treatment with triethylamine, is a very convenient system for converting alkenes into conjugated nitroalkenes. The reaction generally gives good yields and is of broad scope.

$$\underset{}{\bigcirc} \xrightarrow[2)\ NH_4Br]{1)\ NH_4^+NO_3^-\ (CF_3CO)_2O} \underset{\text{"'Br}}{\bigcirc\!\!\!\!\!\nearrow^{NO_2}} \xrightarrow{NEt_3} \underset{}{\bigcirc\!\!\!\!\!\!\nearrow^{NO_2}}$$

65%

4.3.8 Nitroalkenes via Nitro-Mercuration and Nitro-Selenation

Corey[144] developed a nitro-mercuration procedure reacting alkenes, such as cyclohexene, with mercuric chloride and sodium nitrite in aqueous solution. The intermediate nitro-mercurial is then subjected to base-catalyzed elimination to give the nitroalkene

Tomoda,[145] as well as Seebach,[146] has utilized selenium chemistry for preparing nitroalkenes. Cyclohexene, for example, reacts with phenylselenyl bromide, silver nitrite, and mercuric chloride, to give the corresponding nitrophenylselenylalkane. On oxidation with H_2O_2 the nitroalkene is obtained.

4.3.9 Tetranitromethane and Hexanitroethane

The reaction of tetranitromethane with unsaturated hydrocarbons and some of their derivatives was first carried out by Ostromyslenkskii[147a] and by Werner.[147b] They showed qualitatively that tetranitromethane reacts with olefinic bonds.

Titov suggested[147c] that the reactions follow an ionic mechanism with the initial formation of a π-complex. There is, however, insufficient proof to exclude a radical or electron-transfer mechanism.

Schmidt et al.[148a] have carried out a more detailed study of the nitration of arylalkenes with tetranitromethane. Double bonds, conjugated with the aromatic ring, were found to be readily nitrated. Isosafrol for example gave β-nitrosafrol in 72% yield. Hexanitroethane was also used in related nitrations.[148b] Although the method was reviewed[149] it is clear that further studies are needed to gain a better understanding of its scope and mechanism.

4.3.10 Nitronium Salts

The reaction of nitronium salts with olefins depends on the nature of the alkene and the reaction conditions. When excess alkenes react with nitronium salts, a reactive nitrocarbenium ion is formed, which can initiate polymerization.[150]

$$C_2H_5CH=CH_2 + NO_2^+BF_4^- \longrightarrow C_2H_5CHCH_2NO_2BF_4^- \xrightarrow{n(C_2H_5CH=CH_2)}$$

$$C_2H_5CH(CH_2NO_2) - [(CH_2-CH(C_2H_5))] + HBF_4$$

The nitrocarbenium ion can also be stabilized as a result of an intramolecular rearrangement to a nitrocarboxonium ion. The latter reacts with water to form the corresponding ketone.[151]

Olah and Nojima have shown[152] that when alkenes are allowed to react with equimolar $NO_2^+BF_4^-$ in 70% hydrogen fluoride/pyridine nitrofluorination takes place.

$$RCH=CH_2 + NO_2^+BF_4^- \xrightarrow[-BF_3]{\text{pyridinium (HF)}xF^-} RCFH-CH_2-NO_2$$

The nitro-fluorinated adducts can be obtained in good yield (Table 54). β-Nitrofluoroalkanes are useful intermediates in the synthesis of

Table 54. Nitrofluorination of Alkenes with Nitronium Tetrafluoroborate in Pyridinium Polyhydrogen Fluoride

Alkene	Reaction temperature	Reaction time (hr)	Product	Yield (%)
Ethene	20	1	1-Fluoro-2-nitroethane	60
Propene	20	1	2-Fluoro-1-nitropropane	65
2-Butene	20	0.5	2-Fluoro-3-nitrobutane	60
1-Hexene	0	1	2-Fluoro-1-nitrohexane	65
Chloroethene	20	2	1-Chlor-1-fluoro-2-nitroethane	40
1,1-Dichloroethene	20	2	1,1-Dichloro-1-fluoro-2-nitroethane	45
Cyclohexene	0	1	1-Fluoro-2-nitrocyclohexane	70
	20	0.3		80

β-fluoroalkylamines and, via dehydrofluorination with bases, give nitroalkenes.

Chloro- and bromoalkenes, giving less stabilized nitrocarbenium ions, do not require HF for the formation of β-nitrofluoroalkanes. With these alkenes, BF_4^- seems to function as an efficient source of F^-.

Extensive studies of the $NO_2^+BF_4^-$ nitration of alkenes were reported in the Russian literature and an excellent review has appeared,[153] which allows discussion of otherwise difficult to access Soviet work.

If the nitration of olefins by nitronium salts is carried out in acetonitrile, the nitrocarbenium ion intermediates undergo Ritter reaction with the solvent to form nitroacetamides.[154]

$$R^1R^2C=CHR^3 + NO_2^+BF_4^- \xrightarrow[-15°C]{CH_3CN} R^1R^2C^+CHR^3NO_2BF_4^-$$

$$\downarrow CH_3CN$$

$$R^1R^2CCHR^3NO_2 \xleftarrow{H_2O} R^1R^2CCHR^3NO_2$$
$$\quad | \qquad\qquad\qquad\qquad\qquad\quad +|$$
$$NHCOCH_3 \qquad\qquad\qquad\qquad N{\equiv}CCH_3$$

a) $R^1 = CH_3$, $R^2 = R^3 = H$, yield 50%
b) $R^1 = R^2 = CH_3$, $R^3 = H$, yield 23%
c) $R^1 = R^3 = CH_3$, $R^2 = H$, yield 13%

The mode of the reaction between olefins and nitronium salts depends on the nature of the olefin and the reaction conditions. For example, cyclohexene reacts with $NO_2^+BF_4^-$ to form a nitrocarbocation, which leads to the formation of 3-nitrocyclohexene (yield 40%).[155,156]

The nitration of 1-substituted cyclohexenes is accompanied by the formation of 2-fluoro-1-nitro-2-R-cyclohexanes.[157]

(a) R = CH_3, yield 30%[157]; (b) R = Cl, yield 60%[158]

Alkenes with a reduced electron density, e.g., 3-methyl-2,5-dihydrothiophen 1,1-dioxide, react with $NO_2^+BF_4^-$ to give

The reaction of 3,4-dimethyl-2,5-dihydrothiophen 1,1-dioxide, in which the double bond is blocked by the methyl groups, apparently proceeds via an electrophilic substitution mechanism[159]

As mentioned, the reaction of nitronium salts with olefins proceeds via the formation of a nitrocarbocation and the decisive factor that determines the success of the reaction is the stabilization of the nitrocarbocation and the ease of its conversion into the final product. One of the methods applied involves nitration in acetic anhydride. It has been suggested that the complex $NO_2^+BF_4^- \cdot Ac_2O$ reacts and that 2-acetoxy-1-nitroalkane is obtained in 36–60% yield after hydrolysis of the reaction mixture.[160] The reaction involves mainly cis–addition:

$$R^1R^2C=CHR^3 + NO_2^+BF_4^- \xrightarrow{Ac_2O, H_2O} AcOR^1R^2CHR^3NO_2$$

a) $R^1=CH_3$, $R^2=R^3=H$; b) $R^1=R^3=CH_3$ (cis, trans), $R^2=H$

c) $R^1=R^2=CH_3$, $R^3=H$

4-Acetoxy-4-methyl-2-pentanone was isolated together with the nitroacetate in the nitration of isobutylene, which indicates the formation of an acylium cation in the reaction.

The nitration of cycloalkenes is as a rule accompanied by the formation of the β and γ -nitroacetates in 18–29% yields.[161]

The high degree of stabilization of the nitrocarbocation in the nitration of norbornene in acetic anhydride ensures a high (85%) yield of 2-acetoxy-7-nitrobicyclo[2.2.1]heptane.

Overall the nitration of olefins, particularly those containing electron–donating substituents, by nitronium salts is complicated by side reactions and does not always lead to the expected result. In many instances nitration with nitronium tetrafluoroborate takes place with higher yields in the presence of α-picoline.[158,162] In this case the nitrating agent is apparently not the nitronium salt but 1-methyl-N-nitropyridinium tetrafluoroborate, which is formed rapidly when $NO_2^+BF_4^-$ is mixed with α-picoline.[163]

The interaction of nitronium salts with the heteropolar C = N double bond in nitroalkane salts is of great interest. Olsen *et al.* showed that gem-dinitro-compounds are formed in the nitration of 2-nitropropane and nitrocyclohexane salts with $NO_2^+BF_4^-$ in acetonitrile[164]

$$R^1R^2C=NO_2M + NO_2^+BF_4^- \longrightarrow R^1R^2C(NO_2)_2 + MBF_4$$

(a) M = Li, (b) M = Na, (c) M = K

N-Nitropyridinium tetrafluoroborate salts, which are very mild nitrating agents, can also be used for the nitration of mononitroalkane salts.[165,166]

4.3.11 Nitroalkenes via Nitro-Destannylation and Nitro-Desilylation

Nitrocycloalkenes can be obtained in good yield by nitro-destannylation of the corresponding cycloalkenyltrimethylstannanes with tetranitromethane in DMSO.[167,81] Nitro-desilylation and nitro-destannylation of alkenylsilanes and alkenyl stannanes has proved to be relatively difficult to achieve, to contrast with the corresponding reaction of alkyl, allyl, and alkynylsilanes, and alkynylstannes.

So far, there are no reports of a successful nitro–desilylation of vinylsilanes. However, cycloalkenyltrialkylstannanes undergo nitro-destannylation rather readily with tetranitromethane in DMSO and HMPA. This reaction provides a high yield method for the conversion of cycloalkanones to nitrocycloalkenes.

Attempted nitro–destannylation with nitronium salts led to complex mixtures of products. Other milder nitrating agents were not effective.

4.4 Nitration of Alkynes

Acetylene with mixed acid (HNO_3–H_2SO_4) gives tetranitromethane.[168-190] The reaction with HNO_3 leads to trinitromethane which can be further nitrated to tetranitromethane.[171]

A complex mixture of products is obtained by the addition of N_2O_4 to acetylene.[172] N_2O_4 reacts with 3-hexyne to form five products; cis- and trans-3,4-dinitro-3-hexene (4.5 and 13%, respectively) and 4,4-dinitro-3-hexanone (8%) are the only nitro compounds that were found. The other products are propionic acid (6%) and dipropionyl (16%).

Addition of NO_2Cl to acetylenes leads to 1-chloro-2-nitroalkenes.[173,174] Nitroacetylenes are very explosive and little studied, although the nitration of alkynes with N_2O_4 and NO_2Cl has been reviewed.[149]

Britelli and Boswell[175] have shown that $NO_2^+BF_4^-$ reacts with

2-acetoxy-2-ethynyladamantane in an unusual reaction to yield a furoxan in 89% yield.

The acetoxy group is necessary for the reaction to occur. The following mechanism has been suggested for the reaction.

Schmitt et al.[176,177] obtained in high yield nitroacetylenes by the nitro-desilylation of trimethylsilylacetylenes.

$$(CH_3)_3Si-C\equiv C-Si(CH_3)_3 \xrightarrow[-(CH_3)_3SiF]{NO_2^+BF_4^-} (CH_3)_3Si-C\equiv C-NO_2$$

1,2-bis(trialkylsilyl) substituted unsymmetrical acetylenes showed high regioselectivity in the nitro-desilylation reaction. The regioselectivity being determined by the ease of attack of F⁻ at the less hindered silyl substituent. Table 55 summarizes the data.

Nesmeyanov et al.[178] and Jäger et al.[179] described the synthesis of nitroacetylenes by the nitro-destannylation of alkynylstannanes.

$$R_3SnC\equiv C-R' + NO_2^+BF_4^- \longrightarrow O_2N-C\equiv C-R' + (CH_3)_3SnBF_4$$

Table 55. Nitroacetylenes via Nitro–Trimethylsilylation with Nitronium Tetrafluoroborate and Hexafluorophosphate[176]

Nitroacetylene Starting material	Product	Yield (%)
$(CH_3)_3SiC\equiv CSiCH_3$	$(CH_3)_3SiC\equiv CNO_2$	70
$(CH_3)_3SiC\equiv CSi(CH_3)_2-CH(CH_3)_2$	$(CH_3)-CH(CH_3)_2SiC\equiv CNO_2$	34
	$(CH_3)_3SiC\equiv CNO_2$	6
$(CH_3)_3SiC\equiv CSi(CH_3)_2-C(CH_3)_3$	$(CH_3)_3-C-(CH_3)_2SiC\equiv CNO_2$	59
	$(CH_3)_3SiC\equiv CNO_2$	29
$(CH_3)_3SiC\equiv CSi[CH(CH_3)_2]_3$	$[CH(CH_3)_2]_3SiC\equiv CNO_2$	57

Petrov et al. reported the nitro–destannylation of trimethylstannyl-acetylenes with N_2O_4, albeit in low yield

$$R-C\equiv C-Sn(CH_3)_3 + N_2O_4 \longrightarrow R-C\equiv C-NO_2$$

$$R = C_6H_5, C(CH_3)_3, (CH_3)_3Si$$

Nitro–destannylation with N_2O_5 yields the corresponding nitroacetylenes in acceptable yields.

4.5 Nitration at Heteroatoms

Besides C–nitration of hydrocarbons (alkanes, alkenes, and alkynes) heteroatom nitrations such as O–nitration of alcohols, diols, and polyols, N–nitration of amines, amides and other nitrogen compounds, as well as S– and P–nitrations play a significant role in aliphatic nitrations.

4.5.1 Nitration at Oxygen

O–Nitration of alcohols can be considered as an esterification of HNO_3

$$ROH + HNO_3 \rightleftharpoons RONO_2 + H_2O$$

The reaction accordingly can be carried out with nitric acid alone or more frequently with mixed acid (mixture of nitric and sulfuric acids).[180] In the latter case the nitrating mixture consisting of nearly equal volumes of concentrated HNO_3 and H_2SO_4 acid is first treated with urea or urea nitrate to remove any HNO_2 present, which would give alkyl nitrites and render the product unstable. The esterification is generally carried out in the cold by slow addition of the alcohol to the mixed acid. Good stirring and careful temperature control are needed to minimize oxidative side reactions. The alkyl nitrates separate from the acid mixture upon pouring the reaction mixture into cold water. When using HNO_3 alone in the reaction, anhydrous acid is used up as water formed in the reaction dilutes the acid. These methods give satisfactory yields of primary and secondary alkyl nitrates, but not of tertiary ones, which are not stable under the conditions.[181]

O-Nitration of alcohols can also be carried out with acetyl nitrate. The nitration of alcohols is of particular interest for the preparation of polyol esters, such as nitroglycerin (glyceryl trinitrate) and related glycol nitrates, used as explosives.[182,183] Nitrocellulose (cellulose nitrate) is similarly produced.

Alkyl nitrates are also commonly prepared by the nucleophilic metathetic reaction of alkyl halides with silver nitrate

$$RX + AgNO_3 \longrightarrow RONO_2 + AgX$$

$$X = Cl, Br, I$$

The reaction can be carried out under heterogeneous conditions with powdered silver nitrate or in homogeneous acetonitrile solution, in which silver nitrate is soluble. This method also allows preparation of tertiary alkyl and benzylic nitrates.

Concerning the mechanism of the acid-catalyzed nitration of alcohols, Ingold extended his general scheme of aromatic electrophilic C-nitration to the nitration at oxygen (and nitrogen) centers[184]

$$HNO_3 + HNO_3 \underset{-k_1}{\overset{k_1}{\rightleftharpoons}} H_2NO_3^+ + NO_3^- \quad \text{(fast)}$$

$$H_2NO_3^+ \xrightarrow{-k_1} NO_2^+ + H_2O \quad \text{(slow)}$$

$$NO_2^+ + XH \longrightarrow XH \cdot NO_2^+ \quad \text{(slow)}$$

$$XH \cdot NO_2^+ + NO_3^- \longrightarrow XNO_2 + HNO_3 \quad \text{(fast)}$$

$$X = RO^-, R_2N^-, \text{etc.}$$

While studying the O-nitration of alcohols, glycols, and glycerin with excess HNO_3 in nitromethane solution, Ingold et al.[185] found the reactions to be of zero order and identical in rate with one another. For the nitration of methyl alcohol small concentrations of H_2SO_4 increased, whereas nitrate ion decreased the rates. When a sufficient amount of water was added, the kinetics changed to first order. Clearly the formation of the NO_2^+ ion is rate limiting in nitration in the absence of significant amounts of water. O-(and also studied N-)Nitrations thus show close similarity to electrophilic aromatic

C-nitrations with NO_2^+ ion.

Indeed, a significant improvement in the preparation of alkyl nitrates was achieved by Olah et al.[186] who applied stable nitronium salts such as $NO_2^+BF_4^-$ in their preparation.

$$ROH + NO_2^+BF_4^- \longrightarrow RONO_2 + HF + BF_3$$

The reaction gives high (frequently nearly quantitative) yields of primary and secondary alkyl nitrates (Table 56).

When N-nitropyridinium tetrafluoroborate was used,[187] O-nitration of alcohols was achieved under acid-free conditions

$$\text{Py}^+\text{-NO}_2\,BF_4^- + ROH \longrightarrow RONO_2 + \text{Py}^+\text{-NH}\,BF_4^-$$

The method was further improved using N-nitrocollidinium tetrafluoroborate as transfer nitrating agent for alcohols (polyols).[186,188]

Alcohols undergo transfer nitration with N-nitrocollidinium tetrafluoroborate under essentially neutral conditions. Yields were found to be close to quantitative. Thus, ethanol, ethylene glycol, and 1,4-butanediol showed complete conversion to the nitrate esters after 1 hr. Nitration of glycerol was complete in 3 hr. Separation of alkyl nitrates by distillation or crystallization gave good to excellent preparative yields (Table 57).

Table 56. Preparation of Alkyl Nitrates from Alcohols and $NO_2^+BF_4^-$ [186]

R	Percent yield (isolated)
CH_3	87
C_2H_5	92
$n\text{-}C_3H_7$	87
$n\text{-}C_4H_9$	94
$n\text{-}C_9H_{17}$	86
C_2H_4F	88
C_2H_4Cl	85
C_2H_4Br	72
CF_3CH_2	72

Table 57. Nitrate Esters from Alcohols with N–Nitrocollidinium Tetrafluoroborate[188]

Nitrate ester	Yield (%)
$C_2H_5-O-NO_2$	100
(phenyl)–CH_2-O-NO_2	78
$n-C_6H_9-O-NO_2$	51
$H_3C-CH_2OCH(CH_3)-O-NO_2$	38
1-adamantyl–$O-NO_2$	82
(endo-bornyl)–$O-NO_2$	48
(exo-bornyl)–$O-NO_2$	63
1,1'-bicyclohexyl bis(nitrate)	41
$O_2N-O-(CH_2)_2-O-NO_2$	100
$O_2N-O-(CH_2)_4-O-NO_2$	100
$(O_2N-O-CH_2)_2-CH(ONO_2)$	100

Nitration at Heteroatoms

$$\text{R-OH} + \underset{\underset{NO_2}{|}}{\underset{H_3C\,\,\overset{+}{N}\,\,CH_3}{\text{(4-methyl-2,6-dimethylpyridinium)}}}\,\, BF_4^- \longrightarrow \text{R-O-NO}_2 + \underset{\underset{H}{|}}{\underset{H_3C\,\,\overset{+}{N}\,\,CH_3}{\text{(collidinium)}}}\,\, BF_4^-$$

No oxidation of alcohols is observed under the reaction conditions. The N-nitrocollidinium salt is generally less reactive than $NO_2^+BF_4^-$ itself, but gives better control of conditions and superior yields as shown, for example, in the preparation of 1-adamantyl nitrate. Adamantanol gives less than 2% yield of nitrate ester upon treatment with $NO_2^+BF_4^-$ while the N-nitro salt forms the ester in 82% yield.

Optically active nitrate esters are formed with retention of configuration. Furthermore, the reaction of benzyl alcohol shows that O-nitration of alkylaryl alcohols in transfer nitrations is preferred to aromatic C-nitration.

1,2-Diols give dinitrates under the reaction conditions. No pinacolone rearrangement was observed. Glycerol gives trinitroglycerin in quantitative yield.

...e excellent reagents for the conversion of alcohols ... reaction proceeding via the formation of an ...ollowed by proton loss.

$$+ \,\, NO_2^+ \longrightarrow R-\underset{+}{\overset{\overset{NO_2}{|}}{O}}-H \xrightarrow{-H^+} R-O-NO_2$$

...also react with the oxygen of the ether linkage. ... formed in the reaction is readily converted, in ...orresponding aldehyde or ketone.[189]

$$+ \,\, NO_2^+BF_4^- \longrightarrow \underset{\underset{NO_2}{|}}{R-CH_2\overset{+}{O}CH_3}\,\, BF_4^-$$

$$\downarrow -HNO_2$$

$$RCHO \xleftarrow[H_2O]{-CH_3OH} R\overset{+}{CH}=OCH_3 \,\, BF_4^-$$

$R = C_6H_5CH_2, \,\, p\text{-}CH_3C_6H_4CH_2, \,\, p\text{-}NO_2C_6H_4CH_2$

$$R_2CH-O-CH_3 + NO_2^+ \longrightarrow R_2CH-\underset{+}{\overset{NO_2}{\overset{|}{O}}}-CH_3 \xrightarrow{-HNO_2}$$

$$R_2C\!\!=\!\!\overset{+}{O}CH_3 \xrightarrow{H_2O} R_2C\!\!=\!\!O$$

In the presence of other nucleophiles, the nature of the product changes significantly. Thus, methyl ethers of 1-adamantanol, *t*-butanol, and *exo*-norbornan-2-ol react with $NO_2^+BF_4^-$ in acetonitrile solution to yield the products of a formal Ritter reaction.

$$R-O-CH_3 + NO_2^+ \longrightarrow R-\underset{+}{\overset{NO_2}{\overset{|}{O}}}-CH_3 \xrightarrow{CH_3CN}$$

$$R-N^+\!\!\equiv\!\!C-CH_3 \xrightarrow{H_2O} R-NH-\overset{O}{\overset{||}{C}}-CH_3$$

Ethers of secondary alcohols, in general, show poor regioselectivity in the second step, thus yielding mixtures of amides as well as products of oxidation via cleavage with HNO_2 (*vide infra*). In an analogous reaction, treatment of 1-methoxyadamantane with a 1 : 1 mixture of acetyl nitrate and HBF_4 resulted in its conversion to 1-acetoxyadamantane in high yield.

Desilylative nitration of vinyloxysilanes with $NO_2^+BF_4^-$ was studied.

Enol silyl ethers undergo facile nitro-desilylation to yield the corresponding α-nitro ketones in moderate yield.[190]

The reaction probably proceeds via O-nitration, a mechanism similar to the reaction with alkoxysilanes, followed by isomerization of the vinyl nitrate.

$$H_3C-C(OSi(CH_3)_3)=CH_2 + NO_2^+ \longrightarrow H_3C-C(O-NO_2)=CH_2 \longrightarrow H_3C-C(=O)-CH_2-NO_2$$

4.5.2 Nitration at Nitrogen

The nitration of amines and their alkyl and acyl derivatives to give nitramines originally gave rise to interest in the case of aromatic amines,[191-194] but subsequently was extended to aliphatic amines and ammonia itself. The parent nitramine, H_2NNO_2, was isolated by Thiele and Lachman[195] and was also named by them. Wright in an excellent earlier review on nitramines discussed the early chemistry of nitramines and their nomenclature.[196] We follow his suggestion in calling nitramines the N-nitro derivatives of amines. When the nitramino group is bonded to the carbon of a carbonyl group it becomes a nitramide. When the nitramino group is coupled with loss of water with a carbonyl it becomes a nitrimine.

4.5.2.1 Nitric Acid and Derivatives

The nitration of amines and their derivatives to nitramines is of substantial importance. Primary aliphatic amines generally cannot be nitrated directly with HNO_3. This is due both to the relative instability of the monosubstituted nitrammonium ion leading to rapid decomposition with formation of N_2O

$$R-NH_2 + NO_2^+X^- \rightleftharpoons [RNH_2NO_2]^+X^-$$

and the instability of the tautomeric isonitramines in acid

$$RNHNO_2 \longrightarrow R-N=N(O)(OH) \longrightarrow N_2O + ROH$$

Emmons and Freeman[197] found that acetone cyanohydrin nitrate can nitrate primary and secondary amines to their nitramines when the amine is used in excess.

RNH$_2$ (excess) + (CH$_3$)$_2$C(CN)ONO$_2$ $\longrightarrow$ RNHNO$_2$ + HCN + (CH$_3$)$_2$CO

Acetone and HCN produced as by-products react with excess of amine (generally used in five-fold quantity) to give the corresponding aminonitrile. Yield with primary amines in low dielectric solvents is low (20–25%) but increases to 50% in refluxing THF or acetonitrile. Branched aliphatic amines (at the α-carbon) fail to react, probably for steric reasons.

Trifluoroacetyl nitrate in the form of Crivello's nitrating agent, i.e., ammonium nitrate–trifluoroacetic anhydride,[198a] was found by Suri and Chapman[198b] also to be very effective for N-nitration of amines, as well as of amides and imides.

N$_2$O$_5$, NO$_2$Cl, and NO$_2$F can also be used to nitrate primary and secondary amines.[199a,b]

Only weakly basic amines can be converted directly with HNO$_3$ to their nitramines. Because strongly basic amines form ammonium salts with HNO$_3$, apparently with the exclusion of nitronium salts, they cannot be nitrated to give secondary nitramines in satisfactory yield (hence, see Section 4.5.2.2 for nitration with preformed nitronium salts). Chloride ion was found to catalyze the nitration of secondary amines to their nitramines. It is assumed that HNO$_3$ oxidizes the chloride to an electropositive chlorine species leading to chloramines, which in turn are readily nitrated. Dimethylamine was nitrated in 68% yield and morpholine in 93% yield to their nitramines in chloride-catalyzed reactions.[200] Tertiary amines cannot be nitrated, but their nitrolysis is of significance particularly in the production of explosives. Henning, in 1899, obtained a compound, C$_3$H$_6$N$_6$O$_6$, by reacting hexamethylenetetramine with HNO$_3$.[201] Two decades later, Herz[202] recognized that the compound obtained by Henning was 1,3,5-trinitrohexahydro-s- triazine, a powerful, high explosive, later frequently referred to as RDX (cyclonite, hexogen). A detailed report on its preparation with 99.8% HNO$_3$ was given by Hale.[203]

RDX

HMX

The extensive research during World War II on RDX was reviewed[204] including improved preparations in the presence of ammonium nitrate and Ac_2O, allowing complete recycling of the byproducts.

Related polynitramines also gained importance as explosives, particularly the eight-membered ring HMX (a high melting explosive) first found as by-product in the manufacture of RDX. The chemistry of these explosives was reviewed by Urbański.[183]

Amides were the first nitrogen compounds to be nitrated with HNO_3. Thiele nitrated urethane[205] in his preparation of nitramine.

$$H_2NCO_2C_2H_5 \xrightarrow[H_2SO_4; \ 0°C]{HNO_3 \ or \ C_2H_5ONO_2} O_2N-NH-CO_2C_2H_5 \xrightarrow[0°C]{H_2SO_4} H_2NNO_2$$

Subsequently, a large number of linear and cyclic amides were nitrated. Besides urethane, Thiele also nitrated urea[205] via urea nitrate with HNO_3 and studied the nitration of biuret,[206] as well as of guanidine.[207] He considered nitroguanidine a primary nitramine but subsequently[208] it was shown to have a nitrimide structure.

$$H_2N-\overset{\overset{NH}{\|}}{C}-NH_2 \xrightarrow[H_2SO_4]{HNO_3} H_2N-\overset{\overset{NNO_2}{\|}}{C}-NH_2$$

Nitrourea, nitroguanidine, and their alkyl derivatives have been extensively studied.[208]

Secondary nitramides are generally readily prepared by the nitration of the corresponding amides, facilitated by the weakly basic nature of the imino group.

The nitrolysis of dialkylamides (i.e., acylamines) gives nitramines generally in good yield. The method gained particular significance in the preparation of polynitramine explosives. Gilbert et al. described the preparation of RDX by nitrolysis of a variety of 1,3,5-triacylhexahydro-s-triazines.[210]

$$\underset{R}{\text{triazine}} \xrightarrow[\text{or } HNO_3/(CF_3CO_2)_2O]{HNO_3 \ (P_2O_5)} \underset{NO_2}{\text{trinitro triazine}}$$

$R = CH_3CO-, \ C_2H_5CO-, \ C_2H_5OCO-, \ p-CH_3C_6H_4SO_2-$

The mixture of ammonium nitrate and trifluoroacetic anhydride was found by Suri and Chapman[210] to be a convenient and safe reagent for N-nitration in the synthesis of cyclic nitramines, nitramides, and nitrimides.

$$(CF_3CO)_2O + NH_4NO_3 \rightleftharpoons CF_3COONO_2 + NH_4^+CF_3CO_2^-$$

R = H, CHO, CH$_3$CO

$$\overset{X}{\underset{X}{\diagdown}}N-R \longrightarrow \overset{X}{\underset{X}{\diagdown}}N-NO_2$$

The reactions are conveniently carried out in nitromethane solution generally at 0°C. Examples include preparation of 1-nitro-2- pyrrolidone (30% yield), 1,3-dinitro-2-oxotetrahydroimidazole (41%), and 1-nitrohydrantoin (20%). 1,3,5-Triacetylhexahydro-1,3,5-triazine was also successfully nitrated giving predominantly 1-acetyl-3,5-dinitrohexahydro-1,3,5-triazine with some 1,3,5-trinitrohexahydro-1,3,5-triazine (RDX).

4.5.2.2 Nitronium Salts

The first report of the nitration of amines by nitronium salts was by Olah and Kuhn.[211]

Primary and secondary amines are nitrated by $NO_2^+BF_4^-$ in sulfolane or SO_2 solution to yield nitramines.

$$2R_2NH + NO_2^+BF_4^- \longrightarrow R_2N-NO_2 + R_2NH \cdot HBF_4$$

The nitration of amines with NO_2BF_4 to nitramines was studied subsequently by Olsen, Fisch, and Hamel.[212]

Satisfactory yields of nitramines were obtained by reacting two equivalents of secondary aliphatic amines with $NO_2^+BF_4^-$ in methylene chloride solution.

Reactions of primary aliphatic amines, such as *n*-butylamine, however, gave only a 20% yield of *n*-butyl nitramine. In contrast picramide gave N,2,4,6-tetranitroaniline in 85% yield (Table 58).
Extensive studies of N-nitration were carried out by Russian investigators and the topic was reviewed.[153]

Extensive studies reported in Russian literature were reviewed by Guk et al.[153] allowing discussion and credit to some otherwise still difficult to access important work.

It was shown[213] that the nitration of aromatic amines proceeds differently depending on basicity. Amines of moderate to low basicity, such as *bis*(2-cyanoethyl)amine (pK_a = 5.25) and *bis*(2,2,2-trinitroethyl)amine (pK_a = 0.05) are nitrated by $NO_2^+BF_4^-$ to the corresponding N-nitramines in acetonitrile or ethyl acetate in yields of 87–98%. The nitration of highly basic dialkylamines (pK_a=8.70–11.15) is accompanied by the partial reduction of $NO_2^+BF_4^-$ to nitrosonium tetrafluoroborate and the formation of nitrosamines. The content of nitrosamine in the reaction mixture increases with increase of the reaction temperature. Nitronium hexafluorosilicate proved to be a milder nitrating agent; its application makes it possible to reduce greatly the formation of nitroso derivatives.

The nitration of aliphatic–aromatic amines also proceeds smoothly and the low acidity of the medium makes it possible virtually to avoid the N-nitro → C-nitro-rearrangements.[213]

Aromatic methylenebisamines, which are unstable in acid medium, were nitrated for the first time with nitronium salts, and N,N'-diarylmethylene-dinitramines were obtained in a high yield.[214]

$A = BF_4^-, SiF_6^{2-}$; $R = 2-NO_2, 3-NO_2, 4-NO_2, 2,6-$ and $3,5-(NO_2)_2$

Table 58. Nitration of Amines and Their Derivatives with Nitronium Tetralfuoroborate[212]

Compound	Yield of N-nitro-derivative (%)
Di-n-butylamine	54
Morpholine	72
β,β-Bis(cyanoethyl)amine	62
Picramide	85

The reaction of nitronium salts and N_2O_5 with ammonia has been investigated,[215,216,217,218] and it has been shown that nitramine is formed at liquid–nitrogen temperature

$$2NH_3 + NO_2^+A^- \longrightarrow NH_2NO_2 + NH_4^+A^-$$

$$A^- = SO_3F, ClO_4^-, BF_4^-, NO_3^-$$

The formation of nitramine is determined by the nature of the anion; on passing from nitronium chlorosulfate to tetrafluoroborate, the yield increases from 8–43%.

Nitronium salts proved to be particularly useful in the synthesis of N,N-dinitramines, which previously had been difficult to obtain. It was shown[219] that the interaction of primary nitramines (or their salts) with nitronium tetrafluoroborate, pyrosulphate, fluorosulfate, fluorosilicate, as well as other nitronium salts leads to the formation of the N,N-dinitramines.

$$RN(NO_2)X + NO_2^+A^- \longrightarrow RN(NO_2)_2$$

$$X = H, NH_4, K, Li; \quad A = BF_4^-, S_2O_7^{2-}, FSO_3^-, SiF_6^{2-}, ClO_4^-, SbF_6^-, SnF_6^{2-}$$

Best results are obtained in nitrations by nitramine salts in chloroalkanes or acetonitrile. However, the use of more basic solvents, such as ethers and esters, ensures equally high yields in the nitration of both free nitramines and their salts.[220]

Amines (acylamines) and urethanes, as shown by Olsen,[212] gave, with one equivalent of $NO_2^+BF_4^-$ in acetonitrile at −30°C, the corresponding N-nitro-derivatives (Table 59).

$$RCONH_2 + NO_2^+BF_4^- \longrightarrow RCO(NO_2)H + HBF_4$$

Table 59. Nitration of Amides with
Nitronium Tetrafluoroborate[212]

Compound	Yield of N-nitro-derivative (%)	Compound	Yield of N-nitro-compound (%)
Acetamide	13	Ethyl n-butylcarbamate	91
2-Chloroacetamide	55	n-Butylacetamide	40
2,2,2-Trichloroacetamide	62	Succinimide	43
Benzamide	53		

The nitramides of benzoic and chloroacetic acids were obtained in satisfactory yields by the method, but the yield of nitroacetamide was only 12%. This can be explained by the fact that the aliphatic nitroamides are readily hydrolyzed even in the presence of potassium acetate.

Only the use of more basic solvents such as ethyl acetate, 1,4-dioxan, or trimethyl phosphate made it possible to obtain nitramides of different structure in 40–90% yield.[221] Succinimide is nitrated by $NO_2^+BF_4^-$ in ethyl acetate in 43% yield.[212]

It is interesting to note that N-methylsuccinimide does not react with nitronium salts and N-methylphthalimide undergoes nitration in the aromatic ring exclusively.

The most convenient synthesis of sulfonic acid nitroimides consists of the reaction of $NO_2^+BF_4^-$ with imide salts.[222]

$$\begin{array}{c} X \\ \diagdown \\ N^-M^+ \\ \diagup \\ Y \end{array} + NO_2^+BF_4^- \longrightarrow \begin{array}{c} X \\ \diagdown \\ N-NO_2 \\ \diagup \\ Y \end{array}$$

(a) $X = p\text{-}CH_3C_6H_4SO_2$; $Y = CO_2R$

(b) $X = Y = CH_3SO_2$; $C_6H_4SO_2$

As discussed, one of the known methods of synthesis of dialkylnitramines involves the reaction of dialkylamides with HNO_3, which leads to the substitution of the acyl group by the nitro group (nitrolysis). However, when HNO_3 or its mixtures with Ac_2O are used, the yield of nitramines is, as a rule, low and only the use of the $HNO_3-(F_3CCO)_2O$ mixture makes it possible to raise the yield to 90%.[223] For preparative purposes, it is more convenient to nitrate the dialkylamides by nitronium salts.[224] The reaction takes place at 20°C in acetonitrile.

The dialkylnitramines are formed in yields up to 90% and the acyl group is converted into acylium tetrafluoroborate.

$$R_2NC(O)R' + NO_2^+BF_4^- \longrightarrow R_2N-NO_2 + R'CO^+BF_4^-$$

The possibility of the nitrolysis of secondary amides with formation of the corresponding primary N-nitramines (together with 3,3'-azo-1,2,4-triazole) has been demonstrated for an N-acyl derivative of 3-amino-1,2,4-triazole.[225]

[Structure: HN-triazole-NHCOCH₃ + NO₂⁺BF₄⁻ → HN-triazole-NHNO₂ + HN-triazole-N=N-triazole-NH]

Under the reaction conditions, alkyl N-nitramines easily decompose in the presence of the organic acids liberated.[226] Amides containing electron-accepting substituents at the nitrogen atom also undergo nitrolysis by nitronium salts. Thus N-fluoronitramines were obtained from N-fluoroesters.[227]

$$RNFCO_2CH_3 + NO_2^+BF_4^- \longrightarrow RNF-NO_2$$

Alkyl-N,N-dinitramines were formed in high yields from N-alkyl-nitramides.[226]

$$RN(NO_2)C(O)R' + NO_2^+BF_4^- \longrightarrow RN(NO_2)_2 + R'CO^+BF_4^-$$

$$R = CH_3, C_4H_9, R' = CH_3, C_3H_7, CCl_3$$

The reaction of $NO_2^+BF_4^-$ with N-alkylamides has been investigated[212,216] under different conditions, but nevertheless the results permit the conclusion that at a low temperature (−30°C) N-butyl-acetamide and ethyl *n*-butylcarbamate in acetonitrile are nitrated to the N-nitro-derivatives,[212] while at higher temperatures (up to 10°C) nitrolysis takes place with the formation of the corresponding carboxylic acid and alcohol as well as N_2O.[228] The question of the order in which the C–N bond is N–nitrated and cleaved in the reaction of N–alkylamides with $NO_2^+BF_4^-$ requires additional study.

Aliphatic isocyanates react with $NO_2^+BF_4^-$ in ethyl acetate or acetonitrile with formation (after hydrolysis) of alkylnitramides.[229]

$$RNCO + NO_2^+BF_4^- \longrightarrow RNHNO_2 + CO_2 + HBF_4$$

The study of the nitration of a series of N,N–diacylmethylamines showed that the acyl group is substituted by the nitro group and, depending on the conditions and the component ratio, methylnitroacetamide or methyl–N,N–dinitramine is formed.[230]

$$CH_3N(COCH_3)_2 + NO_2^+BF_4^- \longrightarrow \begin{array}{l} CH_3N(NO)_2COCH_3 \\ CH_3N(NO_2)_2 \end{array}$$

The methanesulphonyl group can also be substituted by the nitro group. The tosyl group or an alkoxycarbonyl group do not enter into this kind of reaction.

The mode of reaction of $NO_2^+BF_4^-$ with substituted methylenediamines depends on the nature of the substituents.[230,231] Alkanesulphonyl, arenesulphonyl, or methoxycarbonyl derivatives undergo nitrolysis at the C–N bond with the formation of substituted nitramines. This reaction pathway is favored by the formation of the carbonium-immonium ion stabilized by the amino nitrogen:

$$R'RNCH_2NRR' + NO_2^+BF_4^- \longrightarrow R'NR-NO_2 + R'NR^+CH_2BF_4^-$$

$$R = Alkyl, \quad R' = Alkyl, \ SO_2C_6H_5, \ CO_2CH_3$$

N,N–Diacetylimidazolidine undergoes nitration by nitronium salts to N–acetyl–N'–nitro– or N,N'–dinitro–derivatives:

$$CH_3CO-N\diagup\diagdown N-COCH_3 + NO_2^+BF_4^- \longrightarrow$$

$$O_2N-N\diagup\diagdown N-COCH_3 \ + \ O_2N-N\diagup\diagdown N-NO_2$$

Nitronium salts are convenient nitrating agents for the synthesis of N-nitrimines. 4-Amino-1,2,4-triazole reacts with $NO_2^+BF_4^-$ in acetonitrile to form 4-nitrimino-1,2,4-triazole in 65% yield.[232]

1-Aminobenzimidazole is nitrated analogously and 1,1'-azobenzimidazole is formed as a side product. 3-Amino-1,2,4-triazole reacts with $NO_2^+BF_4^-$ in acetonitrile to form N-nitrotriazole, which rearranges to the nitroaminotriazole.

Both nitronium and nitrosonium salts readily react with azodicarboxylate anions in accordance with

$$NO_2^+ + {}^-OOCN=NCOO^- \longrightarrow N_2O_4 + 2CO_2 + N_2$$

$$NO^+ + {}^-OOCN=NCOO^- \longrightarrow 2NO + 2CO_2 + N_2$$

The reaction of alkyl azodicarboxylates depends on the reactivity of the electrophilic species. $NO_2^+BF_4^-$, for example, reacts with ethyl azodicarboxylate, while the weaker electrophile $NO^+BF_4^-$ does not.

The reaction of $NO_2^+BF_4^-$ with lithium azide in acetonitrile results in the formation of covalent nitronium azide, which is converted into nitrous oxide above $-10°C$.[223]

$$NO_2^+BF_4^- + LiN_3 \longrightarrow LiBF_4 + NO_2N_3 \longrightarrow 2N_2O$$

Nitramines can be also obtained by the nitro-desilylation of silylamines, which can provide a route for the synthesis of certain nitramines heretofore unobtainable by conventional nitration methods. 2-Isopropyl-1,4-dinitroimidazole was synthesized by nitration of the corresponding silylated derivative.[234]

Similarly, N-nitrotriazole was synthesized.

Apparently no report of nitration of silylated aliphatic amines has appeared. The reaction of silylated dialkylamines with $NO_2^+BF_4^-$ should give the corresponding nitramines.

$$R_2NSi(CH_3)_3 \xrightarrow{NO_2^+BF_4^-} R_2N-NO_2 + (CH_3)_3SiF + BF_3$$

N-Methoxycarbonyl and N-methanesulfonylN'-trimethylsilyl-carbodimides undergo nitro-desilylation, resulting in the formation of N-nitrocyanamides.

$$R-N=C=N-Si(CH_3)_3 \xrightarrow{NO_2^+} R-N=C=N-NO_2 \longrightarrow R-N(CN)-NO_2$$

It is not clear if the N-nitrocyanamides are formed via isomerization of the N-nitro derivative or not.

$$R-N=C=N-Si(CH_3)_3 \xrightarrow{NO_2^+} R-N=C=N-NO_2 \longrightarrow R-N(CN)-NO_2$$

However, direct formation of the N-nitrocyanamide seems highly likely.

4.5.3 Nitration of Sulfur

Sulfides react with $NO_2^+PF_6^-$ at $-78°C$ to form S-nitrosulfonium ions, which isomerize to S-nitrosulfonium ion on warming to $-20°C$ and subsequently give the corresponding sulfoxides.[235a]

$$R-S-R + NO_2^+ \longrightarrow \underset{NO_2}{R-\overset{+}{S}-R} \rightleftharpoons \underset{ONO}{R-\overset{+}{S}-R} \longrightarrow \underset{O}{R-\overset{\parallel}{S}-R} + NO^+PF_6^-$$

This reaction demonstrates the ambident reactivity of the NO_2^+ ion, in analogy with the ambident reactivity of NO_2 and NO_2^-. The nitrosulfonium ion intermediate could be observed by ^1H, ^{13}C, and ^{15}N-NMR spectroscopy.

In a similar fashion, sulfoxides were oxidized to sulfones[235b]

$$R-\overset{O}{\underset{}{\overset{\parallel}{S}}}-R + NO_2^+ \longrightarrow \underset{}{R-\overset{ONO_2}{\overset{|}{S^+}}-R} \longrightarrow R-\overset{O}{\underset{ONO}{\overset{\parallel}{S^+}}}-R \longrightarrow R-\overset{O}{\underset{O}{\overset{\parallel}{\underset{\parallel}{S}}}}-R + NO^+$$

S-nitro- or S-nitritosulfonium ions are similar intermediates in the oxidative cleavage of ethylenethioacetals with $NO_2^+BF_4^-$ or sodium nitrate/trifluoroacetic acid.[235c]

$$\underset{R}{\overset{R}{>}}\!\!\!<\!\!\!\underset{S}{\overset{S}{\underset{}{\bigcup}}} \xrightarrow[2)\ H_2O]{1)\ NO_2^+\ \text{or}\ NaNO_3/CF_3CO_2H} \underset{R}{\overset{R}{>}}C=O$$

4.5.4 Nitration at Phosphorus

Reaction of nitronium salts with phosphines yields phosphine oxides[235a] in quantitative yield.

$$R_3P + NO_2^+ \longrightarrow [R_3P^+\text{-}O\text{-}N\text{=}O] \xrightarrow{-NO^+} R_3P\text{→}O$$

Examination of the intermediates by ^{13}C, ^{31}P. and ^{15}N–NMR spectroscopy showed the presence of only the P–nitritophosphonium ion and no P–nitrophosphonium ion (R_3P^+–NO_2) was observed even at very low temperature.

4.6 Nucleophilic Aliphatic Nitration

Aliphatic nitro compounds are prepared, as discussed so far in this chapter, mainly by free–radical and electrophilic nitration methods. Nucleophilic nitration, however, also plays an important role. It is of significance to recall that nitroalkanes were originally prepared from alkyl iodides with silver nitrate[236-238] (by what is now called the Victor Meyer reaction), and similar improved nucleophilic reactions continue to be of substantial importance in the methodology of preparing aliphatic nitro compounds.

4.6.1 Alkyl Halides with Silver Nitrite (Victor Meyer Reaction)

In 1872, Victor Meyer discovered that nitroaliphatic compounds can be prepared by the reaction of alkyl iodides with silver nitrite.[236,237,238]

$$R\text{-}I + AgNO_2 \longrightarrow R\text{-}NO_2 + AgI$$

In the reaction, alkyl nitrites are also formed. For example, amyl iodide and silver nitrite give both nitropentane and amyl nitrite:

$$C_5H_{11}I + AgNO_2 \longrightarrow \begin{cases} C_5H_{11}NO_2 \\ C_5H_{11}ONO \end{cases}$$

Separation of the nitroalkane from the isomeric alkyl nitrites is generally achieved by distillation since the latter usually have a much lower boiling point. Alkyl bromides and chlorides are less convenient or not suitable at all for the reaction. Mercurous nitrite reacts similarly to silver nitrite. The reaction of alkyl iodides with potassium nitrite was reported to yield alkyl nitrites exclusively. Likewise, sodium nitrite was considered in some textbooks to give only alkyl nitrites in the Victor Meyer reaction, although no detailed studies were reported at the time.

The reaction of silver nitrite with primary alkyl iodides or bromides is an excellent method for the preparation of primary nitroalkanes[239,240,241] with the exception of neopentyl iodide.[242] The reaction fails with primary alkyl chlorides. The difference in reactivity of alkyl halides with silver nitrite is of interest, and several α-fluoro- ω-bromoalkanes are converted to α-fluoro- ω-nitroalkanes. The success of the Victor Meyer reaction is closely related to the structure of the halide, and it is apparent from the data in Table 60 that secondary alkyl iodides or bromides are not well suited for the preparation of secondary nitroalkanes.[243,244,245] Branching of the carbon chain is accompanied by a striking decrease in the yield of the nitro compound. With secondary bromides or iodides the products are mainly nitrite esters and olefins, and the small amount of nitro compounds that is produced is contaminated with nitrite esters, which are removable only by a chemical separation attended by loss of product.

Table 60. Preparation of Nitroalkanes from
Alkyl Halides by the Victor Meyer Reaction[243-245]

Nitro compound	Yield (%)	RX X
n-$C_4H_9NO_2$	73	Br
n-$C_5H_{11}NO_2$	67	Br
n-$C_8H_{17}NO_2$	80	Br
$(CH_3)_2CHCH_2CH_2NO_2$	72	Br
$(CH_3)_2CHCH_2NO_2$	59	I
$(CH_3)_2CHNO_2$	19-26	Br
$CH_3CH_2CH(CH_3)NO_2$	19-24	Br
$(CH_3CH_2CH_2)_2CHNO_2$	7-15	Br
$CH_3(CH_2)_5CH(CH_3)NO_2$	17-23	Br
$C_6H_5CH(CH_3)NO_2$	18	Br
$O_2N(CH_2)_{10}NO_2$	50	I
$F(CH_2)_5NO_2$	73	Br
$CH_3CH(NO_2)CO_2C_2H_5$	80	I

The reaction of tertiary halides with silver nitrite for the preparation of nitro compounds is not adaptable.[243] Instead of nitro compounds, alkyl nitrites, olefins, and adducts of olefins with nitrogen oxides are produced.

The reaction has also been adapted for the preparation of certain nitro alcohols, nitro ethers, nitro ketones, and other nitroaliphatic compounds.

α-Nitroesters are obtained only from the corresponding α-iodoesters with silver nitrite. The reaction is slow, but the yields are satisfactory.[246]

The reaction of allyl bromide with silver nitrite leads to the corresponding nitro compound in 55% yield.[50] Epiiodohydrin and silver nitrite yield 1-nitro-2,3-epoxypropane. Similarly, 1-nitro-2,3-epoxybutane and 3-nitro-1,2-epoxybutane are prepared in 70 and 72% yield, respectively.[247]

The reaction of silver nitrite in diethyl ether with optically active 2-bromooctane occurs to give 2-nitrooctane and 2-octyl nitrite with inversion of configuration.[248] The role of the nitrite ion as a nucleophile is important. Failure of neopentyl iodide to react under conditions that facilitate complete reaction with other primary iodides must be attributed to steric conditions that prohibit nitrite ion from participating in the reaction.[242] These features are characteristic of an S_N2 process.

The reaction of silver nitrite with alkyl halides also exhibits features which are of S_N1 character, in accordance with the well-known sequence of reactivity of alkyl halides in nucleophilic reactions, i.e., tertiary>secondary>primary.[242] The study of silver nitrite with a series of benzyl bromides with *para* substituents gave widely different rates, and a reactivity sequence that is expected from a carbocationic process.[242] The difference in the proportion of nitro compounds andnitrite ester isolated is additional evidence for the carbocationic (S_N1) character of the reaction (Table 61).

The mechanism for the nucleophilic displacement reaction of alkyl halides with silver nitrite is one in which the transition state has both S_N1 and S_N2 character, the extent of which is dependent on the structure of the halide. Products of the reaction reflect the variation in nature of the transition state. Nucleophilic attack by nitrite ion on carbon, and electrophilic attack by the silver cation on the halogen of the alkyl halide are essential features of the process. Formation of the silver-halogen bond provides the driving force for the Victor Meyer

Table 61. Reaction of Silver Nitrite with Benzyl Bromides[242]

Bromide	Nitro compound yield (%)	Nitric ester yield (%)
β-Nitrobenzyl	75	5
Benzyl	61	28
β-Methylbenzyl	45	37
β-Methoxybenzyl	26	55

reaction. Sulfonate esters fail to react with silver nitrite.[242]

4.6.2 Alkyl Halides with Alkali Metal Nitrites

Only months after Meyer and Stübers first published an article on the preparation of nitroalkanes, Kolbe reported that primary nitroalkanes are produced by the reaction of α-haloalkanecarboxylic acids with sodium nitrite.[249] The initially formed α-nitroalkanecarboxylic acids immediately decarboxylate to the corresponding nitroalkanes.[250]

$$RCH(Cl)CO_2H + NaNO_2 \longrightarrow RCH(NO_2)CO_2H + NaCl$$

$$RCH(NO_2)CO_2H \longrightarrow RCH_2NO_2 + CO_2$$

Chloroacetic acid gives nitromethane, α-bromopropionic acid nitroethane, and α-bromobutyric acid nitropropane, but the method does not work for the preparation of higher nitroalkanes.[251] The α-carboxylic group clearly activates the C—Cl bond and allows the reaction with sodium nitrite to proceed. In the absence of activating groups sodium nitrite is generally inactive in the Victor Meyer reaction.

4.6.3 Kornblum Modification

Kornblum and co-workers have further investigated the Victor Meyer reaction and extended its scope. They have also developed a useful modification of the Victor Meyer reaction. Primary and secondary alkyl iodides were converted into the corresponding nitroalkanes when reacted with sodium nitrite in DMF solution.[253]

Alkyl nitrites are again significant by-products, but they can be trapped with phloroglucinol (phloroglucinol is nitrosated and the alkyl nitrite is converted into the alcohol). Otherwise, the nitrites react further to give nitro-nitroso compounds.[253] The reaction is preferentially carried out in the presence of urea which apparently increases the solubility of sodium nitrite. As shown in Table 62, the Kornblum modification gives satisfactory yield of nitro compounds.

Use of the alkali metal nitrites in place of silver nitrite for the preparation of nitroalkanes requires that both nitrite and halide be in solution.[254] Although lithium nitrite is more soluble than sodium nitrite in the solvents that are used, sodium nitrite is employed because of its availability. Dimethylformamide was the solvent of choice.

$$CH_2CH_2Br \xrightarrow[DMF]{NaNO_2} RCH_2CH_2NO_2 + RCH_2CH_2ONO$$

Sodium nitrite is more soluble in dimethylsulfoxide (DMSO) than in DMF. Consequently, as Kornblum and Powers reported, in DMSO the reaction can be carried out at a much faster rate with comparable yields.[255] Even α, ω -dinitroalkanes can be prepared this way from the corresponding dibromoalkanes.[256]

Table 62. Preparation of Nitroalkanes in the Reaction of Alkyl and Cycloalkyl Halides with Sodium Nitrite in DMF Solution[252]

Alkyl halide	Percent yield Nitroalkane	Alkyl nitrite
1-Bromoheptane	50	29
1-Iodoheptane	61	31
1-Bromooctane	60	29
1-Iodooctane	60	31
1-Iodododecane	57	25
1-Iodo-3-phenylpropane	58	26
Benzylbromide	55	33
2-Brobooctane	58	–
4-Bromoheptane	61	–
4-Iodoheptane	63	25
Bromocyclopentane	57	25
Iodocyclopentane	55	–
Bromocycloheptane	55	–
Iodocycloheptane	58	–

DMSO is consequently the preferred solvent, whereas ethylene glycol is a much poorer solvent. Primary and secondary alkyl iodides and bromides, as well as sulfonate esters, are satisfactory for the preparation of primary and secondary nitroalkanes. The reaction of alkyl chlorides with sodium nitrite is too slow to be practical. Reaction of tertiary alkyl halides with sodium nitrite for preparing tertiary nitroalkanes is useless even in these systems.

The modified Victor Meyer reaction is applicable to the synthesis of α-nitro esters.[257,258] and β-nitro ketones.[259] Failure of cyclohexyl bromide to react with sodium nitrite to form nitrocyclohexane is a serious limitation[254] that precludes the application of the modified Victor Meyer reaction to the synthesis of nitro steroids.[260]

The reaction of alkyl bromides with nitrite ion is an S_N2 reaction.[242,254] The application of the modified Victor Meyer reaction should succeed where S_N2 reactions may be operative. Failure of *trans*-1,4-cyclohexanediol *bis-p*-toluenesulfonate to react with sodium nitrite in DMF is attributed to the presence of large leaving groups in the equatorial position and consequent crowding by axial hydrogens, which hinder attack by the nitrite ion.[261] A reaction does occur between cyclohexyl iodide and sodium nitrite; cyclohexene is isolated in 57% yield.[254] Ethyl α-bromoisobutyrate and α-bromoisobutyronitrile react to give the nitro compounds in 78% and 50% yields, respectively.[257] A nitrocyclobutene is prepared in 54% yield by the reaction of 1-phenyl-3,3-difluoro-4,4-dichlorocyclobutene with sodium nitrite.[262]

Cyclobutenyl halides generally undergo S_N2 displacement reactions with ease.[263]

4.6.4 ter Meer Reaction

In 1876, ter Meer reported a convenient method for the synthesis of gem-dinitro compounds.[264] He described the reaction of sodium nitrite with 1-bromo-1-nitropropane, which gave 1,1-dinitropropane. He also prepared 1,1-dinitroethane in a similar way.

$$\underset{H}{\overset{Cl}{\underset{|}{\overset{|}{H_3C-C-NO_2}}}} \xrightarrow[2.\ H^+]{1.\ K_2CO_3,\ NaNO_2} CH_3CH(NO_2)_2 \xrightarrow{CH_2O} CH_3C(NO_2)_2CH_2OH$$

Synthesis of α,α, ω, ω-tetranitroalkanes is another useful application of the ter Meer reaction.[265]

$$O_2NCHBr(CH_2)_2CHBrNO_2 \xrightarrow[2.\ H^+]{1.\ KOH,\ KNO_2} (O_2N)_2HC(CH_2)_2CH(NO_2)_2$$

The mechanism of the ter Meer reactions has been established by Hawthorne.[266] The reaction of 1-chloro-1-nitroethane with nitrite ion occurs through its aci-nitro form involving nucleophilic displacement of chloride by nitrite ion. The reaction is inhibited by excess base: therefore the nitronic acid form must be involved.

$$H_3C-\underset{NO_2}{\overset{Cl}{\underset{|}{\overset{|}{C}}}}-H + NO_2^- \rightleftharpoons R-\underset{NO_2}{\overset{Cl}{\underset{|}{\overset{|}{C^-}}}} + HNO_2 \rightleftharpoons$$

$$R-\underset{OH}{\overset{Cl\ \ \ O^-}{\underset{\diagdown}{\overset{\diagup}{C=N^+}}}} + NO_2^- \longrightarrow R-\underset{OH}{\overset{NO_2\ \ \ O^-}{\underset{\diagdown}{\overset{\diagup}{C=N^+}}}} + Cl^-$$

At high nitrite ion concentrations, the reaction shows a first-order dependence on both nitrite ion and chloronitroethane. The rate-determining process is the ionization of chloronitroethane catalyzed by nitrite ion. When 1-chloro-1-nitroethane-1-d was used, a primary kinetic isotope effect of $k_H k_D = 3.3$ was observed.

4.6.5 Kaplan-Schechter Reaction

Preparation of gem-dinitro compounds by the reaction of the aci-salts of primary or secondary nitro compounds with silver nitrate and inorganic nitrites in alkaline or neutral media was carried out by Kaplan and Shechter.[267]

$$R_2C=NO_2^- + 2Ag^+ + NO_2^- \longrightarrow R_2C(NO_2)_2 + 2Ag$$

The reaction is an excellent method for the synthesis of gem-dinitroalkanes. It is also successful with hindered compounds. Yields vary from 60% - 95%. The preparation of primary, secondary, and functionally substituted dinitro compounds were reported. Typical examples are the synthesis of 1,1-dinitropentane, methyl-3,3-dinitropropionate, and 1-cyclopropyl-1,1-dinitroethane. The reaction occurs by formation of a complex salt, the steric nature of which favors decomposition to the dinitro compound.

$$R_2C=NO_2 \xrightarrow[2Ag^+]{NO_2^-} R_2C\begin{pmatrix}N^+-O^-\\ \\N-O\end{pmatrix}Ag^-Ag^+ \longrightarrow R_2C(NO_2)_2 + 2Ag$$

The mechanism for the silver-induced nitration of nitroalkanes to gem-dinitroalkanes, was proposed to involve a redox process via outer-sphere electron transfer[268]

$$(CH_3)_2CNO_2^- + Ag^I \longrightarrow (CH_3)_2CNO_2\cdot + Ag$$

$$(CH_3)_2CNO_2\cdot + NO_2^- \longrightarrow (CH_3)_2C(NO_2)_2^-$$

$$(CH_3)_2C(NO_2)_2^- + Ag^I \longrightarrow (CH_3)_2C(NO_2)_2 + Ag$$

It is interesting to note that besides the oxidative nitration reaction of Shechter and Kaplan, gem-dinitroalkanes were also prepared by reacting salts of secondary nitroalkanes with $NO_2^+BF_4^-$ in acetonitrile at −40°C.[212]

$$R_1R_2C=NO_2M + NO_2^+BF_4^- \longrightarrow R_1R_2C(NO_2)_2$$

The reaction generally gives lower yields than the Shechter process, for pseudonitroles ($R_1R_2C(NO)NO_2$) are formed in significant amounts as by-products. However, it is possible that the $NO_2^+BF_4^-$ used contained $NO^+BF_4^-$ tetrafluoroborate, which was found to react with the salts of nitroalkanes to give pseudonitroles.

4.6.6 Nitrolysis of Dialkylhalonium and Trialkyloxonium Ions

The nitrolysis of diarylhalonium ions with sodium nitrite was discussed in Section 2.3.2. Olah *et al.* have studied[269] the related reaction of dialkylhalonium ions, such as dimethyliodonium hexafluoroantimonate with sodium nitrite and found nitrolysis to yield nitromethane

$$CH_3I^+CH_3SbF_6^- \xrightarrow{NaNO_2} CH_3NO_2 + CH_3I + NaSbF_6$$

Similarly, the reaction of trimethyloxonium hexafluorophosphate gave nitromethane

$$(CH_3)O^+PF_6^- \xrightarrow{NaNO_2} CH_3NO_2 + (CH_3)_2O + NaPF_6$$

These interesting nitrolysis reactions have not yet been fully explored.

Outlook

After 150 years, the study of nitration remains a vibrant and significant field of chemistry. The methods of nitration continue to expand with introduction of new and improved procedures, such as nitronium salt nitrations developed by the Olah group, while the established ones, such as nitric-acid and nitrogen-oxide-based reactions, remain significant. Acid-free and de-organometallative nitrations are gaining importance, particularly in providing enhanced selectivity. Use of solid catalysts promises to minimize environmental problems associated with conventional nitrating acid systems. Methods allowing good recovery and recycling of acid are also going to improve nitration methods. Fundamental chemistry exercises a strong influence on practical industrial applications. This is reflected not only in the active development of new nitration methods, but also in the continuing need for a better understanding of the mechanism of the nitration processes. One cannot help but be impressed by the significant progress made. It is thus with confidence that we predict in years to come an even brighter and more exciting future for nitration chemistry. At the same time we hope that when future investigators of the field look back to the efforts of our generation they will feel that we did our best to extent the knowledge of nitration, one of the most fundamental reactions in chemistry.

References

1. *Faraday's Diary*, Martin, T., ed., London, *1932*, Vol. I, p. 221.

2. Mitscherlich, E. *Ann. Phys. Chem.* **1834**, 625; *Ann. Pherm.* **1834**, *12*, 305.

3. Asinger, F. "*Paraffins: Chemistry and Technology*", Pergamon Press: London, 1968, p. 365.

4. Beilstein, F., Kurbatov, A. *Ber.* **1880**, 13, **1818**, 2029.

5. Konovalov, M. *Ber.* **1893**, *26*, 878.

6. Markownikov, V. *Ber.* **1899**, *32*, 1441, 1445.

7. Hass, H. B. *et al. Ind. Eng. Chem.* **1936**, *28*, 339.

8. Olah, G. A., Lin, H. *J. Am. Chem. Soc.* **1971**, *93*, 1259.

9. Meyer, V., Stüber, O. *Ber.* **1872**, *5*, 203, 399, 514, 1029, 1054.

10. Kornblum, N., Lichtin, N. N., Patton, J. T., Iffland, D. C. *J. Am. Chem. Soc.* **1947**, *69*, 307; Kornblum, N. *et al. Chem. Ind.* **1955**, 443.

11. Baker, P. J., Jr., Bollmeier, A. F., Jr. In: "*Kirk-Othmer Encyclopedia of Chemical Technology*", 3rd ed., Wiley-Interscience: New York, 1981; Vol. **14**, p. 979.

12. Seebach, D., Henning, R., Lehr, F. *Angew. Chem.* **1978**, *90*, 479.

13. Konovalov, M. I. *Zh. Russ. Khim. Obshch.* **1893**, *25*, 389, 472; **1899**, *31*, 255; *Ber.* **1895**, *28*, 1855.

14. For a discussion see: Topchiev, A. V. "*Nitration of Hydrocarbons,*" Pergamon Press: London, New York, 1959.

15. Hopkins, B. U. S. Pat. 1,694,098 (1928).

16. Konovalov, M. I. *Zh. Russ. Khim. Obshch.* **1899**, *31*, 255.

17. Nametkin, S. S. *Zh. Russ. Khim. Obshch.* **1908**, *40*, 184, 1570; **1909**, *41*, 145; **1910**, *42*, 581, 585, 691; *Ber.* **1909**, *42*, 1372.

18. Titov, A. I. *Zh. Obshch. Khim.* **1949**, *19*, 8, 1464.

19. Markownikov, V. V. *Liebigs, Ann.* **1898**, *302*, 15; *Ber.* **1900**, *33*. 1906.

20. Ingold, C. K. *J. Chem. Soc.* **1935**, 244.

21. Titov, A. I. *Tetrahedron* **1963**, *19*, 557, and references therein.

22. Titov, A. I. *Zh. Obshch. Khim.* **1946**, *16*, 1896.

23. Titov, A. I. *Usp. Khim.* **1952**, *21*, 881.

24. Titov, A. I. *Usp. Khim.* **1958**, *27*, 845.

25. Titov, A. I., Matveeva, M. K. *Dokl. Akad. Nauk, CCR* **1952**, *83*, 101.

26. Markownikov, V. *Ann. Chem.* **1899**, *307*, 335.

27. Nametkin, S. S. *J. Russ. Phys. Chem. Soc.* **1911**, *43*, 1607.

28. Shechter, H., Brain, D. K. *J. Am. Chem. Soc.* **1965**, *85*, 1806.

29. Ogata, Y., Tezuka, H., Kamei, T. *J. Org. Chem.* **1969**, *34*, 845.

30. Titov, A. I., Shchitov, N. V. *Dorl. Akad. Nauk, CCR* **1951**, *81*, 1085.

31. Urbański, T., Slon, M. *Rovkn. Chem.* **1936**, *16*, 466; *Chem. Z.* **1937**, *I*, 3626-3627; Urbański, T. *et al. C. R. Acad. Sci. Paris*, **1936**, *203*, 620; **1937**, *204*, 870.

32. Hass, H. B., Dorsky, J., Hodge, E. B. *Ind. Eng. Chem.* **1941**, *33*, 1138.

33. Hass, H. B., *et al. Ind. Eng. Chem.* **1936**, *28*, 339; **1938**, *30*, 67; **1939**, *31*, 648; **1940**, *32*, 427; **1941**, *33*, 1138; Hass, H. B., Riley, E. F. *Chem. Revs.* **1943**, *32*, 380.

34. For reviews see: (a) Asinger, F. *"Paraffins: Chemistry and Technology,"* Eng. ed. Pergamon Press: Oxford, 1968, pp. 365-482; (b) Ogata, Y., Trahanovsky, W. S. *"Oxidation in Organic Chemistry"*, Academic Press: New York, 1973, Part C, pp. 295-342; (c) Ballod, A. P., Shtern, V. Ya., *Russ. Chem. Rev.* **1976**, *45*, 721; (d) Sosnovsky, G. *"Free Radical Reactions in Preparative Organic Chemistry,"* Macmillan: New York, 1964, pp. 282, 355, 387.

35. Matasa, C., Hass, H. B. *Can. J. Chem.* **1971**, *49*, 1284.

36. (a) Bachman, G. B., Hass, H. B., Addison, L. M. *J. Org. Chem.* **1952**, *17*, 906, 914, 928, 935, 942; (b) Hass, H. B., Riley, E. F. *Chem. Rev.* **1943**, *32*, 380.

37. (a) Albright, L. F. In: *"Kirk-Othmer Encyclopedia of Chemical Technology,"* 3rd ed., Wiley-Interscience: New York, 1981, Vol. **15**, p. 841; (b) Albright, L. F. *Chem. Eng.* **1966**, *73*, 149; (c) Topdliev, A. V. *"Nitration of Hydrocarbons,"* Pergamon Press: New York, **1959**, 226-268; Titov, A. I. *Tetrahedron* 1963, **19**, 557, and references therein.

38. Sosnovsky, G. *"Free Radical Reactions in Preparative Organic Reactions,"* Macmillan: New York, 1964, Chapter 7.

39. Ogata, Y. In: *"Oxidation in Organic Chemistry,"* Part C, Academic Press: New York, 1978, Chapter 4.

40. Larson, H. O. In: *"The Chemistry of the Nitro and Nitroso Group,"* Feuer, H. ed., Wiley-Interscience: New York, 1969, Vol. I, Chapter 6.

41. (a) Wislicenus, W., Grutzner, R. *Chem. Ber.* **1909**, *42*, 1930; (b) Wislicenus, W., Waldnuller, M. *Chem. Ber.* **1908**, *41*, 3334.

42. Black, A. P., Babers, F. H. In: *Organic Syntheses*, Coll. Vol. II, (Blatt, A. H. ed.), John Wiley and Sons: New York, 1943, p. 512.

43. Wieland, H., Garbisch, P., Chavan, J. J. *Ann. Chem.* **1928**, *461*, 295.

44. Feuer, H., Shepherd, J. W., Savides, C. *J. Am. Chem. Soc.* **1956**, *78*, 4364.

45. Klager, K. *J. Org. Chem.* **1955**, *20*, 646.

46. Feuer, H., Anderson, R. S. *J. Am. Chem. Soc.* **1961**, *83*, 2960.

47. Feuer, H. Savides, G. *J. Am. Chem. Soc.* **1959**, *81*, 5826.

48. Feuer, H., Vincent, B. F. *J. Org. Chem.* **1964**, *29*, 939.

49. Feuer, H., Pivawer, P. M. *J. Org. Chem.* **1966**, *31*, 3152.

50. Kornblum, N. *Org. React.* **1962**, *12*, 101.

51. Feuer, H., Shepherd, J. W., Savides, C. *J. Am. Chem. Soc.* **1956**, *78*, 4364; Feuer, H., Lawrence, J. P. *J. Org. Chem.* **1972**, *37*, 3662; Feuer, H., Spinicelli, L. F. *J. Org. Chem.* **1976**, *41*, 2981; Feuer, H., Van Buren, II, W. D., Grutzner, J. B. *J. Org. Chem.* **1978**, *43*, 4676; Truce, W. E., Christencen, L. W. *Tetrahedron* **1969**, *25*, 181; March, J. *"Advanced Organic Chemistry,"* 3rd ed., Wiley-Interscience: New York, 1985, p. 638.

52. Matacz, Z., Piotrowska, Z., Urbański, T. *Pol. J. Chem.* **1979**, *53*, 187; Kornblum, N., Singh, H. K., Kelly, W. J. *J. Org. Chem.* **1983**, *48*, 332; Meyer, K. H. In: *Organic Synthesis,* Coll. Vol. I, 1967, 390; Hartman, H. H., Sheppard, O. E. In: *Organic Synthesis,* Coll. Vol. II, 1967, 440; Black, A. P., Babers, F. H. In: *Organic Synthesis,* Coll. Vol. II, 1967, 512.

53. Emmons, W. D., Freeman, J. P. *J. Am. Chem. Soc.* **1955**, *77*, 4391.

54. (a) Emmons, W. D., Freeman, J. P. *J. Am. Chem. Soc.* **1955**, *77*, 4387; (b) Emmons, W. D., Freeman, J. P. *J. Am. Chem. Soc.* **1955**, *77*, 4673.

55. Feuer, H. In: "Chemistry of Functional Groups," (Patai, S., ed.), supplement F, Part 2 *"The Chemistry of Amino, Nitroso and Nitro Compounds and Their Derivatives,"* Wiley: New York, 1982, p. 805.

56. Vanags, G. *Chem. Ber.* **1936**, *69*, 1066; Vanags, G., Dombrowski, A. *Chem. Ber.* **1942**, *75*, 82; Fieser, L. F. *"Experiments in Organic Chemistry,"* 3rd ed., D. G. Heath: Boston, 1957, p. 127.

57. Vanags, G., Vanags, E. *Dok. Akad. Nauk SSSR* **1963**, *90*, 59; *Chem. Abstr.* **1954**, *48*, 3981.

58. Zalukajevs, L., Vanags, E. *Dok. Akad. Nauk SSSR* **1955**, *103*, 619; *Chem. Abstr.* **1956**, *50*, 5603.

59. Gudriniece, E., Neiland, O., Vanags, G. *Zh. Obshch. Khim.* **1954**, *24*, 1863; *Chem. Abstr.* **1955**, *49*, 13128.

60. Neiland, O., Laizane, Z. *Zh. Obshch. Khim.* **1964**, *34*, 2804; *Chem. Abstr.* **1964**, *61*, 14596.

61. Lowry, T. M. *J. Chem. Soc.* **1898**, *73*, 986; Lowry, T. M., Steele, V. *J. Chem. Soc.* **1915**, *107*, 1038.

62. Levand, O., Ph.D. Thesis, University of Hawaii, 1963; Larson, H. O., Levand, O., Barnes, L., Shoolery, J. N. *Australian J. Chem.* **1962**, *15*, 431.

63. Larson, H. O., Wat, E. K. W. *J. Am. Chem. Soc.* **1963**, *85*, 827.

64. Weisblat, D. I., Lyttle, D. A. *J. Am. Chem. Soc.* **1949**, *71*, 3079.

65. Kissinger, L. W., Ungnade, H. E. *J. Org. Chem.* **1958**, *23*, 1340.

66. Wittig, G., Pockels, U. *Chem. Ber.* **1936**, *69*, 790.

67. Titov, A. I. *Obshch. Khim.* **1948**, *18*, 1312; *Chem. Abstr.* **1949**, *43*, 4217.

68. Dox, A. W. In: *Organic Synthesis*, Coll. Vol. I, 2nd ed. (Gilman, H., and Blatt, A. H., eds.), John Wiley and Sons: New York, 1946, p. 266.

69. Sifniades, S. *J. Org. Chem.* **1975**, *40*, 3562.

70. Shevelev, S. A., Semenov, V. V., Fainzil'berg, A. A. *Izv. Akad. Nauk SSSR Ser. Khim.* **1977**, 2530.

71. Shevelev, S. A., Semenov, V. V., Fainzil'berg, A. A. *Izv. Akad. Nauk SSSR Ser. Khim.* **1978**, 1091.

72. (a) Olah, G. A., Lin, C. H. *J. Am. Chem. Soc.* **1971**, *93*, 1259; (b) Olah, G. A., Rao, C. B., Olah, J. A. unpublished results.

73. (a) Olah, G. A. *J. Am. Chem. Soc.* **1972**, *94*, 808; (b) Olah, G. A. "*Carbocations and Electrophilic Reactions*," Verlag Chemie, Weinheim (Germany), Wiley: New York, 1973; (c) *Angew. Chem. Int. Ed.* **1973**, *12*, 173.

74. Olah, G. A., Olah, J. A. *J. Am. Chem. Soc.* **1971**, *93*, 1256.

75. Cremaschi, P., Simonetta, M. *Theor. Chim. Acta* **1974**, *34*, 175.

76. Bach, R. D., Holubka, J. W., Badger, R. C., Rajan, S. Y. *J. Am. Chem. Soc.* **1979**, *101*, 4416.

77. Walborsky, H. M., Baum, M. E., Yossef, A. A. *J. Am. Chem. Soc.* **1961**, *83*, 988.

78. Goering, H. L., Sloan, M. F. *J. Am. Chem. Soc.* **1961**, *83*, 1397.

79. Olah, G. A., Rochin, C. *J. Org. Chem.* **1987**, *52*, 1987.

80. Kashin, A. N., Bumagin, N. A., Bessonova, M. P., Beletskaya, I. P., Reutov, O. A. *J. Org. Chem. USSR* **1980**, *16*, 1153.

81. Corey, E. J., Estreicher, M. *Tetrahedron Lett.* **1980**, 1113 (footnote 9).

82. Wieland, H., Rahn, F. *Ber.* **1921**, *54*, 1775.

83. Michael, A., Carlson, C. H. *J. Am. Chem. Soc.* **1935**, *57*, 1268.

84. Wieland, H., Rahn, F. *Chem. Ber.* **1921**, *54*, 1770.

85. Van der Lee, F. *Rec. Trav. Chim.* **1926**, *45*, 674.

86. Martynov, I. V., Kniglyak, Y. L. *J. Gen. Chem. USSR (Engl. Transl.)* **1965**, *35*, 974.

87. Bull, J. R., Jones, E. R. H., Meakins, G.D. *J. Chem. Soc.* **1965**, 2601.

88. Fieser, L. F., Fieser, M. *"Steroids,"* Reinhold: New York, 1959, pp. 43, 44, 545.

89. (a) Windaus, A. *Chem. Ber.* **1903**, *36*, 3752; (b) Windous, A. *Chem. Ber.* **1920**, *53*, 488; Barton, D. H. R., Rosenfelder, W. *J. J. Chem. Soc.* **1951**, 1048.

90. Bowers, A., Sanchez, M. B., Reingold, H. J. *J. Chem. Soc.* **1959**, *81*, 3702.

91. Anagnostopoulos, C. E., Fieser, L. F. *J. Am. Chem. Soc.* **1954**, *76*, 532.

92. Petrov, A. D., Bulygina, M. A. *Dorl. Akad. Nauk, CCR* **1951**, *77*, 1033.

93. Haitinger, L. *Ann. Chem.* **1878**, *193*, 366.

94. Col, C., Doumani, T. U.S. Pat. 2,478,243 (1950); *Chem. Abstr.* **1950**, *44A*, 1128.

95. Zakharin, L. I. *Izv. Akad. Nauk. CCCR* **1957**, 1064.

96. Topchiev, A. V., Fantalova, E. A. *Dokl. Akad. Nauk CCCR*, **1953**, *88*, 83.

97. (a) Knuyants, I. L., German, L. G., Rozhkov, I. N. *Izv. Akad. Nauk. USSR (Eng. Transl.)* **1962**, 1794; (b) Titov, A. I. *Dokl. Akad. Nauk USSR*, **1963**, *149*, 3301; (c) Coe, P. L., Juges, A. E., Tatlov, J. C. *J. Chem. Soc. [C]* **1966**, 2323; Coe, P. L., Juges, A. E., Tatlov, J. C. *Tetrahedron* **1968**, *24*, 5913.

98. Bordwell, F. G., Garbisch, E. W. *J. Am. Chem. Soc.* **1960**, *82*, 3588.

99. Griswold, A. A., Strarcher, P. S. *J. Org. Chem.* **1966**, *31*, 357.

100. Bordwell, F. G., Garbisch, E. W. *J. Org. Chem.* **1962**, *27*, 2322.

101. Bordwell, F. G., Garbisch, E. W. *J. Org. Chem.* **1962**, *27*, 3049.

102. Bordwell, F. G., Garbisch, E. W. *J. Org. Chem.* **1963**, *28*, 1765.

103. Bachman, G. B., Hokama, T. *J. Org. Chem.* **1960**, *25*, 178.

104. Levy, N., Scaife, C. W. *J. Chem. Soc.* **1946**, 1093, 1096, 1100.

105. Levy, N., Scaife, E. W., Smith, A. E. W. *J. Chem. Soc.* **1948**, 52.

106. (a) Baldock, H., Levy, N., Scaife, C. W. *J. Chem. Soc.* **1949**, 2627; (b) Fokin, A. V., Komarov, V. A., Sorochkin, I. N., Davydova, S. M. *Zh. Vses. Khim.* **1965**, *10*, 354; *Chem. Abstr.* **1965**, *63*, 8182.

107. (a) Seifert, W. K. *J. Org. Chem.* **1963**, *28*, 125; (b) Shechter, H., Conrad, F. *J. Am. Chem. Soc.* **1952**, *74*, 3052.

108. (a) Bonetti, G. A., DeSavigny, C. B., Michalski, C., Rosenthal, R. *J. Org. Chem.* **1968**, *33*, 237; (b) Shechter, H., Conrad, F. *J. Am. Chem. Soc.* **1953**, *75*, 5610.

109. (a) Griswold, A. A., Starcher, P. S. *J. Org. Chem.* **1966**, *31*, 357; (b) Ustavshchikov, B. F., Podgornova, V. A., Dormidontova, N. V., Farberov, M. I. *Neftekjimiya* **1965**, *5*, 873; *Chem. Abstr.* **1966**, *64*, 7981.

110. (a) Shechter, H., Conrad, F. *J. Am. Chem. Soc.* **1953**, *75*, 5610; (b) Shechter, F., *Rec. Chem. Prog.* **1964**, *25*, 55; (c) Hassner, A. *Accts. Chem. Res.* **1971**, *4*, 9.

111. (a) Stevens, T. W., Emmons, W. D. *J. Am. Chem. Soc.* **1958**, *80*, 338; (b) Coffman, D. D., Raasch, M. S., Rigby, G. W., Barnick, P. L., Hanford, W. E. *J. Org. Chem.* **1949**, *14*, 747.

112. Houben-Weyl *"Methoden der Organischen Chemie,"* Vol. **X**, Part 1, Thieme-Stuttgart, 1971, p. 82.

113. Perekalin, V. V., Lerner, O. H. *Dokl. Akad. Nauk. SSSR* **1959**, *129*, 1303.

114. Baldock, H., Levy, N., Scaife, C. W. *J. Chem. Soc.* **1949**, 2627.

115. Brand, J. C. D., Stevens, I. D. R. *J. Chem. Soc.* **1958**, 629.

116. Shechter, H. *Rec. Chem. Prog.* **1964**, *25*, 55.

117. Bonetti, G. A., DeSavigny, C. B., Michalski, C., Rosenthal, R. Prep. Div. Petr. Chem. *Am. Chem. Soc.* **1965**, 135.

118. Sosnovsky, G. *"Free Radical Reactions in Preparative Organic Chemistry,"* MacMillan: New York, 1964, p. 255.

119. Pryor, W. A., Lightsey, J. W., Church, D. F. *J. Am. Chem. Soc.* **1982**, *104*, 6685.

120. Khan, N. A. *J. Chem. Phys.* **1955**, *23*, 2447.

121. Sprung, J. L., Akimoto, H., Pitts, J. N., Jr. *J. Am. Chem. Soc.* **1971**, *93*, 4358.

122. (a) Lachowicz, D. R., Kreuz, K. L. *J. Org. Chem.* **1967**, *32*, 3885; (b) Duynstee, E. F. J., Hennekens, J. L. J. P., van Raayen, W., Voskuil, W. *Tetrahedron Lett.* **1971**, *34*, 3197; (c) Kenley, R. A., Hendry, D. G. *J. Am. Chem. Soc.* **1982**, *104*, 220.

123. (a) Ashmore, P. G., Tyler, B. J. *J. Chem. Soc.* **1961**, 1017; (b) Asquith, P. L., Tyler, B. J. *J. Chem. Soc. D*, **1970**, 744.

124. Anderson, H. R., Jr., Scheraga, H. A., Van Artsdalen, E. R. *J. Chem. Phys.* **1953**, *21*, 1258.

125. Thaler, W. A. In: *"Methods in Free-Radical Chemistry,"* (Huyser, E. S., ed.), Marcel Dekker: New York, 1969, Vol. **2**, pp. 121-127.

126. Vaughan, W. E., Rust, F. F., Evans, T. W. *J. Org. Chem.* **1942**, *7*, 477.

127. Rust, F. F., Vaughan, W. E. *J. Org. Chem.* **1942**, *7*, 491.

128. (a) Sixma, F. L. J., Riem, R. H. *Proc. K. Ned. Akad. Wet., Ser. B.: Phys. Sci.* **1958**, *61B*, 183; (b) McGrath, B. P., Tedder, J. M. *Proc. Chem. Soc. London* **1961**, 80.

129. Wieland, H. *Ann. Chem.* **1903**, *328*, 154.

130. Wieland, H., Bloch S. *Ann. Chem.* **1905**, *340*, 63.

131. Wieland, H. *Chem. Ber.* **1903**, *36*, 2558.

132. Hurd, C. D., Patterson, J. *J. Am. Chem. Soc.* **1953**, *75*, 285.

133. Lipp, P. *Ann. Chem.* **1913**, *399*, 241; Hückel, W., Doll, W., Eskola, S., Weidner, H. *Ann. Chem.* **1941**, *549*, 186; Nickon, A., Lambert, J. B. *J. Am. Chem. Soc.* **1966**, *88*, 1905.

134. Stevens, T. E., Emmons, W. D. *J. Am. Chem. Soc.* **1957**, *79*, 6008.

135. Shechter, H., Conrad, F., Daulton, A. L., Kaplan, R. B. *J. Am. Chem. Soc.* **1952**, *74*, 3052.

136. Bachman, G. B., Hokama, T. *J. Org. Chem.* **1960**, *25*, 178.

137. Ville, J., Dupont, C. *Bull. Soc. Chim. France* **1956**, 804.

138. Steinkopf, W., Kuhnel, M. *Chem. Ber.* **1942**, *75*, 1323.

139. Price, C. C., Sears, G. A. *J. Am. Chem. Soc.* **1953**, *75*, 3276.

140. Tanabe, K., Hayashi, R. *Chem. Pharm. Bull. (Tokyo)* **1962**, *10*, 117.

141. (a) Hassner, K. H., Heathcock, C. *J. Org. Chem.* **1964**, *29*, 1350; (b) Harrison, W. A., Jones, E. R. H., Meakins, G. D., Wilkinson, P. A. *J. Chem. Soc.* **1964**, 3210.

142. Jew, S., Kim, H., Cho, Y., Cook, C. *Chem. Letters (Japan)* **1986**, 1747.

143. Kumaravel, G., Ph.D. Thesis, Indian Inst. Technology, Kanpur, 1988; Vankar, Y. D. personal communication.

144. Corey, E. J., Estreicher, H. *J. Am. Chem. Soc.* **1978**, *100*, 6294; *Tetrahedron Lett.* **1980**, 1113.

145. Hayama, T., Tomoda, S., Takeuchi, Y., Nomura, Y. *Tetrahedron Lett.* **1982**, 1109, 4733.

146. Seebach, D., Calderari, G., Knochel, P. *Tetrahedron* **1985**, *41*, 4861.

147. (a) Ostromyslenskii, I. I. *Chem. Ber.* **1910**, *43*, 97; (b) Werner, A. *Chem. Ber.* **1908**, *42*, 4324; (c) Titov, A. I. *Uspekhi Khimi* **1958**, *27*, 877.

148. (a) Schmidt, E., Fischer, H. *Chem. Ber.* **1920**, *53*, 1537; (b) Allson, F., Kenner, J. *J. Chem. Soc.* **1923**, 2314.

149. Perekalin, V. V. *"Unsaturated Nitro Compounds,"* (Engl. Transl.) Israel Progrm. Sci-Transl. Jerusalem, 1964.

150. Olah, G. A., Quinn, H. W., Kuhn, S. J. *J. Am. Chem. Soc.* **1960**, *82*, 426.

151. Olah, G. A., Schilling, P., Westerman, P. W., Lin, H. C. *J. Am. Chem. Soc.* **1974**, *96*, 3581.

152. Olah, G. A., Nojima, M. *Synthesis* **1973**, 785.

153. Guk, Y. V., Ilyushin, M. A., Golod, E. L., Gidaspov, B. V. *Russ. Chem. Rev.* **1983**, *52*, 284 (Engl. transl.).

154. Scheinbaum, M. L., Dines, M. *J. Org. Chem.* **1971**, *36*, 3641.

155. Smit, V. A., Semenovskii, A. V., Chernova, T. N. *Izv. Akad. Nauk SSSR Ser. Khim.* **1970**, 1681.

156. Smit, W. A., Semenovskii, A. V., Kucherov, V. F., Chernova, T. H., Krimer, M. Z., Lubinskaya, O. V. *Tetrahedron Lett.* **1971**, 3107.

157. Mursakulov, I. G., Zefirov, N. S. *Zhur. Org. Khim* **1977**, *13*, 1121.

158. Mursakulov, I. G., Talybov, A. G. Guseinov, M. M., Smit, V. A. *Zhur. Org. Khim* **1979**, *15*, 95.

159. Speranskii, E. M., Berestovitskaya, V. M., Perekalin, V. V., Gukkel', "*XXVI Gertsenovskie Chteniya, Khimiya, Nauchnye Doklady*" (The XXVIth Gertsen Lectures. Chemistry. Scientific reports), Leningrad, 1973, No. **2**, p. 79.

160. Zlotin, S. G., Krayushkin, M. M., Sevostyanova, V. V., Novikov, S. S. *Izv. Akad. Nauk. SSSR Ser. Khim.* **1977**, 2361.

161. Zlotkin, S. G., Krayushkin, M. M., Sevostyanova, V. V., Novikov, S. S. *Izv. Akad. Nauk. SSSR Ser. Khim.* **1977**, 2286.

162. Talybov, A. G., Mursakulov, I. G. *Azerb. Khim. Zhur.* **1978**, *3*, 64.

163. Olah, G. A., Narang, S. C. Olah, J. A., Pearson, R. L., Cupas, C. A. *J. Am. Chem. Soc.* **1980**, *102*, 3507.

164. Olsen, R. E., Fish, D. W., Hamel, E. E. In: *"Sovremennaya Khimiya Raketnykh Topliv"* (Modern Chemistry of Rocket Fuels), Atomizdat, Moscow, 1972, 55 [*Adv. Chem. Ser.*] **1965**, *54*, 48.

165. Onishchenko, A. A. Shernikova, T. V., Luk'yanov, O. A., Tartakovskii, V. A. *Izv. Akad. Nauk. SSSR Ser. Khim.* **1975**, 2342.

166. Onishchenko, A. A. Shernikova, T. V., Luk'yanov, O. A., Tartakovskii, V. A. *Izv. Akad. Nauk. SSSR Ser. Khim.* **1976**, 2530.

167. Corey, E. J., Estreicher, H. *Tetrahedron Lett.* **1980**, 1113.

168. Orton, K. J. P., McKie, P. V. *J. Chem. Soc. (London)* **1920**, *117*, 783.

169. Hager, K. F. *Ind. Eng. Chem.* **1949**, *41*, 2168.

170. Wetterholm, A. *Tetrahedron Suppl.* **1963**, 155.

171. Schischkoff, L. *Ann.* **1955**, *594*, 59.

172. Schlubach, H. H., Rott, W. *Ann.* **1955**, *594*, 59.

173. Schlubach, H. H., Braum A. *Ann.* **1959**, *627*, 78.

174. Freeman, J., Emmons, W. *J. Am. Chem. Soc.* **1957**, *79*, 1712.

175. Britteli, D. R., Boswell, G. A., Jr. *J. Org. Chem.* **1981**, *46*, 312.

176. Schmitt, R. J., Bedford, C. *Synthesis* **1986**, 493.

177. Schmitt, R. J., Bottaro, J. C., Malhotra, R., Bedford, C. D. *J. Org. Chem.* **1987**, *52*, 2294.

178. Nesmeyanov, A. N., Tolstoya, T. I., Korol'kov, V. V. *Dokl. Akad. Nauk. SSSR* **1978**, *241*, 1103.

179. Jäger, V., Motte, J. C., Viehe, H. G. *Chimica* **1975**, *29*, 516.

180. Boschan, R., Merrow, R. T., van Dolah, R. W. *Chem. Rev.* **1955**, *55*, 485.

181. Marken, C. D., Kristofferson, C. E., Roland, M. M., Manzara, A. P., Barnes, M. W. *Synthesis* **1977**, 484.

182. Lindner, V. In: *"Kirk-Othmer Encyclopedia of Chemical Technology,"* 3d ed., Wiley-Interscience: New York, 1980, Vol. **9**, p. 1561.

183. Urbański, T. *Chemistry and Technology of Explosives,"* Pergamon Press: Oxford, 1965, Vol. **2**.

184. (a) Hughes, E. D., Ingold, C. K., Pearson, R. B. *J. Chem. Soc.* **1958**, 4357; (b) Ingold, C. K. *Substitution at Elements Other Than Carbon,"* Weizmann Science Press: Jerusalem, 1959.

185. Blackall, E. L., Hughes, E. D., Ingold, C., Pearson, R. B. *J. Chem. Soc.* **1958**, 4366.

186. Olah, G. A., Noszko, L., Kuhn, S., Szelke, M. *Chem. Ber.* **1956**, *89*, 2374.

187. Olah, G. A., Olah, J. A., Overchuck, N. A. *J. Org. Chem.* **1965**, *30*, 3373.

188. Olah, G. A.. Narang, S. C., Pearson, R. L. Cupas, C. A. *Synthesis* **1978**, 452.

189. Ho, T. L., Olah, G. A. *J. Org. Chem.* **1977**, *42*, 3097.

190. Shvarts, I. I., Yakovenko, V. N., Krayushkin, M. M., Novikov, S. S., Sevost'yanov, V. V. *Izv. Akad. Nauk. SSSR Ser. Khim.* **1976**, 1674.

191. Romburgh, V. P. *Rev. Trav. Chim.* **1883**, *2*, 13.

192. Bamberger, E. *Chem. Ztg.* **1892**, *16*, 185.

193. Hinsberg, O. *Chem. Ber.* **1892**, *25*, 1092.

194. Hantzsch, A., Dolfuss, F. E. *Chem. Ber.* **1902**, *35*, 226.

195. Thiele, J., Lachmann, A. *Chem. Ber.* **1894**, *27*, 1909; *J. Chem. Soc.* **1894**, *66*, 412.

196. Wright, G. F. In: *"The Chemistry of the Nitro and Nitroso Group,"* (Feuer, H., ed.), Wiley-Interscience: New York, 1969, Vol. I, Chapter 9.

197. Emmons, W. D., Freeman, J. P. *J. Am. Chem. Soc.* **1955**, *77*, 4387.

198. (a) Crivello, J. V. *J. Org. Chem.* **1981**, *46*, 3056; (b) Suri, S. C., Chapman, R. D. *Synthesis* **1988**, 743.

199. Bamberger, E. *Chem. Ber.* **1894**, *27*, 668; *28*, 399; *Ann. Chem.* **1900**, *311*, 91; (b) Mandell, H. C. U. S. Pat. 3,071,438 (1962); *Chem. Abstr.* **1963**, *59*, 447.

200. MacKenzie, J. C., Myers, G. S., Smart, G. N. R., Wright, G. F. *Can. J. Res.* **1948**, *25B*, 138.

201. Henning, G. F. Germ. Pat. 104,280 (1899).

202. Herz, E. V. Swiss Pat. 88,759 (1920); *Chem. Zentr.* **1921**, *92*(IV), 926.

203. Hale, G. C. *J. Am. Chem. Soc.* **1925**, *47*, 2754.

204. Edward, J. T. *J. Chem. Ed.* **1987**, *64*, 599.

205. Thiele, J., Lachman, A. *Ann. Chem.* **1895**, *288*, 275.

206. Thiele, J., Uhlfelder, E. *Ann. Chem.* **1898**, *303*, 93.

207. Thiele, J. *Ann. Chem.* **1892**, *270*, 10.

208. Barton, S. S., Hall, R. H., Wright, G. F. *J. Am. Chem. Soc.* **1951**, *73*, 2201.

209. Davies, T. L., Blanchard, K. C. *J. Am. Chem. Soc.* **1929**, *51*, 1790, 1801.

210. Gilbert, E. E., Leccacorvi, J. R., Warman, M. In: *"Industrial and Laboratory Nitrations,"* (Albright, L. F., ed.), ACS Symposium Series 22; Washington, D.C. 1976; Chapter 23.

211. Olah, G. A., Kuhn, S. J. In: *"Friedel-Crafts and Related Reactions,"* (Olah, G. A. ed.), Wiley-Interscience: New York, 1964, Vol. 3, Chapter 43.

212. Olsen, R. E., Fisch, D. W., Hamel, E. E. In: *"Advanced Propellant Chemistry,"* (Gould, R. E., ed.), Advances in Chemistry Series 54, American Chemical Society: Washingon, D. C. 1966, Chapter 6.

213. Ilyushin, M. A., Golod, E. L., Gidaspov, B. V. *Zhur. Org. Khim.* **1977**, *13*, 11.

214. Ilyushin, M. A., Guk, Yu. V., Golod, E. L., Frolova, G. M., Gidaspov, B. V. *Zhur. Org. Khim.* **1979**, *15*, 103.

215. Vast, P. *Rev. Chim. Miner.* **1970**, *7*, 757.

216. Barbes, H., Vast, P. *Rev. Chim. Miner.* **1971**, *8*, 851.

217. Vast, P., Heubel, J. *Compt. rend.* **1967**, *864c*, 1697.

218. Vast, P., Heubel, J. *Compt. rend.* **1965**, *260*, 5799.

219. Brit. Pat. 1,126,591 (1968); U. S. Pat. 3,428,667 (1969); *Chem. Abstr.* **1969**, *70*, 67584.

220. Lebedev, B. A., Ilyshin, M. A., Andreev, S. A., Gidaspov, B. V. *Zhur. Org. Khim.* **1978**, *14*, 2205.

221. Andreev, S. A., Lebedev, B. A. Tselinskii, I. V. *Zhur. Org. Khim.* **1978**, *14*, 2513.

222. Luk'yanov, O. A., Kozlova, I. K., Tartakovskii, V. A. *"Tezisy VI Vsesoyuznogo Soveshchaniya po Khimii Nitrosoedinenii"* (Abstracts of Reports at the VIth All-Union Conference on the Chemistry of Nitro-Compounds), Moscow, 1977, p. 68 (quoted in reference 153).

223. Robson, J. H. *J. Am. Chem. Soc.* **1955**, *77*, 107.

224. Andreev, S. A., Gidaspov, B. V. *Zhur. Org. Khim.* **1978**, *14*, 240.

225. Pevzner, M. S., Kulibabina, T. N., Povarova, N. A., Killina, L. V. *Khim. Geterostsikl. Soed.* **1979**, 1132 (quoted in reference 153).

226. Andreev, S. A., Lebedev, B. V. *Zhur. Org. Khim.* **1978**, *14*, 907.

227. Grakauskas, V. U. S. Pat. 3300181 (1968); *Chem. Abstr.* **1968**, *69*, 35406.

228. Andreev, S. A., Lebedev, B. A., Tselinskii, I. V. *Zhur. Org. Khim.* **1978**, *14*, 909.

229. Cherednichenko, L. V., Lebedev, B. A., Gidaspov, B. V. *Zhur. Org. Khim.* **1978**, *14*, 735.

230. Luk'yanov, O. A., Mel'nikova, T. G., Seregina, N. M., Tartakovskii, V. A. "*Tezisy VI Vsesoyuznogo Soveshchaniya po Khimii Nitrosoedinenii*" (Abstracts of Reports at the VIth All-Union Conference on the Chemistry of Nitro-Compounds), Moscow, 1977, p. 33 (quoted in reference 153).

231. Luk'yanov, O. A., Seregina, N. M., Tartakovskii, V. A. *Izv. Akad. Nauk SSSR Ser. Khim.* **1976**, 225.

232. Katritzky, A. R., Mitchell, J. W. *J. Chem. Soc. Perkin Trans. (I)* **1973**, 2624.

233. Doyle, M. P., Whitefleet, J. A. L., Debruyn, D. J., Wierenge, W. *J. Am. Chem. Soc.* **1977**, *99*, 494.

234. Glass, C. R. S., Blount, J. F., Batler, D. *Can. J. Chem.* **1972**, *50*, 3472.

235. (a) Olah, G. A., Gupta, B. G. B., Narang, S. C. *J. Am. Chem. Soc.* **1979**, *101*, 5317; (b) Olah, G. A., Gupta, B. G. B. *J. Org. Chem.* **1983**, *48*, 3585; (c) Olah, G. A., Narang, S. C., Salem, G. F., Gupta, B. G. B. *Synthesis* **1979**, 273.

236. Meyer, V., Stüber, O. *Chem. Ber.* **1872**, *5*, 203, 399, 514.

237. Meyer, V., Chojnacki, C. *Chem. Ber.* **1872**, *5*, 1034.

238. Meyer, V. *Ann. Chem.* **1874**, *171*, 1.

239. Plummer, C. W., Drake, N. L. *J. Am. Chem. Soc.* **1954**, *76*, 2720.

240. Kornblum, N., Taub, B., Ungnade, H. E. *J. Am. Chem. Soc.* **1954**, *76*, 3209.

241. Kornblum, N., Ungnade, H. E. In: *Organic Synthesis*, Coll. Vol. IV, 1963, p. 724.

242. Kornblum, N., Smiley, R. A., Blackwood, R. K., Iffand, D. C. *J. Am. Chem. Soc.* **1955**, *77*, 6269.

243. Kornblum, N., Smiley, R. A., Ungnade, H. E., White, A. M., Taub, B., Herbert, S. A. *J. Am. Chem. Soc.* **1955**, *77*, 5528.

244. Feuer, H., Leston, G. In: *Organic Synthesis,* Coll. Vol. IV, (Rabjohn, E., ed.), Wiley: New York, 1963, p. 368.

245. Kornblum, N., Ungnade, H. E. *In: "Organic Synthesis, Coll. Vol.* IV, Wiley: New York, 1963, 724.

246. Kornblum, N., Chalmers, M., Daniels, R. *J. Am. Chem. Soc.* **1955**, *77*, 6654.

247. Sokovishina, I. F., Perekdin, V. V., Lerner, O. M., Andreeva, L. M. *Zh. Org. Khim.* **1965**, *1*, 636.

248. Kornblum, N., Fischbein, L., Smiley, R. A. *J. Am. Chem. Soc.* **1955**, *77*, 6261.

249. Kolbe, H. *J. Prakt. Chem.* **1872**, *5*, 427.

250. Whitmore, F. C., Whitmore, M. G. In: *Organic Synthesis*, Coll. Vol. I, 1951, p. 401.

251. Treibs, W., Reinhenkel, H. *Chem. Ber.* **1954**, *87*, 341.

252. Kornblum, N., Blackwood, R. K. *J. Am. Chem. Soc.* **1956**, *78*, 4037.

253. Kornblum, N., Blackwood, R. K., Mooberry, D. D. *J. Am. Chem. Soc.* **1956**, *78*, 1501.

254. Kornblum, N., Larson, H. O. Blackwood, R. K., Mooberry, D. D., Oliveto, E. P., Graham, G. E. *Chem. Ind. (London)* **1955**, 433; Kornblum, N., Larson, H. O. Blackwood, R. K., Mooberry, D. D., Oliveto, E. P., Graham, G. E. *J. Am. Chem. Soc.* **1956**, *78*, 1497.

255. Kornblum, N., Powers, J. W. *J. Org. Chem.* **1957**, *22*, 455; Kornblum, N. *et al. J. Am. Chem. Soc.* **1956**, *78*, 1497; Kornblum, N., Blackwood, R. K., Powers, J. W. *J. Am. Chem. Soc.* **1957**, *79*, 2507.

256. Stille, J. K., Vessell, E. D. *J. Org. Chem.* **1960**, *25*, 479.

257. Kornblum, N., Blackwood, R. K., Powers, J. W. *J. Am. Chem. Soc.* **1957**, *79*, 2507.

258. Kornblum, N., Blackwood, R. K. In: *Organic Synthesis*, Coll. Vol. IV (Rabjohn, N., ed.), John Wiley and Sons: New York, 1963, p. 454.

259. Fusco, R., Rossi, S. *Chem. Ind. (London)* **1957**, 1650.

260. Bull, J. R., Jones, E. R. H., Meakins, G. D. *J. Chem. Soc.* **1965**, 2601.

261. Nielsen, A. T. *J. Org. Chem.* **1962**, *27*, 1993.

262. Breslow, R., Kivelebich, D., Mitchell, M. J., Fabian, W., Cendel, K. *J. Am. Chem. Soc.* **1965**, *87*, 5132.

263. Caserio, M. C., Simmons, H. E., Johnson, A. E., Roberts, J. D. *J. Am. Chem. Soc.* **1960**, *82*, 3102.

264. ter Meer, E. *Ann. Chem.* **1876**, *181*, 1.

265. Feuer, H., Colwell, G. E., Leston, G., Nielsen, A. T. *J. Org. Chem.* **1962**, *27*, 3598.

266. Hawthorne, M. F. *J. Am. Chem. Soc.* **1956**, *78*, 4980.

267. Kaplan, R. B., Shechter, H. *J. Am. Chem. Soc.* **1961**, *83*, 3535.

268. Kornblum, N., cited by H. Feuer, *Tetrahedron Suppl.* **1964**, *1*, 107.

269. Olah, G. A., Farnia, M., unpublished results.

Author Index

Achord, J. M., 167
Adams, R. C., 44, 53
Adams, R., 204
Addison, C. C., 93, 94
Addison, L. M., 228
Adkins, H., 55
Akimoto, H., 252
Alazard, J. P., 13
Albright, F., 3
Albright, L., 228
Albright, L. F., 1, 5, 15, 56, 60, 145, 157, 228, 277, 278
Allson, F., 260
Al-Omran, F., 132
Amos, D. W., 94
Anagnostopoulos, C. E., 245
Anderson, H. R., Jr., 255
Anderson, R. S, 230
Ando, W., 84, 90
Andreev, S. A., 280, 281, 282, 283
Andreeva, L. M., 289
Andrews, L. J., 136, 137, 141
Andrisano, R., 38
Angus, W. R., 129
Anzilotti, W. F., 20
Archer, S., 19, 86
Arzamaskova, L. N., 123
Asensio, G., 98, 101
Ashmore, P. G., 255
Asinger, F., 3, 219, 227
Asquith, P. L, 255.
Attina, M., 157, 177, 179
Ausloos, P., 174, 177, 178
Avanesov, D., 89
Aynsley, E. E., 58
Babers, F. H., 229, 231

Bach, R. D., 239
Bachman, G. B., 49, 51, 52, 55, 228, 249, 257
Badger, R. C., 239
Baines, D. A., 94
Baker, P. J., Jr., 220
Baldock, H., 250, 252
Ballod, A. P., 227
Bamberger, E., 275, 276
Banyshnikova, A. N., 50
Barbes, H., 280, 282
Barlas, H., 90
Barnes, C. E., 205
Barnes, L., 232
Barnes, M. W., 266, 269
Barnett, J. W., 22
Barnick, P. L., 251
Barton, D. H. R., 245
Barton, S. S., 277
Batler, D., 66, 285
Baum, M. E., 240
Bayliss, N. S., 129
Bazant, V., 82
Bedford, C. D., 266, 267
Beilstein, F., 2, 219
Beletskaya, I. P., 243
Benezra, S. A., 173, 178
Benford, G. A., 164, 120, 130
Benkeser, R. A, 82, 96
Benkesser, R. H., 96
Berestovitskaya, V. M., 263
Berger, K., 117, 121
Berkheimer, H. E., 122
Bernardi, F., 179, 180
Bessonova, M. P., 243
Beug, M., 159

Black, A. P., 229, 231
Blackall, E. L., 26, 270
Blackstock, D. J., 162, 163
Blackwood, R. K., 288, 289, 290, 291, 292
Blatt, A. H. ed., 98
Bloch S., 256
Blount, J. F., 66, 82, 285
Blucher, W., 22, 23, 24, 25, 26, 92, 132
Bluestein, B. R., 51, 52
Bodor, N., 44
Bodtker, E., 38
Boer, P., 146
Boeters, O., 76
Bohm, W., 98
Bollmeier, A. F., Jr., 220
Bonetti, G. A., 251, 252
Bonner, T. G., 84
Bontempelli, G., 192
Boorman, P. M., 94
Bordwell, F. G., 44, 248, 249
Boschan, R., 266, 269
Boswell, G. A., Jr., 266
Bottaro, J. C., 266, 267
Boughriet, A., 93, 187, 196
Bourne, E. J., 21
Bowers, A., 245
Boyer, J. H., 65, 66
Brady, J. D., 118
Brain, D. K., 224, 225
Brand, J. C. D., 252, 255
Braum A., 266
Breslow, R., 292
Britteli, D. R., 266
Brouwer, D. M., 137, 139
Brown, H. C., 21, 118, 137, 142, 143, 164, 180
Bruce, M. R., 148
Bruice, T. C., 88
Brumfield, P. E., 82, 96
Brunton, G., 85
Bucher, W. G., 59, 67
Bull, J. R., 245, 292
Bulygina, M. A., 246
Bumagin, N. A., 243
Bunton, C. A., 117, 118, 130, 164
Bursey, M. M., 173, 178
Butler, D., 82
Buttrill, S. E., Jr., 175, 176, 177

Cacace, F., 157, 177, 179, 180
Caesar, G. V., 55, 56
Calderari, G., 260
Carlson, C. H., 244
Carr, R. V. C., 12
Carr, R. V. S., 12
Caserio, M. C., 292
Cendel, K., 292
Cerfontain, H., 73, 141
Chadwick, D. H., 15
Challis, B. C., 132
Chalmers, M., 289
Chapman, R. D., 276
Chavan, J. J., 229
Chedin, J., 124
Chénevert, R., 69, 142, 145
Cherednichenko, L. V., 283
Chernova, T. H., 263
Cho, Y., 259
Chojnacki, C., 287
Christencen, L. W., 231
Christy, P. F., 153
Church, D. F., 252, 253, 254
Chvalovsky, V., 82
Ciaccio, L. L., 11, 22
Cimino, G. G., 124, 129, 186
Claes, P., 32, 44
Clarke, H. T., 64
Clemens, A. H., 189, 194, 196, 199
Cleveland, T. H., 15
Clouet, F. L., 148
Coe, P. L., 247
Coffman, D. D, 251
Col, C., 246
Collis, M. J., 45
Colonna, H., 38
Colwell, G. E., 293
Conrad, F., 251, 252, 257, 258
Cook, C., 259
Coombes, R. D., 153, 154
Coombes, R. G., 95, 123, 131, 141
Coon, C. L., 22, 23, 24, 25, 26, 59, 67
Cooper, K. E., 55
Corey, E. J., 243, 260, 265, 266
Cornélis, A., 33, 34
Cox, E. G., 19
Craig, L. C., 98
Cram, D. J., 143
Cremaschi, P., 239

Author Index

Cretney, J. R., 162
Crivello, J. V., 32, 34, 35, 37, 276
Cruse, H. W., 85
Csürös, Z., 21
Cupas, C. A., 65, 69, 265, 266, 271, 272
Dachlauer, K., 46
Dallinga, G., 137, 139
Danforth, R. H., 79
Daniels, R., 289
Daulton, A. L., 257, 258
Davies, B., 79, 84
Davis, T. L., 76, 77
Davydova, S. M., 250
Deans, F. B., 81
de la Mare, P., 118
de Lange, M. P., 160
Delaude, L., 34
Deno, N. C., 122
de Petris, G., 179
DePuy, C. H., 44
Derendyaev, B. G., 161
DeSavigny, C. B., 251, 252
Desvergnes, L., 15, 64
Detsina, A. N., 161
Dever, J. L., 55
Dewar, M. J. S., 34, 44, 136, 137, 141
Di Giaims, M. P., 38
DiGiaimo, M. P., 16, 17, 18
Dinçtürk, S., 96, 171, 172
Dines, M., 262
Dingle, R., 129
Doering, W. v. E., 44
Dolfuss, F. E., 275
Doll, W., 256
Dolphin, D., 86
Dombrowski, A., 231
Donalson, W. J., 38, 64
Dormidontova, N. V., 250, 251
Dorsky, J., 226
Doumani, T., 246
Doumani, T. F., 158
Downs, W. J., 78
Dox, A. W., 232
Doxsee, K. M., 143
Drake, N. L., 76, 77, 287, 288
Draper, M. R., 12, 95, 171
Drummond, A. A., 64
Dunbar, R. C., 174, 175, 178

Dupont, C., 258
Duynstee, E. F. J., 255
Eaborn, C., 81, 82
Eberson, L., 4, 84, 85, 88, 166, 167, 168, 187, 189, 191, 194, 195
Edward, J. T., 276, 277
Effenberger, F., 59, 84
Elliott, R. J., 119, 181
Elsenbaumer, R. J., 60, 69, 142, 145
Emmons, W., 266
Emmons, W. D., 40, 231, 251, 257, 275
Eskola, S., 256
Estreicher, H., 260, 243, 265, 266
Euler, E., 11, 117
Evans, J. C., 14, 51, 55
Evans, T. W., 238, 255
Evans, W. L., 122
Even, C., 32, 44
Eyring, H., 117
Fabian, W., 292
Fainzil'berg, A. A., 233
Fantalova, E. A., 246
Farberov, M. I., 250, 251
Fauquenoit, C., 32, 44
Feldman, K. S., 205
Feng, J., 182
Fenichel, L., 21
Feuer, H., 3, 51, 52, 229, 229, 230, 231, 288, 293
Field, B. O., 94
Fieser, L. F., 245
Fieser, M., 245
Finar, I. L., 38
Fisch, D. W., 279, 280, 281, 282, 294
Fischbein, L., 289
Fischer, A., 44, 162
Fischer, H., 88, 89, 260
Fischer, J. C., 187, 196
Fischer, J-C., 93
Fischer, P., 84
Fish, D. W., 265, 266
Fisher, A., 162, 163
Fleming, I., 119, 181
Flewett, G. W., 94
Flood, S. H., 14, 60, 135, 136, 137, 138, 144, 178
Fokin, A. V., 250
Forsyth, D. A., 161
Foster, P. W., 144, 146

Frazer, M. G., 18
Frazer, V. S., 18
Freeman, J., 266
Freeman, J. P., 40, 231, 275
Freiser, B. S., 176
Fridman, A. L., 86
Frolova, G. M., 279
Fujiwara, K., 132
Fukunaga, K., 33
Fukuzumi, S., 139, 184, 166
Fung, A. P., 30, 31, 72, 73
Fusco, R., 292
Garbisch, E. W., 44, 229, 248, 249
Garner, D. C., 94
Garrett, A. B., 30
German, L. G., 247
Gerstmans, A., 34
Gidaspov, B. V., 65, 66, 82, 262, 279, 280, 281, 283
Giffney, J. C., 132
Gilbert, E. E., 277, 278
Gillespie, R. J., 19, 50, 117, 118
Gintz, F. P., 45
Glass, C. R. S., 285
Glass, R. S., 66, 82
Goddard, D. R., 19, 45, 57, 117, 118
Goering, H. L., 240
Goke, J., 59
Golad, E. L., 65, 66
Gold, V., 44, 142, 143
Golod, E. L., 82, 262, 279
Goug, L. C., 86
Gould, R. E., 279, 280, 281, 282, 294
Goulden, J. D, 129.
Goulden, J. D. S., 50
Grabovskaya, Zh. E., 123
Graham, G. E., 291, 292
Graham, J., 50, 117, 118
Grakauskas, V., 282
Grandinetti, F., 179, 180
Gregory, M. J., 88
Griswold, A. A., 248, 249, 250, 251
Groggin, P. H., 5
Gross, M. L., 137, 138, 178, 179
Grutzner, J. B., 231
Grutzner, R., 229
Gudriniece, E., 232, 250
Guk, Y. V., 82, 262, 279

Guk, Yu. V., 65, 66, 279
Gupta, B. G. B., 286
Guseinov, M. M., 263, 265
Haas, H. B., 2
Habib, R. M., 32
Hager, K. F., 266
Haines, L. B., 55
Haitinger, L., 246
Hale, G. C., 276
Halevi, E. A., 117, 118, 140
Hall, R. H., 277
Halvarson, K., 44
Hamel, E. E.,265, 266, 279, 280, 281, 282, 294
Hammett, L., 137
Hancock, R. A., 84
Haney, W. A., 184, 185
Hanford, W. E., 251
Hanna, S. B., 133, 145, 146, 153, 186
Hanson, C., 60, 145, 157
Hantzsch, A., 18, 57, 117, 121, 275
Hardy, C. J., 94
Harrar, J. E., 12
Harris, C. R., 158
Harrison, A. G., 179
Harrison, W. A., 258
Hartman, H. H., 231
Hartman, W. W., 64
Hartshorn, S. R., 44
Hartshoth, M. R., 162
Hass, H. B., 219, 226, 227, 228
Hassel, O., 137, 138
Hassner, A., 251, 252
Hassner, K. H., 258
Hawkins, J. G., 98, 99
Hawthorne, M. F., 293
Hayama, T., 260
Hayashi, E., 43
Hayashi, R., 258, 259
Heathcock, C., 258
Hebdon, E. A., 45
Heimkanys, R. W., 76, 77
Helsby, P., 194
Henderson, G. N., 162, 163
Hendry, D. G., 255
Hennekens, J. L. J. P., 255
Henning, G. F., 276
Henning, R., 220

Hennion, G. F., 20
Henry, P. M., 78
Herbert, S. A, 288
Herz, E. V., 276
Hetherington, G., 58
Hetherington, O., 48
Heubel, J., 280
Heubel, J., 280
Hey, D. H., 202
Heyworth, F., 98, 99
Hilinski, E. F., 185
Hill, M. E., 22, 23, 24, 25, 26, 59, 67
Hinsberg, O., 275
Ho, T. L., 266, 273
Hodge, E. B., 226
Hodgson, H. H., 9, 97, 98, 99
Hoffmann, M. K., 173, 178
Hofstra, A., 137, 139
Hoggett, J. G., 3, 44, 141, 153, 154, 172, 178
Hokama, T., 49, 249, 257
Holbenstadt, E. S., 164
Holleman, A. F., 117
Holman, R. W., 137, 138, 178, 179
Holubka, J. W., 239
Hood, G. C., 21
Hopkins, B., 221
Horita, H., 32
Houben-Weyl., 14, 20, 43, 97, 251
Hoyano, F., 77
Huang, G. F., 9
Hückel, W., 256
Hughes, E. D., 19, 57, 44, 50, 117, 118, 121, 130, 164, 266, 270
Hum, G. P., 132
Hunziker, E., 133, 134, 140, 144, 145, 146, 153, 186
Hurd, C. D., 256
Hussey, C. L., 167
Huyser, E. S., ed., 255
Ichikawa, K., 78, 79
Iffland, D. C., 220, 288, 289, 290, 292
Ihyushin, M. A., 82
Ikarigawa, T., 90
Ilyshin, M. A., 280
Ilyushin, M. A., 65, 66, 262, 279
Inana, K., 43
Ingold, C., 266, 270
Ingold, C. J., 19, 57

Ingold, C. K., 3, 6, 9, 14, 44, 50, 55, 59, 68, 117, 118, 120, 121, 130, 144, 164, 202, 222, 266, 270
Ioffe, S. L., 82
Isaeva, L. S., 98
Ishikawa, H., 90
Ishikawa, T., 43
Iwata, E., 77
Iyer, L. M., 162, 163
Jacobs, D. I. H., 130
Jäger, V., 266, 267
Jaruzelski, J. J., 122
Jayasuriya, K., 157, 181
Jeffrey, G. A., 19
Jew, S., 259
Johnson, A. E., 292
Johnson, G. D., 98, 99
Johnson, M. W., 205
Johnson, R. M., 92, 93, 94
Johnston, J. F., 194
Jones, E. R. H., 292, 245, 258
Jones J., 64
Jones, M. H., 130
Jonsson, L., 167
Juges, A. E., 247
Kagan, H. B., 13
Kalmár, A., 21
Kamei, T., 225
Kameo, T., 27, 28
Kanno, S., 90
Kaplan, E. P., 95
Kaplan, R. B., 257, 258, 293
Kapustina, N. I., 95
Kashin, A. N., 243
Katritzky, A. R., 129, 284
Keefer, R. M., 136, 137, 141
Kelly, W. J., 231
Kenley, R. A., 255
Kenner, J., 164, 260
Kesten, E. M., 86
Khan, N. A., 252
Kharash, M. S., 118, 121
Kharkats, , Yu I., 188, 192
Killina, L. V., 282
Kim, E. K., 75
Kim, E. K., 185, 195
Kim, H., 259
Kimura, M., 33
Kirpal, A., 98

Kissinger, L. W., 232
Kivelebich, D., 292
Klager, K., 229
Klemenc, A., 55
Klimenko, V. I., 86
Klopman, G., 119, 181
Kniglyak, Y. L., 245, 247
Knochel, P., 260
Knowles, J. R., 137
Knuyants, I. L., 247
Kobe, K. A., 158
Kochi, J. K., 75, 139, 166, 184, 185, 186, 195
Kolbe, H., 290
Komarov, V. A., 250
Komoto, H., 77
König, W., 43
Konovalov, M., 219
Konovalov, M. I., 221
Koptyug, V. A., 161
Kornblum, N., 220, 231, 288, 289, 290, 291, 292, 294
Korol'kov, V. V., 266, 267
Korte, F., 90
Kotzias, D., 90
Kozlova, I. K., 281
Krayushkin, M. M., 264, 266, 274
Kreuz, K. L., 255
Krimer, M. Z., 263
Krishnamurthy, V. V., 77, 78
Kristofferson, C. E., 266, 269
Kucherov, V. F., 263
Kudav, N. A., 32
Kuhlmann, K. F., 124, 125, 126, 127
Kuhn, L. P., 5
Kuhn, S., 3, 14, 20, 38, 45, 46, 50, 51, 55, 56, 57, 58, 59, 60, 61, 62, 63, 65, 135, 136, 137, 138, 139, 144, 146, 159, 178, 261, 266, 278
Kuhnel, M., 258
Kulibabina, T. N., 82, 282
Kumaravel, G., 259
Kurbatov, A., 2, 219
Kurtz, W., 84
Kurz, M. E., 44, 53, 204
Lachman, A., 275, 276, 277
Lachowicz, D. R., 255
Laizane, Z., 232
Lambert, J. B., 256

Lammertsma, K., 145
Landesman, H., 82
Landesman, H., 96
Larson, H. O., 229, 231, 232, 291, 292
Laszlo, P., 33, 34
Laurence, P. R., 157, 181
Lawrence, J. P., 231
Lawson, A. J., 132
Lawsson, S. O. 164, 181
Lazlo, P., 33
Lebedev, B. A., 280, 281, 282, 283, 282
Leccacorvi, J. R., 277, 278
Leckie, A. H., 129
Lee, H. W. H., 205
Lehr, F., 220
Lerner, O. H., 252
Lerner, O. M., 289
Leston, G., 288, 293
Levand, O., 232
Levy, N., 250, 251, 252
Lias, S. G., 174, 177, 178
Lichtin, N. N., 220
Lien, A. P., 139
Lightsey, J. W., 252, 253, 254
Likhomanova, G. I., 65
Lin, C. H., 233, 235
Lin, H., 219
Lin, H. C., 26, 39, 40, 45, 47, 64, 65, 71, 74, 97, 138, 142, 143, 145, 149, 150, 153, 161, 261
Lindner, V., 266, 270
Lipp, P., 256
Llewellyn, D. R., 117
Logan, N., 94
Lotz, F., 86
Lowry, T. H., 173, 205, 232
Lubinknowski, J. K., 98
Lubinskaya, O. V., 263
Luk'yanov, O. A., 265, 266, 281, 283
Lütgert, H., 96
Lyttle, D. A., 232
MacKenzie, J. C., 276
Mackor, E. L., 137, 139
MacLean, C., 137, 139
Macsi, B., 38
Magno, F., 192
Mahadevan, A. P., 97
Mahadevan, A. P., 98, 99
Main, L., 133

Majumdar, M. P., 32
Maletina, I. I., 59
Malhotra, R., 27, 28, 38, 39, 40, 41, 42, 52, 53, 54, 73, 74, 92, 93, 94, 124, 125, 126, 127, 133, 167, 266, 267
Mamatyuk, V. I., 161
Manabe, O., 27, 28
Mandell, H. C., 276
Manzara, A. P., 266, 269
March, J., 231
Marcus, R. A., 11, 22, 167
Marken, C. D., 266, 269
Markovnik, A. S., 169, 193
Markownikov, V., 219, 222, 224
Marsden, E. J., 98, 99
Martin T., 2, 219, 222
Martynov, I. V., 245, 247
Marziano, N., 124, 129, 186
Masci, B., 142, 145
Masci, B. J., 69
Maslina, I. A., 82
Masnovi, J. M., 185
Matacz, Z., 231
Matasa, C., 227
Matveeva, M. K., 223
Mazzocchin, G. A., 192
McCaulay, D. A., 139
McEwen, W. E., 98
McGrath, B. P., 255
McKie, P. V., 266
Mckillop, A., 79
Meakins, G. D., 258, 292
Meakins, G.D., 245
Meisenheimer, J., 202
Mel'nikova, T. G., 283
Melander, L., 44, 123
Menke, J. B., 32
Merrow, R.T., 266, 269
Meyer, K. H., 231
Meyer, V., 220, 287
Michael, A., 244
Michalski, C., 251, 252
Millen, D. J., 19, 49, 50, 117, 118, 129, 144
Miller, D. G., 28
Milligan, B., 28, 84, 133
Minkoff, G. J., 130, 164
Misumi, S., 32

Mitchell, J. W., 284
Mitchell, M. J., 292
Mitscherlich, E., 2, 219
Mlinko, A., 20, 57, 58, 60, 135, 137, 138, 144,
Mo, Y. K., 74, 161
Mole, T., 34
Molinari, E., 86
Mooberry, D. D., 290, 291, 292
Moodie, R. B., 3, 16, 18, 21, 22, 44, 123, 126, 128, 131, 133, 141, 153, 154, 172, 178, 186,
Moran, K. D., 133
Motte, J. C., 266, 267
Mulliken, R. S., 137, 138, 180
Mursakulov, I. G., 263, 265, 266
Mustafa, H. T., 32
Myers, G. S., 276
Myhre, P. C., 73, 133, 134, 140, 144, 159, 163, 205
Nabeshima, T., 87
Nakagura, S., 164, 180
Nakaoka, I., 84
Namba, K., 77, 84
Nametkin, S. S., 222, 224
Narang, S. C., 27, 28, 30, 31, 38, 39, 40, 41, 42, 45, 52, 53, 54, 65, 69, 71, 72, 73, 73, 74, 77, 78, 82, 89, 145, 149, 150, 153, 167, 168, 169, 198, 265, 266, 271, 272, 286
Neiland, O., 232, 250
Nesmeyanov, A. N., 98, 266, 267
Nickon, A., 256
Nielsen, A. T., 292, 293
Nightingale, D. V., 160
Nikishin, G. I., 95
Nishimura, S., 27, 28
Nogradi, J., 58
Nojima, K., 90
Nojima, M., 261
Nomura, Y., 260
Norman, R. O. C., 79, 44, 137
Noszko, L., 266, 271
Novikov, S. S., 264, 266, 274
O'Brien, D. H., 147
Oae, S., 87
Ogata, Y., 225, 227, 228
Okada, T., 78, 79

Okamoto, Y., 137
Okano, M., 32, 79, 80, 82, 84
Olah, G. A, 3, 13, 14, 16, 20, 22, 26, 27, 28, 30, 31, 38, 39, 41, 42, 45, 50, 51, 52, 53, 54, 55, 56, 57, 58, 60, 61, 62, 63, 64, 65, 68, 69, 71, 72, 73, 74, 77, 78, 82, 89, 97, 98, 101, 118, 119, 126, 135, 136, 137, 138, 139, 140, 142, 143, 144, 145, 146, 147, 148, 149, 150, 152, 153, 154, 159, 161, 167, 168, 169, 174, 175, 178, 198, 203, 204, 219, 233, 234, 235, 237, 241, 261, 265, 266, 271, 272, 273, 278, 286, 295
Olah, J. A., 30, 31, 38, 45, 64, 65, 69, 71, 73, 74, 89, 145, 149, 150, 153, 167, 168, 169, 198, 233, 234, 237, 266, 271
Oliveto, E. P., 291, 292
Olsen, R. E., 265, 266, 279, 280, 281, 282, 294
Omran, F., 194
Onishchenko, A. A., 265, 266
Ono, Y., 54
Orda, V. V., 59
Orton, K. J. P., 266
Osawa, T., 77
Osei-Gymah, P., 86
Ostromyslenskii, I. I., 260
Ott, R. J., 153
Overchuck, N. A., 38, 64, 54, 65, 69, 89, 152, 153, 154, 203, 204, 266, 271
Oxford, A. E., 43
Parker, V., 188, 191
Parker, V. D.,165, 166
Parlar, H., 90
Parr, W. J. E., 79
Passerini, A., 124, 129, 186
Passino, M. J., 19
Patai, S., ed., 3
Patterson, J., 256
Patton, J. T., 220
Paul, A. P., 16, 17, 18, 38, 45
Pavlath, A., 139
Pawellek, D., 9, 14
Pearson, D. E., 18, 266, 270
Pearson, R. K., 12, 65, 69, 265, 266

Pearson, R. L., 266, 271, 272
Pederson, E. B., 164, 181
Peeling, E. R. A., 50, 117, 118
Pennetreau, P., 33
Penton, J. R., 3, 44, 133, 134, 140, 144, 141, 153, 178, 186
Perekalin, V. V., 252, 261, 263, 266
Perekdin, V. V., 289
Perrin, C. L., 159, 164, 167
Person, W. B., 180
Petersen, T. E., 164, 181
Peterson, H. J., 122
Petrov, A. D., 246
Pevzner, M. S., 65, 66, 82, 282
Pfeiffer, P., 139
Pigeon-Gosselin, M., 69, 142, 145
Pinck, J. A., 50
Pinck, L. A., 130
Pinna, F., 124, 129, 186
Piotrowska, Z., 231
Pitts, J. N., Jr., 252
Pivawer, P. M., 230
Plato, A., 86
Plazak, E., 38
Plummer, C. W., 287, 288
Pockels, U., 232
Podgornova, V. A., 250, 251
Politzer, P., 157, 181
Pollitt, A. A., 64
Poole, H. C., 117, 118, 144
Povarova, N. A., 282
Powers, J. W., 291, 292
Pradier, J. C., 124
Prakash, G. K. S., 22, 26, 126, 145
Price, C. C., 45, 117, 258
Pross, A., 5, 166, 194
Pryor, W. A., 252, 253, 254
Quinn, H. W., 261
Raasch, M. S., 251
Rabjohn, E., 288
Rabjohn, N., ed., 292
Radcliffe, L. G., 64
Radda, G. K., 44, 137
Radner, F., 84, 85, 88, 167, 168, 191, 189, 194, 195,
Rahn, F., 244
Rajan, S. Y., 239
Ram, M. S., 187, 193, 196

Rao, C. B., 233, 237
Raudnitz, H., 38
Read, A. J., 44
Redlich, O., 21
Reed, R. E., 164
Reed, R. I., 117, 118, 130
Reents, W. D., Jr., 176
Reid, D. H, 88
Reingold, H. J., 245
Reinhenkel, H., 290
Rentzepis, P. M., 185
Reutov, O. A., 243
Ri, T., 117
Richards, K. E., 162
Richards, W. G., 119, 181
Richardson, K. S., 173, 205
Riches, K. M., 85
Ridd, H. J., 60
Ridd, J. H., 12, 44, 95, 96, 118, 132, 136, 153, 171, 172, 189, 194, 196, 198, 199
Riem, R. H., 255
Rigby, G. W., 251
Riham, T. I., 32
Riley, E. F., 228
Rinn, H. W., 51, 55
Riordan, J. F., 88
Ritchie, C. D., 152
Roberts, J. D., 292
Robinson, F. L., 48
Robinson, P. L., 58
Robinson, R., 6, 18
Robson, J. H., 281, 284
Rochester, C. H., 15
Rochin, C., 241
Rodrigue, A., 69, 142, 145
Roland, M. M., 266, 269
Rolle, F. R., 84
Romburgh, V. P.,275
Rosenfelder, W. J., 245
Rosenthal, R., 251, 252
Ross, D. S., 12, 92, 93, 94, 124, 125, 126, 127, 132, 133, 167, 175, 176, 177
Rossi, S. , 292
Rott, W., 266
Roupuszynski, 38
Rozhkov, I. N., 247
Russell, L. W., 95

Rust, F. F., 238, 255
Rys, P., 15, 140, 153
Sackwild, V., 119, 181
Saheki, K., 84
Saito, T., 133, 153, 145, 146, 186
Sakakibara, T., 98, 101
Sakata, Y., 32
Salem, G. F., 286
Salem, L., 4
Salih, Z. S., 82
Sampoli, M., 124, 129, 186
Sanchez, M. B., 245
Sandall, J. P. B., 189, 194, 196, 199
Sankararaman, S., 184, 185
Sastry, S., 32
Savides, C., 229, 230, 231
Savoie, R., 69, 142, 145
Scaife, C. W., 250, 251, 252
Schaarschmidt, A., 51, 52
Scheinbaum, M. L., 44, 262
Scheraga, H. A., 255
Schilling, P., 261
Schischkoff, L., 266
Schleyer, P. v. R., eds., 137, 139
Schlubach, H. H., 266
Schmeisser, H., 55, 58
Schmidt, E., 88, 89, 260
Schmitt, R., 167, 175, 176, 177, 266, 267
Schofield, K., 3, 11, 14, 16, 18, 21, 22, 44, 55, 123, 126, 128, 129, 131, 133, 141, 153, 154, 159, 160, 163, 172, 178, 186, 160
Schramm, R., 76
Schramm, R. F., 78
Schrieshiem, A., 122
Schumacher, I., 54, 55
Scriven, E. V., 129
Sears, C., 45
Sears, G. A., 258
Seebach, D., 220, 260
Seel, F., 58
Segel, E., 76
Seidenfaden, W., 9, 14
Seifert, W. K., 251
Semenov, V. V., 233
Semenovskii, A. V., 263
Seregina, N. M., 283
Setton, R., 13

Sevostyanova, V. V., 264, 266
Shaik, S. S., 5, 166
Shani, A., 32
Shchitov, N. V., 225
Sheats, G. F., 128
Shechter, F., 251, 252
Shechter, H., 224, 225, 251, 252, 257, 258, 293
Shen, J., 174, 175, 178
Shepherd, J. W., 229
Shepherd, J. W., 231
Sheppard, O. E., 231
Shernikova, T. V., 265, 266
Shevelev, S. A., 233
Shih-Hucé, W., 96
Shilling, P. , 148
Shoolery, J. N., 232
Shorygin, P. P., 86
Shtern, V. Ya., 227
Shusherina, N. P., 65
Shvarts, I. I., 266, 274
Sifniades, S., 232
Sigh, D., 45
Silverman, R. S., 84
Simmons, H. E., 292
Simonetta, M., 239
Simons, J. H., 19
Simpson, W. B., 94
Singer, K., 129
Singh, H. K., 231
Sisler, H. H., 30
Sitkin, A. I., 86
Sixma, F. L. J., 255
Sjoberg, P., 157, 181
Skinner, G. A., 159
Slavinskaya, R. A., 38
Sloan, M. F., 240
Slon, M., 225
Smart, G. N. R., 276
Smiley, R. A., 288, 289, 290, 292
Smit, V. A., 263, 265
Smith, A. E. W., 250, 251
Smithy, I. W., 158
Sokolovsky, M., 88
Sokovishina, I. F., 289
Sommer, J., 22
Sondheimer, F., 32
Sorochkin, I. N., 250
Sosnovsky, G., 227, 228, 252

Speckles, E., 55
Speier, J. L., 82
Speranskii, E. M., 263
Spinicelli, L. F., 231
Spitzer, U. A., 21
Sprague, R. W., 30
Sprung, J. L., 252
Ssigba-Aullin, N. C., 39
Stacey, M., 21
Stafford, W. H., 88
Stafford, W. L., 88
Stanburg, D. M., 187, 193, 196
Starcher, P. S., 250, 251
Starkey, E. B., 98
Stears, N. S., 153
Stedman, G., 118
Steele, V., 232
Stein, R., 122
Steinkopf, W., 258
Stevens, I. D. R., 252, 255
Stevens, T. E., 257
Stevens, T. W., 251
Stewart, J., 21
Stille, J. K., 291
Stock, L. M., 77, 142, 143, 157
Strachan, A. N., 128
Strarcher, P. S., 248, 249
Stromme, K. O., 137, 138
Stüber, O. , 220, 287
Suri, S. C., 276
Sutton, D., 94
Suzuki, E., 54
Suzuki, J., 169
Szelke, M., 266, 271
Takahashi, K., 84
Takami, T., 77
Takata, T., 87
Takeuchi, Y., 260
Talybov, A. G., 263, 265, 266
Tanabe, K., 258, 259
Tanaka, J., 164, 180
Tanaka, K., 96
Tapia, R., 27
Tartakovskii, V. A., 65, 66, 82, 265, 266, 281, 283
Tatlov, J. C., 247
Tatlow, J. C., 21
Taub, B., 288
Taube, H., 193

Author Index

Taylor, E. C., 79
Taylor, N. H., 5
Taylor, P. G., 22
Taylor, R., 126, 137, 186
Tedder, J. M., 13, 21, 255
Teipel, J., 27
Telder, A., 73, 141
ter Meer, E., 292
Terem, B., 129
Tezuka, H., 225
Thaler, W. A., 255
Thiele, J., 275, 276, 277
Thoennes, D., 27
Thomas, C. B., 79, 84
Thomas, P. N., 18
Thomas, R. J., 20
Thompson, M. J., 40, 44
Tisue, T., 78
Titov, A. I., 38, 50, 88, 89, 201, 203, 222, 223, 225, 227, 228, 232, 260
Tobin, G. D., 21, 153, 154
Todres, Z. V., 192, 193
Tohmori, K., 54
Tolgyesi, W. S., 152
Tolstaya, T. P., 98
Tolstoya, T. I., 266, 267
Tomoda, S., 260
Topchiev, A., 86
Topchiev, A. V., 29, 38, 75, 221, 246, 228
Török, L., 21
Torres, G., 27
Torsell, K., 164, 181
Toseland, B. A., 12
Toshimitsu, A., 32, 79, 80, 82, 84
Trahanovsky, W. S., 227
Traverso, P. G., 124, 129, 186
Treibs, W., 290
Tronow, B. W., 39
Truce, W. E., 231
Truter, M. R., 19
Tsang, S. M., 16, 17, 18, 38
Tselinskii, I. V., 281, 283
Tsuno, Y., 139
Tsutsumi, S., 77
Tyler, B. J., 255
Uemura, S., 32, 78, 79, 80, 82, 84
Ugi-ie, K., 96
Uhlfelder, E., 276, 277

Underwood, G. R., 84
Ungnade, H., 288
Ungnade, H. E., 232, 288
Urbański, T., 3, 225, 231, 266, 270, 277
Urch, D. S., 34
Ustavshchikov, B. F., 251, 250
Valderrama, J. A., 27
Vallee, B. L., 88
Vamplew, P. A., 129
Van Artsdalen, E. R., 255
Van Buren, II, W. D., 231
Van der Lee, F., 244
van der Waals, J. H., 137, 139
van Dolah, R. W., 266, 269
van Raayen, W., 255
Vanags, E., 231, 232
Vanags, G., 231, 232 , 250
Vanderwalde, A., 84
Vankar, Y. D., 259
Vast, P., 280, 282
Vaughan, J., 44, 162
Vaughan, W. E., 238, 255
Veibel, S., 131, 133
Vessell, E. D., 291
Viehe, H. G., 266, 267
Ville, J., 258
Vincent, B. F., 230
Vinnik, M. I., 123
Vogt, C. B., 51, 52
Vogt, C. M., 51, 52
Voskuil, W., 255
Vyatskin, I., 89
Wait, A. R., 16
Wakabayashi, T., 84
Walborsky, H. M., 240
Walden, P., 117
Waldnuller, M., 229
Wallwork, S. C., 94
Walters, S. L., 88
Walton, D. R. M., 82
Wang, K. B., 54, 55
Ward, E. R., 2, 97, 98, 99
Warford, E. W. T., 34
Warman, M., 277, 278
Wartel, M., 93, 187, 196
Washburn, L. C., 18
Wasserman, E., 69, 142, 145
Wat, E. K. W., 232

Watanabe, K., 90
Watson, D., 129
Watts, D. W., 129
Wayland, B. B., 78
Weidner, H., 256
Weisblat, D. I., 232
Weiss, J., 164
Werner, A., 260
Westerman, P. W., 261
Westheimer, F. F., 118, 121
Westheimer, F. H., 76
Weston, J. B., 153, 154
Wetterholm, A., 266
Wheeler, A. S., 158
Wheland, G. W., 118, 136, 139
White, A. M., 147, 288
Whitmore, F. C., 290
Whitmore, M. G., 290
Whittle, A., 85
Wibaut, J. P., 117
Wieland, H., 244, 255, 256
Wilkie, R. J., 129
Wilkinson, J. H., 38
Wilkinson, P. A., 258
Williams, G. H., 202
Win, H., 152
Windaus, A., 245
Wirkkala, R. A., 21
Wislicenus, W., 229
Wittig, G., 232

Wizinger, R., 139
Woeffenstein, R., 76
Worrall, D. E., 76, 77
Wright, G. F., 275, 276, 277
Wright, G. J., 162
Wright, H. R., 38, 64
Wright, O. L., 27
Wright, T. L., 77
Yagupolskii, L. M., 59
Yakovenko, V. N., 266, 274
Yamato, S., 77
Yanez, M., 177
Yang, L. t. A., 44, 53, 204
Yoshida, T., 60, 77, 84
Yossef, A. A. 240
Young, A. M., 76, 77
Yousif, G., 84
Yukawa, T., 139
Zahora, E. P., 44, 53, 204
Zakharin, L. I., 246
Zalukajevs, L., 232
Zamen, M. B., 124
Zefirov, N. S., 263
Zerner, M. C., 182
Zheng, Z., 182
Zlotin, S. G., 264, 266
Zollinger, H., 133, 134, 140, 144, 133, 186, 145, 146, 153
Zoseland, B. A., 12

Subject Index

Acetone cyanohydrin nitrate, 40, 275
Acetyl nitrate, 12, 33, 43, 54, 247
   preparation of, 43, 248
Acetylacetonates, 93
Acetylium ion, 146
Acyl nitrates, 43
Aliphatic nitration, 219
   of alkenes, 243
   of alkynes, 267
   electrophilic, 229
   free-radical, 220
   at heteroatoms, 269
   industrial methods, 219
   nucleophilic, 287
Alkanes, nitration of,
   electrophilic, 229
   free-radical, 220
   Hass reaction, 227
   with nitronium salts, 233
Alkenes, nitration of, 243
   with acetyl nitrate, 247
   by HNO_3, 244
   with HNO_3/H_2SO_4, 247
   with HNO_3/HF, 247
   with hexanitroethane, 260
   mechanism by NO_2^+ salts, 262
   mechanism with N_2O_4, 251
   with N_2O_3, 255
   with N_2O_4, 250
   with N_2O_5, 257
   via nitro-desilylation, 267
   via nitro-destannylation, 267
   via nitro-mercuration, 260
   via nitro-selenation, 260
   with nitronium salts, 261
   with nitryl halides, 257
   with tetranitromethane, 260
Alkyl nitrates, 36, 231
   preparation of, 271

Alkynes, nitration of, 267
   regioselectivity, 267
Amides,
   formation of, with NO_2^+ salts, 240
   nitration of, 280
   preparation of, 283
Ammonium nitrate, 34
   with trifluoroacetic anhydride, 34
Arenediazonium ion salts, 97
Arenium ions, 139
Aromatic nitration,
   acid–catalyzed, 9
   effect of NO^+, 195
   electron transfer mechanism, 164
   electrophilic, 117
   ipso attack, 158
   kinetics, 11
mechanism of,
   electrophilic nitration, 117
   nucleophilic nitration, 205
   free-radical nitration, 201
methods of,
   acid–catalyzed, 9
   free–radical, 83
   nitric acid, 11
   nitronium salts, 57
   nucleophilic nitration, 96
   nucleophilic, 205
   radical, 201
reagents,
   acetone cyanohydrin nitrate, 40
   acyl nitrates, 43
   alkyl nitrates, 36
   ammonium nitrate/trifluoroacetic anhydride, 34
   arene diazonium salts, 97
   bidentate metal nitrates, 94
   metal complex, 93
   monodentate metal nitrates, 29

Aromatic nitration (cont'd)
  nitric acid, 11
  nitric acid-graphite, 13
  nitric acid-HF, 19
  nitric acid-HF-BF$_3$, 20
  nitric acid-oleum, 15
  nitric acid-solid-acid catalysts, 26
  nitric acid-supported acid catalysts, 28
  nitric methanesulfonic acid, 22
  nitric-fluorosulfuric acid, 25
  nitric-Magic acid, 26
  nitric-perchloric acid, 18
  nitric-phosphoric acid, 15
  nitric-sulfuric acid, 13
  nitric-triflic acid, 22
  nitric-trifluoracetic acid, 21
  nitrogen dioxide, 83
  nitrogen oxides, 89
  nitronium salts, 57
  nitronium tetrafluoroborate, 61
  nitryl halides, 45
  NO/O$_2$/N$_2$O$_4$, 92
  oxidative nitration with NO$^+$ salts, 75
  sodium nitrite, 96
  transfer nitrating agents, 68
  trimethylsilyl nitrate, 42
  via diarylhalonium ions, 98
  via metallation, 75
  solvent effects, 12
  theoretical calculations, 180
Azeotropic nitration, 27
Benzoyl nitrate, 44
Celite-545, 28
Ceric ammonium nitrate, 95, 169
Charge-transfer complex, 180
Charge-transfer studies, 184
  kinetics of, 185
Chemical oxidation, 169
CIDNIP studies, 189
π-Complex, 137, 140
σ-Complex, 118, 137
Crown ethers, 142
Crown-ether complexes, 68
Desilylative nitration, 80
  mechanism of, 81
Diarylhalonium ions, 98
Dimethylsulfoxide-NO$^+$BF$_4^-$, 70

Dinitrogen pentoxide, 55, 225, 257
  preparation of, 55
Dinitrogen tetroxide, 50, 83
Dinitrogen trioxide, 49, 255
Electrochemical oxidation, 167
Electron transfer mechanism, 164
  charge-transfer complex, 180
  charge-transfer studies, 184
  chemical oxidation, 169
  CIDNIP studies, 189
  early suggestions, 164
  electrochemical studies, 167
  energetic considerations, 193
  ICR studies, 173
  isomer ratios, 168
  naphthalene nitration, 165
  other examples of, 166
  oxidation potentials, 192
  Perrin's mechanism, 165
  radiolytic studies, 177
  reaction rates, 168
  summary of, 197
  theoretical calculations, 180
  use of Marcus theory, 187
Electrophilic nitration, 117, 229
  of active CH$_2$ compounds, 229
  with alkyl nitrate, 232
  bond reactivity, 237
  mechanism of, 234
  via nitro-desilylation, 241
  via nitro-destannylation, 241
  with nitronium salts, 233
Encounter pair, 154
Ethyl nitrate, 38
Explosives, 6, 220, 277
Fluorosulfuric acid, 25
Free-radical nitration, 201, 220
  energetics of, 202
  gas-phase with N$_2$O$_4$, 227
  Konovalov reaction, 221
  mechanism of, 202, 222
  via N$_2$O$_5$, 225
  with nitric acid, 221
  regioselectivity, 203
Friedel-Crafts nitration,
  with N$_2$O$_4$, 52
  with N$_2$O$_5$, 56
Frontier orbital theory, 181
Gas-phase nitration, 88, 173, 227

Subject Index

gem-Dinitroalkanes,
  mechanism of formation, 292
  preparation of, 292
Graphite nitrate, 13
Heteroatom nitration, 269
Hexamethylbenzene, 74
Hexanitroethane, 260
Ingold-Hughes mechanism, 117
  acidity dependence, 122, 124
  effect of HNO_3 concentration, 120
  effect of nitrous acid, 129
  enthalpies of activation, 127
  isomer distribution, 119
  kinetic isotope studies, 123
  kinetics of, 117, 120
  kinetics of ^{18}O exchange, 121
  ^{14}N NMR studies, 124
  medium effects, 123
  substituent effect, 119
Inner-sphere electron transfer, 193
Ipso attack, 158
Ipso nitration,
  definition of, 159
  mechanism of, 159
  1,2-migrations, 163
  positional reactivity, 160
Kinetic isotope studies, 123
Kinetics,
  of charge-transfer studies, 184
  of gas-phase nitration, 177
  of Ingold-Hughes mechanism, 120
  of Schofield mechanism, 154
Konovalov reaction, 221
Kornblum modification, 290
  mechanism of, 290
  scope of, 291
Liquid-phase nitration, reactivity order, 225
Magic acid, 26
Marcus theory, 187
  bond reorganization, 192
  rate constant for, 187
Mechanism
  of electron transfer, 164
  of electrophilic nitration, with NO_2^+ salts, 233

  of free-radical nitration, 222, 228
  of Kornblum modification, 290
  of Meyer reaction, 287
  of N_2O_4 nitration, 227
  of N_2O_5 nitration, 226
  of nitration via nitrosation, 131
  of nitration at oxygen, 271
  of nitro-desilylation, 241
  of nitronium ion nitration, 117
  of nucleophilic nitration, 205
  of ter Meer reaction, 292
Medium effects, Ingold-Hughes mechanism, 123
Meisenheimer complex, 205
Mercuration, 76
  mechanism of, 76
  regioselectivity, 77
Metal nitrates, 29, 94
  mechanism of, 94
  reactivity, 29
Methanesulfonic acid, 22
Methyl nitrate, 38, 178
Meyer reaction, 287
  mechanism of, 289
  scope of, 287
Mixed acid, 13
Monodentate metal nitrates, 29
n-Butyl nitrate, 38
N-nitrocollidinium BF_4^-, 271
N-nitro quinolinium salts, 69
N-nitropyridinium salts, 65
Nafion-H, 27, 41, 53, 54, 77
Nitracidium ion, 19, 120, 145
Nitracidium perchlorate, 19
Nitramines, preparation of, 276, 279
Nitrate esters, preparation of, 271
Nitration,
  aliphatic, 219
  at heteroatoms, 269
  with N-nitrocollidinium BF_4^-, 271
  with HNO_3, 275
  at nitrogen, with nitronium salts, 280
  with $NO_2^+BF_4^-$, 271
  at oxygen, 269
  at phosphorus, 287
  at sulfur, 286
  of alkenes, 219, 243

Nitration
  of alkenes (cont'd)
    with acetyl nitrate, 247
    by HNO_3, 244
    with HNO_3/H_2SO_4, 247
    with HNO_3/HF, 247
    with hexanitroethane, 260
    mechanism by HNO_3, 244, 246
    mechanism by NO_2^+ salts, 262
    with N_2O_3, 255
    with N_2O_4, 250
    with N_2O_5, 257
    via nitro-desilylation, 267
    via nitro-destannylation, 267
    via nitro-mercuration, 260
    via nitro-selenation, 260
    with nitronium salts, 261
    with nitryl halides, 257
    with tetranitromethane, 260
  with alkyl nitrates, 231
  of alkynes, 267
    regioselectivity, 267
  of amides, 280
  aromatic, 9
  of arylalkanes, 220
  of cycloalkanes, 220
  desilylative, 80
  effects of crown ethers, 142
  electron transfer mechanism, 164
  Friedel-Crafts with N_2O_4, 52
  Friedel-Crafts with N_2O_5, 56
  gas-phase, 88, 227
  historical, 1
  industrial use, 5
  ipso attack, 158
  mechanisms of, 117
  using $NH_4NO_3(CF_3CO)_2O$, 35
  via nitro-desilylation, 241
  via nitro-destannylation, 241
  with nitronium salts, 233
  via nitrosation, 131
  with NO_2^+ salts, effect of superacid, 233, 237
  with $NO_2^+BF_4^-$, 61
  nucleophilic, 96, 287
  photochemical, 89
  of polymers, 37
  of small hydrocarbons, 227
  of steroids, 245, 258
  theoretical calculations, 180
  of toluene, 16
Nitric acid, 11, 242
Nitric acid-BF_3, 20
Nitric acid,-HF, 19
Nitric acid-oleum, 15
Nitric acid-solid acid catalysts, 26
Nitric-fluorosulfuric acid, 25
Nitric-Magic acid, 26
Nitric-methanesulfonic acid, 22
Nitric-phosphoric acid, 15
Nitric-sulfuric acid, 13
Nitric-triflic acid, 22
Nitric-trifluoroacetic acid, 21
9-Nitroanthracene, 73
9-Nitroanthracenium ion, 73
Nitro-dediazoniation, 97
Nitro-dehalogenation, 96
Nitro-demetallation, 80
Nitro-desilylation, 241, 267
  of vinyloxysilanes, 275
Nitro-destannylation, 241, 266
Nitro-mercuration, 260
Nitro-selenation, 260
Nitrocarbenium ion, 261
Nitrofluorination, 261
Nitrogen dioxide, 50, 83, 225, 250
Nitrogen oxides, 48, 89, 131
Nitrohexamethylbenzenium ion, 74, 161
Nitronium hexafluorophosphate, 67
  preparation of, 67
Nitronium ion,
  electrophilic nature of, 144
  geometry of, 144
  as hydride abstractor, 239
  oxidation potential, 192
  oxidizing ability, 199
  proto-solvated, 237, 239
Nitronium perchlorate, 19
Nitronium salts, 57, 233, 261, 280
  characterization of, 60
  ion-pairing of, 144
  nitration with, 57, 233, 261
  preparation of, 57
  reactivity of, 64
  solubility of, 60
  steric requirements, 237
Nitronium tetrafluoroborate, 56, 61

Nitronium tetrafluoroborate (cont'd)
    crown-ether complex, 68
    nitration with, 57, 234, 261
    preparation of, 61
Nitronium tetratriflatoborate, 25
Nitronium triflate, 25, 67, 68
Nitronium trifluoromethanesulfonate, 67
Nitroonium salts, 71
    preparation of, 71
Nitrosation, 131
Nitrosonium ion, oxidation potential, 192
Nitrosonium salts, 75
    oxidative nitration with, 75
Nitrosonium tetrafluoroborate, 70
Nitrosylsulfuric acid, 51
Nitrous acid, 129
    dissociation of in HNO_3, 129
Nitryl bromide, 47
Nitryl chloride, 45
    nitration with, 45
    preparation of, 46
Nitryl fluoride, 48
Nitryl halides, 45, 257
    in situ formation, 259
$NO_2^+BF_4^-$-crown ether complexes, 68
Nucleophilic aliphatic nitration, 287
  methods of,
    alkyl halides, with $AgNO_2$, 289
    alkyl halides, with alkali metal nitrates, 290
    Kaplan/Shechter reaction, 293
    Kornblum reaction, 290
    ter Meer reaction, 292
    V. Meyer reaction, 287
Nucleophilic nitration, 96
    mechanism of, 205
    Meisenheimer complex, 205
    regioselectivity of, 205
Olah mechanism, 134
    π-complex, 137, 140
    σ-complex, 118, 137
    criticisms of, 152
    effect of mixing, 153
    energy profile, 141
    isomer ratios, 136
    isotope effect study, 140

    nature of intermediates, 136
    positional selectivity, 137
    rates of nitration, 135
    use of nitronium salts, 135
    Wheland intermediate, 136
Oleum, 15
Outer-sphere electron transfer, 187
Oxidation potential,
    of aromatics, 193
    nitronium ion, 192
Palladation, 78
    mechanism of, 79
Palladium salts, 78
Perchloric acid, 18
Perfluorosulfonic acid, 27
Perrin mechanism, 165
Phosphine oxides, preparation of, 286
Phosphoric acid polymer, 16
Photochemical nitration, 89
Polyphosphoric acid, 15
Polystyrenesulfonic acid, 27
Protosolvation, 237, 239
Pyridine-N-oxide, 70
Pyridine-N-oxide-$NO^+BF_4^-$, 70
Raman studies, acidity dependence, 124
Regioselectivity,
    of alkyne nitration, 267
    using charge-transfer formulation, 184
    with crown-ethers, 68
    of diarylhalonium ions, 100
    effect of crown ethers, 142
    of free-radical nitration, 202
    of mercuration, 77
    of metal nitrates, 94
    of metallation, 75
    of mixed acid, 14
    of N-nitropyridium ions, 70
    of nitric-phosphoric acid, 16
    of nitric-triflic acid, 23
    of nitronium salts, 135
    of nucleophilic nitration, 205
    of $NO/O_2/N_2O_4$, 92
    with N_2O_4, 89
    solvent effects, 11, 17
    of triflic acid, 22
Ritter reaction, 240, 262

Schofield mechanism,
  encounter pair, 154
  kinetic studies, 154
  positional selectivity, 156
  substrate selectivity, 156
  Wheland intermediate, 155
Silver nitrate, 30
Single—electron transfer,
  see electron transfer
Sodium nitrite, 96, 97
Solid acid catalysts, 26
Solution nitration, 83
Solvent effects,
  on aromatic nitration, 11
  on polyphosphoric acid nitration, 17
Steroids, nitration of, 245, 258
Sulfones, preparation of, 286
Sulfonic acid nitroimides,
  preparation of, 283
Sulfoxides, preparation of, 286
Sulfuric acid, 12
Supported solid catalysts, 54

ter Meer reaction, 292
  mechanism of, 293
Tetranitromethane, 260, 266
Thallation, 79
Theoretical calculations, 180
  ab initio approach, 181
  frontier orbital approach, 181
  MNDO approach, 182
Transfer nitrating agents, 68
Transfer nitration, definition of, 68
Triflatoboric acid, 25
Triflic acid, 22
Triflic anhydride, 68
Trifluoroacetic acid, 13, 21
Trifluoroacetic anhydride, 34
Trifluoroacetyl nitrate, 44
Trifluoromethanesulfonic acid, 22
Trimethylsilyl nitrate, 42
Wheland intermediate, 118, 136, 155, 194
Zeolites, 54

RETURN TO →	CHEMISTRY LIBRARY 100 Hildebrand Hall	642-3753
LOAN PERIOD 1 **7 DAYS**	2 1 MONTH	3
4	5	6

ALL BOOKS MAY BE RECALLED AFTER 7 DAYS
Renewable by telephone

DUE AS STAMPED BELOW

NON-CIRCULATING UNTIL: MAR 15 '93		MAY 18
JUN 01 1993	DEC 19 1996	MAY 18
JAN 03 1994	DEC 16 1999	NOV 15
JUN 01 1995	AUG 16 2002	AUG 14
SEP 01 1995		
APR 29 REC'D AUG 16 1996	MAY 22 2004	
JUN 03 REC'D	JAN 17 2006	
DEC 19 1996		

UNIVERSITY OF CALIFORNIA, BERKELEY
FORM NO. DD5, 3m, 12/80   BERKELEY, CA 94720

JAN 21 '93